ENCYCLOPAEDIA OF

CLASSICAL INDIAN SCIENCES

Natural Science	Technology	Medicine

ENCYCLOPAEDIA OF CLASSICAL INDIAN SCIENCES

Natural Science	Technology	Medicine

Edited by
HELAINE SELIN and RODDAM NARASIMHA

Universities Press

Universities Press (India) Private Limited

Registered Office
3-6-747/1/A & 3-6-754/1 Himayatnagar, Hyderabad 500 029 (A.P.), India
Email: info@universitiespress.com

Distributed by
Orient Longman Private Limited

Registered Office
3-6-752 Himayatnagar, Hyderabad 500 029 (A.P.), India

Other Offices
Bangalore / Bhopal / Bhubaneshwar / Chennai
Ernakulam / Guwahati / Hyderabad / Jaipur / Kolkata
Lucknow / Mumbai / New Delhi / Patna

First Published 2007

ISBN 13: 978-81-7371-555-6
ISBN 10: 81-7371-555-6

Cover and book design

Typeset in Palatino 10/12 by
Techastra Solutions (P) Ltd.,
Hyderabad 500 482

Printed in India at
Graphica Printers
Hyderabad 500 013

Published by
Universities Press (India) Private Limited
3-6-747/1/A & 3-6-754/1 Himayatnagar, Hyderabad 500 029

Cover visual: *Secular and sacred mathematics*. In the background is a fragment of the Bakhshālī manuscript (written probably around 800 CE), a secular manual devoted to arithmetic and algebra, arraying numbers in a calculation and using special notations for writing equations of various kinds.

In the foreground is a figure representing the geometrical devices called yantras used in the practice of meditation. Some yantras are quite complex: a *śrī-yantra*, for example, consists of nine interwoven triangles generating 43 subsidiary triangles, with sides that may have up to six (often multiple) intersections with each other. The outer figure, with four 'gates', separates the world of order within from the world of chaos outside.

Contents

Preface

Given the extraordinary and widespread curiosity in India about its own scientific heritage, it is remarkable that there are so few books that respond to that curiosity for the interested non-specialist reader. India's scientific and technological accomplishments are among the oldest in the world. Indians have used mathematics for sacred and secular purposes for some 3000 years, probably longer. The decimal place value system with zero as a numera – a system that has been universally adopted today – had its origin in Hindu mathematics.

Indian brick technology is more than 4000 years old. Indians were experienced creators and users of metal products, especially with iron and zinc. People of the Indus valley civilization, ca. 2000 BCE, lived in planned cities and towns, with advanced public drainage systems. All over India people devised elaborate methods of irrigation with rivers, canals and reservoirs. They were excellent sailors and boat builders.

In medicine, Āyurveḍa, a system of living well to promote a long and healthy life, has been practised for millennia. It is now experiencing a global revival. Āyurvedic medicines can be purchased in shops anywhere in the world, and the basic premise that people of different personalities and temperaments should be treated differently is increasingly becoming a part of medical treatment everywhere. Ancient herbal remedies keep getting converted to modern drugs. More people practise yoga in the rest of the world now than Indians do. But its history is long, and the first mention of yoga occurs in the *Upaniṣads.*

All of these accomplishments were supplemented by a philosophy and a view of consciousness that is touched on in many of the articles in the Encyclopaedia. The *Vēdas,* dating from the second millennium BCE or earlier, were the basis of much Indian thought and early science. Together with the *Upaniṣads* that followed them, they set out a philosophical basis on which Indian sciences were built. But there was a vigorous protestant tradition as well, nourished for example by Buddhist schools of thought in ancient universities like Nālanda; the Jains made important contributions of their own, as did Muslims in more recent times.

This *Encyclopaedia of Classical Indian Sciences* contains articles on all these topics, as well as biographical articles on many ancient Indian scientists. There are also some more polemical essays, such as those on *Colonialism and science* or *Values and science.*

Our objective in this Encyclopaedia is to provide as authentic an account as possible of what is actually known and widely accepted today about Indic science by scholars across the world. There is, of course, a great deal that is not yet known. David Pingree's catalogue of Indian astronomical works (in seven volumes) lists an enormous number of manuscripts that have not yet been examined at all. It is only in recent decades that we have begun to appreciate the brilliant mathematical work of the Kerala school (12th to 16th century CE). Some of the greatest classics of Indic science have not yet had complete and modern English translations; e.g. Brahmagupta's *Brahma-sphuṭa-siddhānta,* or Bhāskara's *Siddhānta-śiromaṇi.* Nor has classical Indic science yet attracted a scholar of the breadth and depth of Joseph Needham, whose monumental work on Chinese science and technology has altered our notions of the contributions of eastern civilizations. As his studies progressed Needham raised the important question about 'the failure of China and India to give rise to distinctively modern science while being ahead of Europe for fourteen previous centuries', and went on to ask '... how [it was that] Galilean science could come to birth in Pisa but not in Patna or Peking'. This question, in some form or other, has vexed many Indian leaders and thinkers too, beginning with Raja Rammohan Roy. This Encyclopaedia may help in suggesting answers to such questions.

As Takao Hayashi says at the beginning of his 1995 volume on *The Bakhshālī Manuscript* (which was discovered near Peshawar in 1881), the history of that manuscript, 'just like the study of the history of Indian mathematics itself ... has been suffering from two kinds of preoccupation, namely, Indian nationalism and Hellenism'. Because of exaggerations on both sides a fair and authentic account of the history of Indic science has been difficult to find. (An honourable exception, although not written explicitly for the intelligent lay reader, is *A Concise History of Indian Science* by Sen, Bag and Subbarayappa.) By drawing on scholarship from all over the world, we have attempted to provide here as unbiased an account as seemed feasible.

In 1997 Helaine Selin compiled an *Encyclopaedia of the History of Science, Technology and Medicine in Non-Western Cultures* (Kluwer, Dordrecht). The volume was received with critical acclaim; it received the *Choice Magazine* Outstanding Academic Book award, and a review that called it 'a landmark in the history of science'.

The editors had the good fortune of meeting each other in the year 2000, when HS attended the International Conference on Knowledge and East-West Transitions, organized at the National Institute of Advanced Studies jointly by Dr Susantha Goonatilake and RN. As the discussions veered around to how generally ignorant we still are about the history of science and technology in India, it seemed suddenly clear to RN that HS had already done much of the ground work necessary in her earlier encyclopaedia. The present volume is the outcome of that thought, and the grant of a Fulbright fellowship to HS in December 2004 that enabled us to meet again and make concrete plans for the present volume (which incidentally includes much new material not present in the earlier encyclopaedia).

It only remains to explain the title of the present work. First of all we have (somewhat reluctantly) preferred the characterization *Indian* to *Indic*. India is now a political entity with a certain well-defined frontier, and *Indian* might be construed as limited to this political state. At various times, however, contributions to the same science have come from areas now outside the present political frontiers of India. *Indic* therefore seemed in many ways a more appropriate adjective for the wider cultural area that shared in the creation of a specific kind of science. In the end, however, we preferred a word that was more familiar to most people.

We also owe an explanation for the word *classical*. Much debate has taken place about what periods in history could be included within the scope of that adjective. We have used the word here as a convenient designation of the kind of science that was done in India before the advent of modern European science around the turn of the 18th century to the 19th. But even more than this chronological interpretation we use the word to describe all science that was done in the spirit of a distinctive Indic approach. Thus some *pañcāṅga-kāras* and āyurvedic physicians are certainly practising, and perhaps even creating, classical Indic science *today* – in the sense in which we use those words to define the present work.

We hope that this book can be a reference accessible to every interested Indian reader. We would be delighted if specialists will find it a useful starting point for information on fields outside their areas of specialisation.

The editors welcome suggestions and comments on this work. We are aware that the present work is by no means complete, and would be happy to expand it should the response to this attempt indicate a need for doing so.

Helaine Selin
Hampshire College, Amherst, Massachusetts, USA

Roddam Narasimha
Jawaharlal Nehru Centre for Advanced Scientific Research/
National Institute of Advanced Studies, Bangalore, India

Acknowledgements

My thanks go first to Prof. Roddam Narasimha. It was his inspiration that encouraged him to unite my work with his vision. Prof. Narasimha also helped with my getting the Fulbright award which was a wonderful gift of a month in India and the opportunity to meet with so many scholars at the National Institute of Advanced Studies in Bangalore. At NIAS and in Bangalore, I would like to thank Hamsa Kalyani, Prof. D P Sengupta, Dr. Neelima Narasimha, Dr. Sharada Srinivasan, Prof. Dilip Ahuja, Dr. A R Vasavi, and Dr. Sangeetha Menon for going out of their way to make my stay enjoyable and productive. Thanks to Mr. K S Ramakrishna, for making my Macintosh work in a climate that does not support apples. I would especially like to thank K Nagarathna, who spent many hours of her own time making sure that all the edited work passed smoothly between us. It turns out that, even with email, the simplest transactions are not always so simple on both continents. I was lucky to be able to work with her and make friends with her. Madhu Reddy and his colleagues at Universities Press have been enthusiastic and supportive since our first discussion, and I thank them for their interest and good care. Finally, I would like to thank India. I am so fortunate to have been able to visit, to partake of the colours, the food and the kindness of strangers. I am honoured to have had this opportunity.

H.S.

I must first of all thank Ms. Helaine Selin for so enthusiastically agreeing to pursue the project of this encyclopaedia. Her Fulbright visit to Bangalore was essential for the success of the project, so I must thank the Fulbright Foundation for making her visit possible. The month she spent here in 2005 was a great pleasure, and her cheerfulness even when things were difficult was infectious. I am grateful to Dr. K Kasturirangan, Director of NIAS, for his enthusiastic support of this project and Ms. Selin's visit to Bangalore. Without the ever-cheerful support and the untiring devotion of Ms. K Nagarathna this work would not have seen the light of day for quite some time, and I am as always grateful to her for her efficiency and commitment. Mr. Madhu Reddy and Jebah, of Universities Press, have been most helpful on all matters connected with the production of the volume.

R.N.

A

Agriculture

The Harappan culture related to the earliest agricultural settlements in the Indian subcontinent is dated between 2300 and 1700 BCE. The crops of the Harappan period were chiefly of West Asian origin. They included wheat, barley, and peas. Of indigenous Indian origin were rice, tree cotton, and probably sesame. Rice first appeared in Gujarat and Bihar, not in the centre of the Harappan culture in the Indus Valley. There is some rather doubtful evidence that African crops were also grown by the Harappans. There is a record of sorghum (*jowar*) from Sind and *Pennisetum* (*bajra*) from Gujarat. The earliest record of the African cereal, *Eleusine coracana* (*ragi*) is from Mysore, about 1899 BCE. The Southeast Asian crops of importance to India are sugar cane and banana, and they both appear in the early literary record. Crops of American origin include maize, grain amaranths, and potato. The dating of the introduction of maize is uncertain, the characteristics and distribution of some forms being such as to lend support to the view that they reached India in pre-Columbian times. Crops of the Indian subcontinent have influenced the agricultural development of ancient Egyptian, Assyrian, Sumerian, and Hittite civilisations through their early spread to these regions of the Old World. The Buddhists took several Indian crops and plants to South-east Asian countries, and there was much early exchange of plant material with Africa. The Arabs distributed crops such as cotton, jute, and rice to the Mediterranean region in the eighth to tenth centuries CE. There was also a reciprocal exchange of several New World domesticates.

Agriculture Today

India is characterised by a wide variety of climates, soils, and topographies. It is rich in biodiversity and a seat of origin and diversification for several crop plants such as rice, millets, pigeon pea, okra, eggplant, loofah, gourds, pumpkin, ginger, turmeric, citrus, banana, tamarind, coconut, and black pepper. Because India is ethnically diverse, traditional agriculture is still practiced in many places. There are about 100 million operational holdings, and the country has over 20% of the world's farming population.

India has different ecosystems such as irrigated, rain fed, lowland, upland, semi-deep/deep water, and wasteland. Agriculture is primarily rain fed (rain dependent); it supports 40% of the human population, 60% of cattle, and contributes 44% to the total food production. Owing to differences in latitude, altitude, variation in rainfall, temperature and edaphic diversity, great variety exists in crops and cropping patterns.

There are two important growing seasons in India: the *Kharif* or the summer season, especially important for rice; and the *Rabi* or the winter season in which the major crop grown is wheat. The *Kharif* crop is primarily rain dependent, and the *Rabi* is relatively more reliant on irrigation. The *Kharif*/rainy season cropping patterns include major crops such as rice, sorghum, pearl millet, maize, groundnut, and cotton. The *Rabi*/winter season cropping patterns include important crops like wheat, barley and to some extent oats, sorghum, and gram/chickpea. Mixed cropping is also practiced, especially during the *Kharif* season. Pulses, grain legumes, and oilseeds are grown with maize, sorghum, and pearl millet. Brassica and safflower are grown mixed with gram or even with wheat. Under subsistence farming, on small holdings, mixed cropping provides food security and is consumption oriented.

India is the major producer of a number of agricultural commodities including rice, groundnut, sugar cane, and tea. Food grains constitute roughly two-thirds of the total agricultural output. These consist of cereals, principally rice, wheat, maize, sorghum, and minor millets. India is the second largest producer of vegetables next to China. Mango accounts for almost half of the area and over a third of production; banana is the second largest and is followed by citrus fruits, apple, guava, pineapple, grape, and papaya. Of non-food cash crops, the most important are oilseeds especially groundnut, short staple cotton, jute, sugar cane, and tea.

Owing to improvements in recent years there has been widening of inter-regional disparities in agricultural production and productivity. Regions such as the north and north-west and the delta regions of peninsular India have prospered under assured irrigation, but dryland and semi-arid regions have not done so well.

R. K. Arora

REFERENCES

Arora, R.K. "Plant Diversity in the Indian Gene Centre." In *Plant Genetic Resources: Conservation and Management*. Ed. R.S. Paroda and R.K. Arora. New Delhi: International Board for Plant Genetic Resources. 1992. pp. 25–54.

Forty Years of Agricultural Research and Education in India. New Delhi: Indian Council of Agricultural Research. 1989.

Handbook of Agriculture. New Delhi: Indian Council of Agricultural Research. 1992.

"India." In *Regional Surveys of the World. The Far East and Australasia.* 24th ed. London: Europa Publications. 1993. pp. 275–335.

Alchemy

Alchemy was an art practiced in ancient India. This is evident from the description in *Artha-śāstra,* a monumental work on state craft by Kauṭilya (400 BCE), of a type of gold which was then being prepared by the transmutation of base metals. It was more recently (in 1941–1942) demonstrated in New Delhi in the presence of renowned national leaders. Two marble slabs with the inscription of these two events still adorn the *yajña-vedī* (altar for the Vedic sacrifice) behind the *Lakṣmī-Nārāyaṇa temple* (popularly known as Birla temple after the name of the donor). The English translation of the first inscription is as follows:

> On the first day of *śukla pakṣa* (bright fortnight) in the month of *Jyeṣṭha* (May–June) of the sambat 1998, i.e. 27th May 1941, Pandit Kṛṣṇalāla Śarmā, in our presence ... prepared one *tolā* (12 grams approximately) of gold from out of one *tolā* of mercury in Birla house, New Delhi. The mercury was kept inside a fruit of *riṭhā* (bot. *Sapindus trifoliatus Linn*). Inside this, a white powder of some herbs and a yellow powder which were perhaps one and half *ratti* [one *ratti* is equal to approximately 125 milligrams] in weight were added. Thereafter, the fruit of *riṭhā* was smeared with mud and kept over a charcoal fire for about forty-five minutes. When the charcoal became ash, water was sprinkled over it. From inside the fruit which originally contained mercury, gold came out. In weight, the gold was one or two *rattis* less than one *tolā.* It was pure gold. We could not ascertain the mystery behind the performance. The nature as well as the identity of both the powders which were added were not disclosed to us. During the whole experiment, Pandit Kṛṣṇalāla Śarmā was standing ten to fifteen feet away from us.... We were all surprised to witness this performance..."

The second plaque tells the story of a similar event in 1942, in which mercury was mixed with an unnamed drug, kept over the fire for half an hour, and transformed into gold.

The primary aim of giving the above inscriptions is to show that such alchemical practices are prevalent even now, and they are not mere myths or superstitious beliefs as some people claim.

According to Indian tradition, alchemy is not an end in itself. Mercury, when processed through eighteen different steps which are called *saṁskāras* in Ayurvedic parlance, helps a person to attain positive health and to prevent disease. It also cures several obstinate and otherwise incurable diseases. Prior to administering this processed mercury to human beings, at the seventeenth step, it must be tested on ordinary mercury or other base metals. Depending upon its potency, these base metals become transmutated into gold or silver. Thus alchemy is only a step to test the effectiveness of the recipe before it is administered to human beings to improve their physical and mental health. These are necessary for attaining spiritual perfection in the form of *jīvan mukti* (salvation while remaining alive). These methods are always kept secret and disclosed only to trusted disciples. It is also ensured that the knowledge does not fall into the hands of untrustworthy people, who by amassing wealth may create social problems. In fact, acquiring wealth by alchemical methods is considered a great sin. Such wealth should never be used for personal benefit, but for charitable purposes. This is why saints adept in this technique are reluctant to demonstrate alchemy in public. While doing so, only the end result is shown without disclosing the details of the technique. Pandit Kṛṣṇalāla Śarmā, who demonstrated the method described above, learnt it from a saint of Haradwar named Nārāyaṇa Swāmī. However, he did not teach the detailed technique to anybody because he could not find a worthy disciple.

Because of the secrecy involved, many manuscripts describing this alchemical technique have perished. According to anecdotes, Nāgārjuna, the Buddhist monk perfected this technique and wrote several books on it. All these are unfortunately no longer extant. Some of the extant works dealing with both *deha-siddhi* (attaining perfection of the body by rejuvenation) and *lauha-siddhi* (transmutation of base metals into gold, etc) are as follows:

(1) *Rasahṛdaya-tantra* by Govinda Bhagavat-pāda, fl. ninth century;

(2) *Rasendra-cūdāmaṇi* by Somadeva, fl. twelfth century;

(3) *Rasaprakāśa-sudhākara* by Yaśodhara, fl. thirteenth century;

(4) *Rasasāra* by Govindācārya, fl. thirteenth century;

(5) *Rasendra-cintāmaṇi* by Dhuṇḍukanātha, fl. fourteenth century;

(6) *Rasa-paddhati* by Bindu, fl. fifteenth century;

(7) *Āyurveda-saukhya* by Ṭoḍara Malla, fl. sixteenth century;

(8) *Āyurveda-prakāśa* by Mādhava Upādhyāya, fl. seventeenth century;

(9) *Rasāyana-sāra* by Śyāma Sundarācārya, fl. twentieth century;

In Āyurveda, in addition to mercury, several other metals, minerals, gems, and costly stones are used for therapeutic purposes. Before these ingredients are added to recipes, special processes are required to make them non-toxic and therapeutically potent. The branch of Āyurveda describing such details is called *Rasa-śāstra*. All the books mentioned above belong to this branch. In addition to the processing of mercury both for *deha-siddhi* and *lauha-siddhi*, the technique of processing other metals is described.

Mercury is processed through eighteen different steps, both for *deha-siddhi* and *lauha-siddhi* Although there are some minor differences in different texts, these saṁskāras are (1) *svedana* or fomentation, (2) *mardana* or trituration, (3) *mūrchana* or causing disintegration of particles, (4) *utthāpana* or revival of the natural physical properties of mercury, (5) *pātana* or distillation and sublimation, (6) *bodhana* or potentisation, (7) *niyāmana* or regulation of physical properties, (8) *dīpana* or enhancing the power of digestion (of other metals), (9) *grāsa-māna* or determination of the quantity of other metals to be added, (10) *cāraṇa* or impregnation with *bīja* (preparations used as seed), (11) *garbha-dṛti* or internal digestion, (12) *vāhya-dṛti* or external digestion, (13) *jāraṇa* or assimilation, (14) *rañjana* or colouration, (15) *sāraṇa* or excessive potentisation, (16) *krāmaṇa* or enhancing the power of penetration, (17) *vedha* or testing the efficacy and potency of mercury by way of transmutating base metals into gold and silver, and (18) *śarīrayoga* or administration of processed mercury to human beings for the purpose of rejuvenating the body.

Even though the details of all the above-mentioned steps are described in books, the description is cryptic and some vital techniques are kept secret. Many people, in their enthusiasm to practice alchemy on the basis of the description in books, have lost lots of energy and money. They are unsuccessful, because the secret codes and hidden techniques can be learned only from a guru or master, and such adept masters disclose these techniques only to worthy disciples who are absolutely free from worldly attachments.

Bhagwan Dash

References

Acarya, Yadavji Trikamji. *Rasāmṛtam.* Banārasa: Motilala Banarasidasa. 1951.

Ārya, Satyendrakumāra. Āyurveda Rasaśāstra kā udbhava evaṃ-vikāsa. Varanasi: Krsnadas Academy. 1984.

Dash, Bhagwan. *Alchemy and Metallic Medicines in Ayurveda*. New Delhi: Concept. 1986.

Dvivedī, Vāsudeva Mūlaśaṅkara. *Pāradavijñānīyam*. Datiyā: Śarmā Āyurveda Mandira. 1969.

Govinda Bhagavatpāda. *Rasahṛdayatantra.* Kaleda: Kṛṣṇa Gopāla Āyurveda Bhavana. 1958.

Kangle, R.P. *The Kauṭilīya Arthaśāstra.* Delhi: Motilal Banarasidass. 1992.

Mookerjee, Bhudeb. *Rasa-jala-nidhi.* Vārāṇasi: Śrīgokul Mudraṇalaya. 1984.

Nityanātha Siddha. *Rasāyanakhaṇḍa of Rasaratnākara.* Varanasi: Chaukhamba Amarabharati Prakasan. 1982

Panta, Tārādatta. Ed. *Rasārṇava or Rasatantram.* Benaras: Chowkhamba Sanskrit Series Office. 1939.

Ray, P.C. *History of Chemistry in Ancient and Medieval India.* Calcutta: Indian Chemical Society. 1956.

Seal, B.N. *Positive Sciences of Ancient Hindus.* Delhi: Motilal Banarasi Dass. 1958.

See also: Medicine: Āyurveda

Algebra: *Bījagaṇita*

Bījagaṇita, which literally means "mathematics (*gaṇita*) by means of seeds (*bīja*)", is the name of one of the two main fields of medieval Indian mathematics, the other being *pāṭīgaṇita* or "mathematics by means of algorithms". *Bījagaṇita* is so-called because it employs algebraic equations (*samīkaraṇa*) which are compared to seeds (*bīja*) of plants since they have the potential to generate solutions to mathematical problems. *Bījagaṇita* deals with unknown numbers expressed by symbols. It is therefore also called *avyaktagaṇita* or "mathematics of invisible (or unknown) [numbers]". Algebraic analyses are also employed for generating algorithms for many types of mathematical problems, and the algorithms obtained are included in a book of *pāṭī. Bījagaṇita* therefore also means "mathematics as a seed [that generates *pāṭi* (algorithms)]".

Extant works in *bījagaṇita* include Chapter 18 (*kuṭṭaka* only) of Āryabhaṭa's *Mahāsiddhānta* (ca. CE 950 or 1500), Chapter 14 (*avyaktaganita*) of Śrīpati's *Siddhāntaśekhara* (ca. CE 1050), Bhāskara's *Bījagaṇita* (CE 1150), Nārāyaṇa's *Bījagaṇitāvataṃsa* (before CE 1356, incomplete), and Jñānarāja's *Bījādhyāya* (ca. CE 1500). Srīdhara's work (ca. CE 750), from which Bhāskara quotes a verse for the solution of quadratic equations, is lost. Chapter 18 (*kuṭṭaka*) of Brahmagupta's *Brāhmasphuṭasiddhānta* (CE 628) has many topics in common with later works of *bījagaṇita*, but the arrangement of its contents is not so systematic as that of the later works, and an unusual stress is placed on *kuṭṭaka* as the title of the chapter suggests. *Kuṭṭaka* (lit. pulveriser) is a solution to the linear indeterminate equation: $y = (ax + c)/b$.

The symbols used for unknown numbers in *bījagaṇita* are the initial letters (syllables) of the word *yāvattāvat* (as much as) and of the colour names such as *kālaka* (black), *nīlaka* (blue), *pīta* (yellow), etc. The use of the colour names may be related to Āryabhaṭa's *gulikā* (see below). Powers of an unknown number are expressed by combination of the initials of the words *varga* (square), *ghana* (cube), and *ghāta* (product). A coefficient is placed next (right) to the symbol(s) to be affected by it, and the two sides of an equation are placed one below the other. A dot (or a small circle) is placed above negative numbers. Thus, for example, our equation,

$$5x^5 - 4x^4 + 3x^3 - 2x^2 + x = x^2 + 1,$$

would be expressed as:

yāvaghaghā 5 *yāvava* $\dot{4}$ *yāgha* 3 *yāva* $\dot{2}$ *yā* 1 *rū* 0

yāvaghaghā 0 *yāvava* 0 *yāgha* 0 *yāva* 1 *yā* 0 *rū* 1

where *rū* is an abbreviation of *rūpa* meaning an integer or an absolute term. The product of two (or more) different unknowns is indicated by the initial letter of the word *bhāvita* (produced): e.g. *yākābhā* 3 for 3 *xy*.

These tools for algebra had been fully developed by the twelfth century, when Bhāskara wrote his book *Bījagaṇita*, the main topics of which are "four seeds" (*bījacatuṣṭaya*), namely, (1) *ekavarṇasamīkaraṇa* or equations with one colour (i.e. in one unknown), (2) *maddhyamāharaṇa* or elimination of the middle term (solution of quadratic equations), (3) *anekavarṇasamīkaraṇa* or equations with more than one colour, and (4) *bhāvitakasamīkaraṇa* or equations with "the product" (i.e. of the type $ax + by + c = dxy$).

At least part of this algebraic notation was known to Brahmagupta. He uses the words *avyakta* (invisible) and *varṇa* (colour) for denoting unknown numbers, when he gives his rules concerning the same four seeds as Bhāskara's, in Chapter 18 (*kuṭṭaka*) of his *Brāhmasphuṭasiddhānta.* The details of Brahmagupta's algebraic notation are, however, not known to us.

Bhāskara, a contemporary of Brahmagupta, did know the word *yāvattāvat* meaning an unknown number, but it is not certain if he used it in equations, because he expresses the equation, $7x + 7 = 2x+12$, without the symbol *yā* as:

7	7
2	12

in his commentary (CE 629) on the *Āryabhaṭīya.* In the same work he refers to four seeds which are said to generate "mathematics of practical problems" (*vyavahāragaṇita*) having eightfold of names beginning with "mixture", but what kinds of seeds he mentioned by the names *yāvattāvat, vargāvarga* (square?), *ghanāghana* (cube?), and *viṣama* (odd), are not exactly known. Similar terms (*yāvattāvat, varga, ghana,* and *vargāvarga*) occur in a list of ten mathematical topics given in a Jaina canon, *Sthānāṅga* (Sūtra 747), which is ascribed to the third century BCE.

Āryabhaṭa used the term *gulikā* (a bead) for an unknown number when he gave his rule for linear equations of the type $ax + b = cx + d$ in his *Āryabhaṭīya* (CE 499). All the equations to which he gave solutions (including *kuṭṭaka*) are linear, although his rules for the interest and for the period of an arithmetical progression pre-suppose the solution of quadratic equations.

Brahmagupta gave many theorems for *vargaprakṛti* (lit. square nature), that is, the indeterminate equation of the second degree: $Px^2 + t = y^2$, but it is Jayadeva (the eleventh century or before) that gave a complete solution for the case $t = 1$ (the so-called Pell's equation).

Bījagaṇita reached its culmination in the twelfth century, when Bhāskara gave solutions to various types of equations of higher degrees by means of *kuṭṭaka* and *vargaprakṛti.* After him significant developments in the field of *bījagaṇita* are not known.

Takao Hayashi

REFERENCES

Bag, A.K. *Mathematics in Ancient and Medieval India.* Varanasi: Chaukhambha Orientalia. 1979.

Colebrooke, H.T. *Algebra with Arithmetic and Mensuration from the Sanscrit of Brahmegupta and Bháscara.* London: Murray. 1817. Reprinted, Wiesbaden: Dr. Martin Söndig oHG. 1973.

Datta, B. and Singh, A.N. *History of Hindu Mathematics.* 2 vols. Lahore: Motilal. 1935/38. Reprinted in one vol., Bombay: Asia Publishing House. 1962.

Ganguli, S. "Indian Contribution to the Theory of Indeterminate Equations of the First Degree." *Journal of the Indian Mathematical Society, Notes and Questions* 19: 110–120, 129–142, 153–168. 1931/32.

Lal, R. "Integral Solutions of the Equation $Nx^2 + 1 = y^2$ in Ancient Indian Mathematics (*Cakravāla* or the Cyclic Method)." *Gaṇita Bhāratī* 15:41–54. 1993.

Selenius, C.-O. "Rationale of the Cakravāla Process of Jayadeva and Bhāskara II." *Historia Mathematica* 2:167–184. 1975.

Sen, S.N. "Mathematics," In *A Concise History of Science in India.* Ed. D.M. Bose et al. New Delhi: Indian National Science Academy. 1971. pp. 136–212.

Shukla, K.S. and Sarma. K.V. Eds. and trans. *Āryabhaṭīya of Āryabhaṭa.* New Delhi: Indian National Science Academy. 1976.

Sinha, K.N. "Algebra of Śrīpati: An Eleventh Century Indian Mathematician." *Gaṇita Bhārati* 8:27–34. 1986.

Srinivasiengar, C.N. *The History of Ancient Indian Mathematics*. Calcutta: The World Press. 1967.

See also: Arithmetic: *Pāṭīgaṇita* – Āryabhaṭa – Śrīpati – Bhāskara – Nārāyaṇa – Brahmagupta – Śrīdhara – Jayadeva

Arithmetic: *Pāṭīgaṇita*

Pāṭīgaṇita, which literally means "mathematics *(gaṇita)* by means of algorithms *(pāṭī)*", is the name of one of the two main fields of medieval Indian mathematics, the other being *bījagaṇita* or "mathematics by means of seeds". The two fields roughly correspond to arithmetic (including mensuration) and algebra respectively.

The compound *pāṭīgaṇita* seems to have come into use in relatively later times. In older works, the expressions, *gaṇitapāṭī* and *gaṇitasyapāṭī* (mathematical procedure, i.e. algorithm), are common, and sometimes the word *pāṭī* occurs independently. *Pāṭīgaṇita* is also called *vyaktagaṇita* or "mathematics of visible (or known) [numbers]", while *bījagaṇita* is called *avyaktagaṇita* or "mathematics of invisible (or unknown) [numbers]". Some scholars maintain that the word *pāṭī* originated from the word *paṭṭa* or *paṭa* meaning the calculating board, but its origin seems to be still open to question.

The division of mathematics *(gaṇita)* into the two fields was not practiced in the *Āryabhaṭīya* (CE 499), which has a single chapter called *gaṇita*, but it existed in the seventh century, when Brahmagupta included two chapters on mathematics in his astronomical work, *Brāhmasphuṭasiddhānta* (CE 628). Neither the word *pāṭī* nor *bījagaṇita* occurs in the book, but Chapter 12 (simply called *gaṇita*) deals with almost the same topics as later books of *pāṭī*, and Chapter 18, though named *kuṭṭaka* or the pulveriser (solution of a linear indeterminate equation), has many topics in common with later books of *bījagaṇita*. Śrīdhara (ca. CE 750) is known to have written several textbooks of *pāṭī* and at least one of *bījagaṇita*.

Extant works of *pāṭī* include Śrīdhara's *Pāṭīgaṇita* (incomplete) and *Triśatikā* (and *Gaṇitapañcaviṃśī*?), Mahāvīra's *Gaṇitasārasaṃgraha* (ca. CE 850), Chapter 15 (*pāṭī*) of Āryabhaṭa's *Mahāsiddhānta* (ca. CE 950? or 1500?), Śrīpati's *Gaṇitatilaka* (incomplete) and Chapter 13 (*vyaktagaṇita*) of his *Siddhāntaśekhara* (ca. CE 1050), Bhāskara's *Līlāvatī* (CE 1150), and Nārāyaṇa's *Gaṇitakaumudī* (CE 1356).

A book (or chapter) of *pāṭī* consists of two main parts, namely, fundamental operations (*parikarmāṇi*) and "practical problems" (*vyavahārāḥ*). The former usually comprises six or eight arithmetical computations (addition, subtraction, multiplication, division, squaring, extraction of the square root, cubing, and extraction of the cube root) of integers, fractions, and zero, several types of reductions of fractions, and rules concerning proportion including the so-called rule of three (*trairāśika*). The latter originally consisted of eight chapters (or sections), i.e. those on mixture (*miśraka*), mathematical series (*śreḍhī*), plane figures (*kṣetra*), ditches (*khāta*), stacking [of bricks] (*citi*), sawing [of timbers] (*krākacika*), piling [of grain] (*rāśi*), and on the shadow (*chāyā*).

To this list of the practical problems, *Śrīdhara* added in his *Pāṭīgaṇita* one named "truth of zero" (*śūnyatattva*). A large portion of the work including that chapter is, however, missing in the only extant manuscript. The way the *Gaṇitasārasamgraha* of Mahāvīra divides its contents into chapters is unusual, but it can still be characterised as a book of *pāṭī*. It is quite rich in mathematical rules and problems.

In his *Līlāvatī*, Bhāskara separated the rules on proportion from the arithmetical computations, and created with them a new chapter named *prakīrṇaka* (miscellaneous [rules]), in which he also included the *regula falsi*, the rule of inverse operations, the rule of sum and difference, etc. After the ordinary topics of practical problems, he treated *kuṭṭaka* as well as *aṅkapāśa* or the nets of numerical figures (combinatorics). Written in elegant but plain Sanskrit and organised well, the *Līlāvatī* became the most popular textbook of *pāṭī* in India.

In his *Gaṇitakaumudī*, Nārāyaṇa included in the practical problems not only *kuṭṭaka* and *aṅkapāśa*, but also *vargaprakṛti* or the square nature (indeterminate equations of the second degree including the so-called Pell's equation), *bhāgādāna* or the acquisition of parts (factorisation), *aṃśāvatāra* or manifestation of fractions (partitioning), and *bhadragaṇita* or mathematics of magic squares. These topics had already been dealt with to a certain extent by his predecessors, but he developed them considerably. He also investigated new mathematical progressions, some of which turned out to be useful when Mādhava (ca. CE 1400) and his successors obtained power series for the circumference of a circle (or π), sine, cosine, arctangent, etc.

Takao Hayashi

References

Bag, A.K. *Mathematics in Ancient and Medieval India.* Varanasi: Chaukhambha Orientalia. 1979.

Colebrooke, H.T. *Algebra with Arithmetic and Mensuration from the Sanscrit of Brahmegupta and Bháscara*. London: Murray, 1817. Reprinted, Wiesbaden: Dr Martin Söndig oHG. 1973.

Datta, B. and Singh, A.N. *History of Hindu Mathematics*. 2 vols. Lahore: Motilal. 1935–38. Reprinted in one volume. Bombay: Asia Publishing House. 1962.

Ramanujacharia and Kaye, G.R., trans. "The *Triśatikā* of Śrīdharācārya." *Bibliotheca Mathematica* Series 3. 13: 203–17. 1912–13.

Raṅgācārya, M. Ed. and trans. *Gaṇitasārasaṃgraha of Mahāvīra*. Madras: Government Press. 1912.

Sen, S.N. "Mathematics." In *A Concise History of Science in India*, Ed. D.M. Bose et al. New Delhi: Indian National Science Academy. 1971. pp. 136–212.

Shukla, K.S., ed. and trans. *Pāṭīgaṇita of Śrīdhara*. Lucknow: Lucknow University Press. 1959.

Sinha, K.N. "Śrīpati's *Gaṇitatilaka*: English Translation with Introduction". *Gaṇita Bhāratī* 4: 112–133. 1982.

Sinha, K.N. "Śrīpati: An Eleventh-Century Indian Mathematician." *Historia Mathematica* 12: 25–44. 1985.

Sinha, K.N. "Vyaktagaṇitādhayāya of Śrīpati's *Siddhāntaśekhara*." *Gaṇita Bhāratī* 10: 40–50. 1988.

Srinivasiengar, C.N. *The History of Ancient Indian Mathematics*. Calcutta: The World Press. 1967.

See also: Āryabhaṭa – Śrīdhara – Mahāvīra – Bhāskara – Nārāyaṇa – Combinatorics – Magic Squares – Mādhava

Armillary Spheres

The armillary sphere, known in Hindu astronomy by the terms *Golabandha* and *Gola-yantra* (globe instrument), was constructed from early times for study, demonstration, and observation. Among texts and commentaries which have either a brief mention or a detailed treatment of the armillary sphere, the following might be mentioned: *Āryabhaṭīya* of Āryabhaṭa (b. 476), *Pañcasiddhāntikā* of Varāhamihira (505), *Śiṣyadhīvṛddhida* of Lalla (eighth century), *Brāhmasphuṭasiddhānta* of Brahmagupta (b. 628), *Siddhāntaśekhara* of Śripati (1039), *Sūrya-siddhānta, Siddhāntaśiromaṇi (Golabandhādhikāra)* of Bhāskara II (b. 1114), and *Goladīpikā* of Parameśvara (1380–1460).

The movable and immovable circles which form parts of the instrument are made out of thin bamboo strips, the earth and the celestial bodies are of wood or clay, and the lines are connected by means of strings.

The axis is made of iron and is mounted on two vertical posts, so that it is possible to rotate the sphere as needed.

The *Goladīpikā* describes the construction of a simple *Golabandha* with two spheres, the inner one representing the *Bhagola* (starry sphere) moving inside an outer sphere which represents the *Khagola* (celestial sphere), both fitted on the same central axis. The movements of the planets, etc. are projected and measured on these spheres. A circular loop made of a thin bamboo strip kept vertically in the north–south direction represents the solsticial colure (*Dakṣiṇottara*). Another similar circle fixed to the former in the east–west direction would be the equinoctial or celestial equator (*Ghaṭikā-maṇḍala*). Still another circle fixed around these, cutting them at right angles and making crosses at the four cardinal points, represents the equinoctial colure. The celestial equator is graduated into 60 equal parts and the other two into 360 equal parts. Another circle is fixed passing through the east and west crosses and inclined at 24° north and south of the zenith and the nadir – this would be the ecliptic (*Apama-vṛtta*). Several smaller circles are now constructed across the solsticial colure on either side of the celestial equator and parallel to it, at the required declinations – these would be the diurnal circles (*Ahorātra-vṛttas*), of different magnitudes. The orbits of the Moon and other planets are now constructed, crossing the ecliptic (the band of the zodiac through which the Sun apparently moves in its yearly course) at the nodes (*pātas*) of the respective planets and diverging from it, north and south, by their maximum latitudes at 90° from the nodes. A metal rod is inserted through the north and south crosses to form the central axis. This figuration is called the starry sphere (*Bhagola*).

Three circles are constructed outside the *Kha-gola* (celestial sphere). The horizontal circle is called the horizon (*Kṣitija*), the east–west circle is called prime vertical (*Samamaṇḍala*), and the north–south circle is called the meridian (*Dakṣiṇottara*). A model of the Earth in spherical form is then fixed to the centre of the axis. This would be the figuration at zero latitude.

If the armillary sphere is to be used in any other place, two holes are made in the celestial sphere at a distance equal to the latitude of the place, below and above the south and north crosses, and the axis of the starry sphere is made to pass through them. It is also necessary, in this case, to construct a circle called the equinoctial colure (*Unmaṇḍala*), which passes through the two ends of the axis and the east and west crosses. To keep the two sets of circles in position, wooden pieces are fixed to the axis in between them, so that the spheres do not get displaced.

The inner starry sphere revolves constantly, while the celestial sphere remains stationary, and directions are reckoned therefrom. A diurnal circle with its radius equal to the sine latitude is also constructed with a point on the central axis as the centre, just touching the horizon. The sine and cosine of the place can be measured on this circle.

While the armillary sphere described above is more for study and demonstration, certain other texts speak of more circles and also enjoin observation of celestial bodies. Thus *Brāhmasphuṭasiddhānta* (XXI. 49–69) prescribes the construction of 51 movable circles, *Śiṣyadhīvṛddhida* speaks of diagonal circles (*Koṇa-vṛttas*) and spheres for each planet, and *Siddhānta-śiromaṇi* adds a third circle called *Dṛggola* outside the *Khagola* (celestial sphere). According to *Śiṣyadhīvṛddhida*, the *lagna* (Orient ecliptic point) and time are also found by means of the armillary sphere.

K.V. Sarma

References

Dikshit, Sankar Balakrishna. *Bharatiya Jyotish Sastra* (*History of Indian Astronomy*), Pt. II. New Delhi: Indian Meteorological Department. 1981. pp. 224–25.

The Goladīpikā by Parameśvara. Ed. and trans. K.V. Sarma. Madras: Adyar Library and Research Centre. 1957.

"Golam Keṭṭal (Golabandham)" (in Malayalam). In *Bhāratīya Śāstra-manjūṣa*. Ed. M.S. Sreedharan. Trivandrum: Bharatiya Sastra-manjusha Publications. 1987. Vol. II. pp. 40–55.

Ohashi, Yukio. *A History of Astronomical Instruments in India*. Ph.D. Thesis. Lucknow University. 1990.

See also: Astronomical Instruments

Āryabhaṭa

Āryabhaṭa (b. CE 476) was a celebrated astronomer and mathematician of the classical period of the Gupta dynasty (CE 320 to ca. 600). This era is called the Golden Age in the history of India, during which Indian intellect reached its high water mark in most branches of art, science, and literature, and Indian culture and civilisation reached a unique stage of development which left its deep impression upon succeeding ages. Āryabhaṭa played an important role in shaping scientific astronomy in India. He is designated as Āryabhaṭa I to differentiate him from Āryabhaṭa II, who flourished much later (ca. CE 950–1100) and who wrote the *Mahāsiddhānta*.

Āryabhaṭa I was born in CE 476. This conclusion is reached from his own statement in the *Āryabhaṭīya*: "When sixty times sixty years and three quarters of the *yuga* (now *Mahā*) had elapsed, twenty three years had then passed since my birth" (III, 10).

Since the present *Kaliyuga* (the last quarter of the *Mahāyuga*) started in 3102 BCE, Āryabhaṭa was 23 years old in 3600 minus 3101, that is, in CE 499. The exact date of birth comes out to be March 21st, when the Mean Sun entered the zodiac sign of Aries in CE 476. The significance of mentioning CE 499 is that the precession of equinoxes was zero at the time, so that the given planetary mean positions did not require any correction. According to some commentators, CE 499 was also the year of composition of the *Āryabhaṭīya*.

We have no knowledge about his parents or teachers, or even about his native place. Āryabhaṭa composed the *Āryabhaṭīya* while living at Kusumapura, which has been identified as Pāṭaliputra (modern Patna in Bihar State), the imperial capital of the Gupta empire. It is possible that Āryabhaṭa headed an astronomical school there.

The association of Patna, where Āryabhaṭa taught and wrote on mathematics and astronomy, with his professional career, does not settle the question of his birthplace, but it may have been a place where he was educated.

Āryabhaṭa's fame rests mainly on his *Āryabhaṭīya*, but from the writings of Varāhamihira (sixth century CE), Bhāskara I, and Brahmagupta (seventh century), it is clear that earlier he composed an *Āryabhaṭa Siddhānta*. Although voluminous, the *Āryabhaṭa Siddhānta* is not extant. It is also called *Ārdharātrika Tantra*, because in it the civil days were reckoned from one midnight to the next. Its basic parameters are preserved by Bhāskara I in his *Mahābhāskarīya* (Chapter VII). Rāmakṛṣṇa Ārādhya (CE 1472) has quoted 34 verses on astronomical instruments from the *Āryabhaṭa Siddhānta*, of which some were devised by Āryabhaṭa himself.

The *Āryabhaṭīya* is an improved work and the product of a mature intellect. Considering the genius of Āryabhaṭa, it is easy to agree with the view that he composed it at the age of 23. The date is also in fair agreement with the recent research and analysis by Roger Billard. Unlike the *Āryabhaṭa Siddhānta*, the civil days in the *Āryabhaṭīya* are reckoned from one sunrise to the next – a practice which is still prevalent among the followers of the Hindu calendar. The *Āryabhaṭīya* consists of four sections or *pādās* (fourth parts):

1. *Daśagītikā* (10 + 3 couplets in Gīti meter);
2. *Gaṇitapāda* (33 verses on mathematics);
3. *Kāla-kriyāpāda* (25 verses on time-reckoning); and
4. *Golapāda* (50 verses on spherical astronomy).

That the *Āryabhaṭīya* was quite popular is shown by the large number of commentaries written on it, from Prabhākara (ca. CE 525) through Nīlakaṇṭha Somayāji (ca. 1502) to Kodaṇḍarāma (ca. 1850).

An Arabic translation of the *Āryabhaṭīya* entitled *Zīj al-Ārjabhar* was made in about 800, possibly by al-Ahwāzī. In spite of the *Āryabhaṭīya*'s popularity, H.T. Colebrooke failed to trace any work of Āryabhaṭa anywhere in India.

The use of modern scientific methodology, as described by Roger Billard in his *L'astronomie indienne*, along with new ephemerides, clearly shows that both of Āryabhaṭa's major works were based on accurate planetary observations. In fact, the use of better planetary parameters, the innovations in astronomical methods, and the concise style of exposition rendered the *Āryabhaṭīya* an excellent textbook in astronomy. In opposition to the earlier geostationary theory, Āryabhaṭa held the view that the earth rotates on its axis. His estimate of the period of the sidereal rotation of earth was 23 hours, 56 minutes, and 4.1 seconds, which is quite close to the actual value.

Āryabhaṭa has also been considered the father of Indian epicyclic astronomy. The resulting new planetary theory enabled Indians to determine more accurately the true positions and distances of the planets (including the Sun and the Moon). He was the first Indian to provide a method of finding celestial latitudes. He also propounded the true scientific cause of eclipses (instead of crediting the mythological demon Rāhu). In fact his new ideas gave rise to the formation of a new school of Indian astronomy: the Āryabhaṭa School or *Āryapakṣa*, for which the basic text was the *Āryabhaṭīya*.

Exposition and computation based on the new astronomical theories were made easy by Āryabhaṭa, because of the development of some mathematical tools. One of them was his own peculiar system of alphabetic numerals. The 33 consonants of the Sanskrit alphabet (Nāgarī script) denoted various numbers in conjunction with vowels which themselves stood for no numerical value. For example, the expression *khyughṛ* (=*khu* + *yu* + *ghṛ*) denoted

$$2 \times 100^2 + 30 \times 100^2 + 4 \times 10^3 = 4{,}300{,}000,$$

which is the number of revolutions of the Sun in a Yuga.
The development of Indian trigonometry (based on sine instead of chord, as the Greeks had done) was another of Āryabhaṭa's achievements which was necessary for astronomical calculations.

Because of his own concise notation, he could express the full sine table in just one couplet, which students could easily remember. For preparing the table of sines, he gave two methods, one of which was based on the property that the second order sine differences were proportional to sines themselves.

Āryabhaṭa seems to have been the first to give a general method for solving indeterminate equations of the first degree. He dealt with the subject in connection with the problem of finding an integral number N which will give a remainder r when divided by an integer a, and s when divided by b. This amounts to solving the equations

$$N = ax + r = by + s.$$

Although at present the topic of indeterminate analysis comes under pure mathematics, in ancient times it arose and was used for practical and astronomical problems. In fact, Āryabhaṭa successfully used his theory of indeterminate analysis to determine a mean conjunction of all planets at the zero mean longitude at the start of the *Kaliyuga* (3102 BCE). Recently it has been shown that his algorithm solves more general problems than the Chinese remainder theorem, and works irrespective of the sign of numbers.

The solution of a general quadratic equation and the summation of certain series were some other algebraic topics dealt with by Āryabhaṭa. The methods of adding an arithmetical progression were known in all ancient cultures, but he was perhaps the first to supply a general rule for finding the number of terms (n) when the first term (a), the common difference (d) and the sum (s) were given. His solution is a root of the quadratic equation

$$dn^2 + (2a - d)n = 2s,$$

which comes from the usual formula for the sum of an arithmetical progression.

In geometry, his greatest achievement was an accurate value of π. His rule amounts to the statement

$$\pi = 62832/20000 \text{ nearly.}$$

This implies the approximation 3.1416 which is correct to its last decimal place. How he arrived at this is not known.

From what we know about Āryabhaṭa, it is clear that he was an outstanding astronomer and mathematician. His scientific attitude, rational approach, and mathematical methodology ushered in a new era in the history of the exact sciences in India. It was quite befitting that the first Indian satellite launched on 19th April, 1975 was named Āryabhaṭa.

R. C. Gupta

REFERENCES

Ayyangar, A.A. Krishnaswami. "The Mathematics of Āryabhaṭa." *Quarterly Journal of the Mythic Society* 16: 158–179. 1925–26.

Billard, Roger, "Āryabhaṭa and Indian Astronomy: An Outline of an Unexpected Insight." *Indian Journal of History of Science* 12(2): 207–224. 1977.

Elfering, Kurt. *Die Mathematik des Āryabhaṭa I*. Munich: Wilhelm Fink Verlag. 1975.

Jha, Parmeshwar. *Āryabhaṭa I and his Contribution to Mathematics*. Patna: Bihar Research Society. 1988.

Kak, Subhash, "Computation Aspects of the Āryabhaṭa Algorithm." *Indian Journal of History of Science* 21(1): 62–71. 1986.

Sen, S.N. "Āryabhaṭa's Mathematics." *Bulletin of the National Institute of Sciences of India* 21: 27–319. 1963.

Sengupta, P.C. "Āryabhaṭa, the Father of Indian Epicyclic Astronomy." *Journal of the Department of Letters, Calcutta University* 18: 1–56. 1929.

Shukla, K.S., and K.V. Sarma, eds. and trans. *Āryabhaṭiya of Āryabhaṭa*. New Delhi: Indian National Science Academy. 1976.

Shukla, K.S. *Āryabhaṭa: Indian Mathematician and Astronomer*. New Delhi: Indian National Science Academy. 1976.

Shukla, K.S., ed. *Āryabhaṭīya, with the Commentary of Bhāskara I and Someśvara*. New Delhi: Indian National Science Academy. 1976.

See also: Mathematics – Astronomy – Trigonometry

Astrology

In India, astrology, *Jyotiṣa*, is defined as *Jyotiṣām sūryādi grahāṇām bodhakam śāstram*, the system which explains the influences of the Sun, Moon, and planets.

Indian astrology came explicitly to light around 1200 BCE, when the monk Lagadha compiled the *Vedānga-Jyotiṣa* on the basis of *Vedas*, in which lunar and solar months are described, with their adjustment by *Adhimāsa* (lunar leap month). *Ṛtus* (seasons), years, and *yugas* (eras) are also described. Twenty-seven constellations, eclipses, seven planets, and twelve signs of the zodiac were also known at that time.

In the period from 500 BCE to the beginning of the Christian era some texts were written on the subject of astrology. Nineteen famous sages composed their *Siddhāntās* (texts). *Candra-prajnapti, Surya-prajnapti* and *Jyotiṣakaraṇḍaka* were written. The *Sūryasiddhānta*, the ancient text of Indian astrology, was composed around 200 BCE.

In the first five centuries of the Christian era, there were some important contributions by Jain writers. *Angavijjā* is a large collection about *Śakuna* (omens). *Kālaka* and *Ṛsiputra* also contributed around this time. At the end of the fifth century, Āryabhaṭa I mentioned in his text *Āryabhaṭīya* that the Sun and stars are constant and that day and night are based on the movement of the Earth.

The period CE 500–1000 was very productive. Lallācārya, the disciple of Āryabhaṭa, composed two texts -- *Śiṣyadhīvṛddhida* and *Ratnakoṣa* — dealing with mathematical theories. The astrologer Varāhamihira composed several texts, and his son Pṛthuyaśā composed a brief horary called *Ṣat-Pañcāśikā*. Bhāskarācārya I wrote a commentary on the *Āryabhaṭīya* in the seventh century, and Brahmagupta composed the *Brāhmasphuṭasiddhānta* and the *Khaṇḍakhādyaka* around CE 635. Other scholars wrote commentaries on the texts of their predecessors and independent texts of their own.

In 1000–1500, there was a great deal of enhancement to the literature concerning the construction of astronomical instruments for observation. In the twelfth century, Bhāskara composed the famous text *Siddhāntaśiromaṇi*. The *Līlāvatī* of Rājāditya is another of the texts of that century. In the fifteenth century, Keśava wrote more than ten books, and his son Gaṇeśa composed the *Grahalāghava* at the age of thirteen.

Many more texts and commentaries were written from the sixteenth century onwards. A few noteworthy ones are: *Tājikanīlakaṇṭhī* of Nīlakaṇṭha (sixteenth century), *Meghamahodaya* by Meghavijayagaṇi (seventeenth century), *Janmapatrīpaddhati* by Lābhacandragaṇi (eighteenth century), and the nineteenth century works of astrologer Bāpūdeva Śāstri.

A knowledge of *pañcānga* is a prerequisite to understanding the subject of astrology. This is the fivefold system of *tithi* (lunar day), *vāra* (weekday), *nakṣatra* (asterism), *yoga* (sum of the solar and lunar longitudes) and *karaṇa* (half lunar day). *Tithi*, the lunar date, is the duration of time in which the Moon moves 12°. The fifteen *tithis* of the white fortnight (from new moon to full moon) are:

1. *Pratipadā*; 2. *Dvitīyā*; 3. *Tṛtīyā*; 4. *Caturthī*; 5. *Pañcamī*; 6. *Ṣaṣṭhī*; 7. *Saptamī*; 8. *Aṣṭamī*; 9. *Navamī*; 10. *Daśamī*; 11. *Ekādaśī*; 12. *Dvādaśī*; 13. *Trayodaśī*; 14. *Caturdaśī*; 15. *Purṇimā* ($15 \times 12^\circ = 180^\circ$).

In the black fortnight (from full moon to new moon), the fifteenth day is called *Amāvasyā* and the remainder are the same as above. *Tithis* are classified into five groups: *Nandā* (*tithis* 1,6,11), *Bhadrā* (2,7,12), *Jayā* (3,8,13), *Riktā* (4,9,14), and *Pūrṇā* (5,10,15).

The seven *vāras* (weekdays) are based on the names of the *grahas*: Sun, Moon, Mars, Mercury, Jupiter, Venus, and Saturn. The position of the *grahas* at Varanasi on 21 March 1994 is shown in Chart 1.

Chart 1 *Positions of Grahas (planets) on 21 March 1994 at 6:02 am at Varanasi*

Grahas (Planets)	Sun	Moon	Mars	Mercury	Jupiter	Venus	Saturn	*Rāhu*	*Ketu*
Rāśi (sign)	11	2	10	10	6	11	10	7	1
Anśa (degree)	6	15	16	9	23	22	8	3	3
Kalā	21	7	57	51	19	13	21	3	3
Vikalā	27	45	18	17	40	8	21	44	44

There are twenty-seven *nakṣatras* (asterisms) bifurcating the ecliptic into twenty-seven parts, each of 13.33°. These are mentioned in Table 1.

Table 1 *Twenty-seven nakṣatras (asterisms)*

Kṛttikā	Rohiṇī
Mṛgaśiras	Ārdrā
Punarvasu	Puṣya
Āśleṣā	Maghā
Pūrvāphālgunī	Uttarāphālgunī
Hasta	Citrā
Svātī	Viśākhā
Anurādhā	Jyeṣṭhā
Mūla	Pūrvāṣāḍhā
Uttarāṣāḍhā	Śroṇā
Śraviṣṭhā	Śatabhiṣaj
Pūrva-Bhādrapada	Uttara-Bhādrapada
Revatī	Aśvinī
Bharaṇī	

The ecliptic is again bifurcated into twelve parts through *Rāśis* (signs, each of 30°). The twelve signs are equal to twenty-seven *nakṣatras*, or 1 sign = 2.25 constellations. For example, *Aśvīnī, Bharaṇī*, and one quarter of *Kṛttikā* make the sign *Meṣa* (Aries). The remaining three quarters of *Kṛttikā, Rohiṇī*, and half of *Mṛgaśira* make the sign *Vṛṣa* (Taurus). The same pattern holds true for the other signs: *Mithuna* (Gemini), *Karka* (Cancer), *Simha* (Leo), *Kanyā* (Virgo), *Tulā* (Libra), *Vṛścika* (Scorpio), *Dhanu* (Sagittarius), *Makara* (Capricorn), *Kumbha* (Aquarius), and *Mīna* (Pisces). Thus twenty-seven constellations represent twelve signs.

Yoga is the sum of the solar and lunar longitudes. If the sum of their degrees is between 0 and 13.33°, that is called *Viṣkambha Yoga* – from there until 26.66° it is *Prīti* – up to 40° it is *Āyuṣmāna*. The remaining yogas are *Saubhāgya, Śobhana, Atigaṇḍa, Sukarmā, Dhṛti, Śūla, Gaṇḍa, Vṛdhi, Dhruva, Vyāghāta, Harṣaṇa, Vajra, Siddhi, Vyatīpāta, Varīyāna, Parigha, Śiva,*

Siddha, Sādhya, Śubha, Śukla, Brahma, Aindra, and *Vaidhṛti* (13.33° × 27 = 360°).

Karaṇa (constant or moveable) is the half part of the *tithi.* Constant *Karaṇa Śakuna* belongs to the second half of *caturdaśī Catuṣpada* and *Nāga* to that of *Āmāvasyā* in the black fortnight, while *Kistughna* exists in the first half of the *Pratipada* of the white fortnight in every lunar month. The remaining 14.5 *tithis* of the white and 13.5 *tithis* of the black fortnight contain eight rounds of seven moveable *Karaṇas*: *Bava, Bālava, Kaulava, Taitila, Gara, Vaṇija,* and *Viṣṭi.*

The subject matter of astrology may be divided into five groups: *Saṃhitā, Siddhānta, Jātaka, Praśna,* and *Śakuna*. In ancient India, *Saṃhitā* was the miscellaneous collection of astrological materials out of which the remaining four grew.

Siddhānta or *Gaṇita* refers to mathematical calculations about time, distance, and position of the planets. On the basis of the proper positions of twelve signs and nine planets, a chart containing twelve chambers may be sketched. In northern, southern, and eastern India, astrologers sketch charts such as Charts 2, 3, and 4 which are called *Janmāṅga* or ascendant.

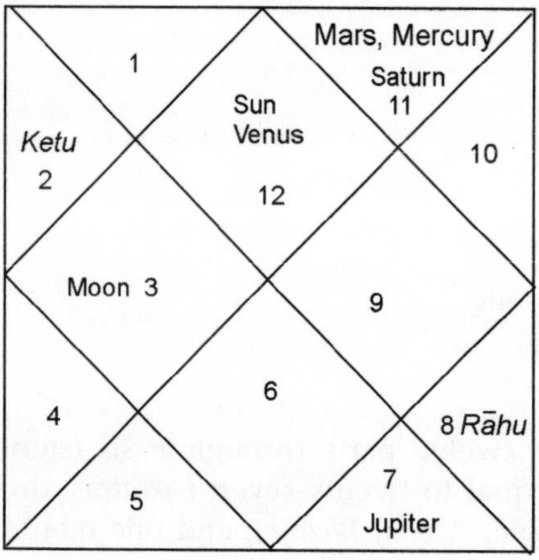

Chart 2 *Ascendant as sketched in in northern India*

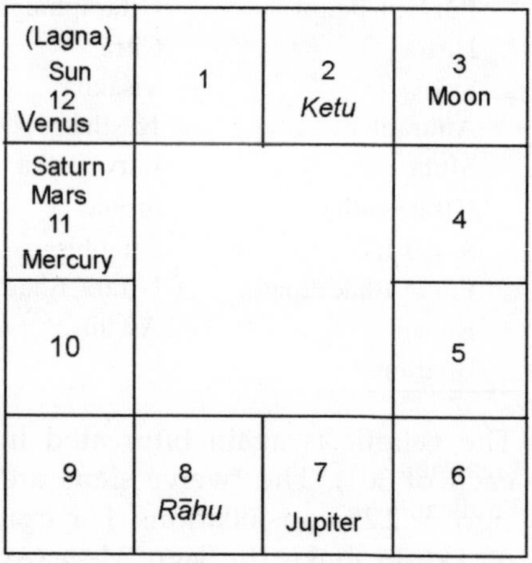

Chart 3 *Ascendant as sketched southern India*

Jātaka (native) is the person about whom a prediction is made on the basis of a birth chart. Twelve houses represent the health, wealth, brother/sister, mother, offspring, diseases/enemies, wife/husband, death, fate, father, income, and expenses, as in Chart 2. *Daśās* (periods) are defined in numerous ways. The period of any planet becomes favourable or harmful according to its position and power in the horoscope.

There are many other astrological methods in India. As an example, in the Kerala-system, numbers are assigned to alphabets, and the astrologer advises the person to say the names of a flower, river, or god on which the calculation depends.

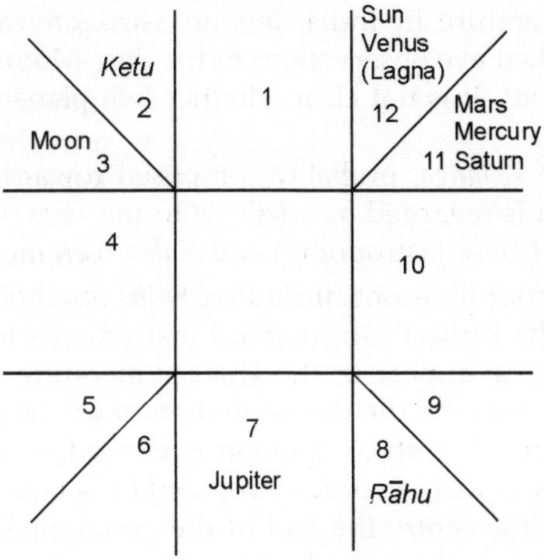

Chart 4 *Ascendant as sketched in eastern India (West Bengal and Orissa)*

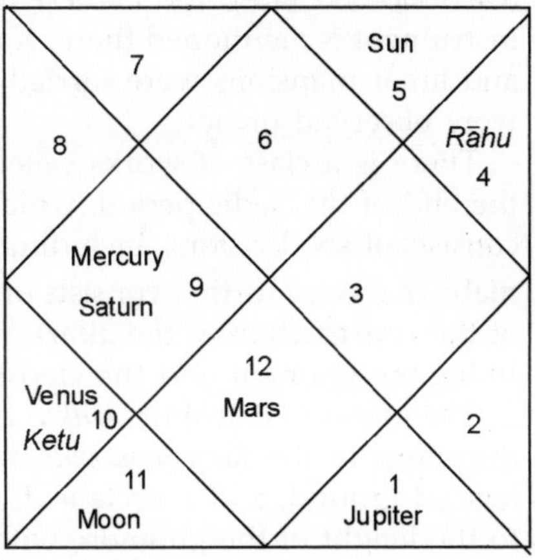

Chart 5 *Navamānśa chart*

Astrology is applied to many aspects of Indian life. There are rules concerning times for travelling, planting, and building. Favourable times for the preparation of medicines and treatment are also prescribed.

Vijaya Narayan Tripathi

References

Jagannātha, R. *Principles and Practice of Medical Astrology*. New Delhi: Sagar Publications. 1994.

Lokamani, D. *Bhāratīyajyotiṣaśāstrasyetihāsaḥ* (History of Indian Astrology). Varanasi: Chaukhambha Surabharti. 1990.

Raghunandanaprasāda, G. *Ādhunika-Jyotiṣa* (Modern Astrology). Delhi: Ankur Publications. 1994.

Raman, B. V. *A Manual of Hindu Astrology*. Delhi and London: UBS Publishers. 1992.

Sureṣacandra, M. *Jyotiśa-Sarvasva* (Complete Hindu Astrology). New Delhi: Ranjan Publications. 1994.

Astronomical Instruments

Astronomical knowledge in India can be traced back to the Vedic literature (ca. 1500–500 BCE), the earliest literature in India, but no astronomical instrument is mentioned there. Naked eye observations of the Sun, Moon, and lunar mansions were carried out. It is not clear whether five planets were observed or not.

There is a class of works called *Vedāṅga*, probably composed towards the end of the Vedic period, which is regarded as auxiliary to the *Veda*. It consists of six divisions, including *Jyotiṣa* (astronomy) and *Kalpa* (ceremonial). The *Kalpa* further consists of four divisions, including *Śulba* (method of the construction of the altar). The earliest astronomical instruments in India, the gnomon and the clepsydra, appear in the *Vedāṅga* literature.

The gnomon (Sanskrit: *śaṅku*) is used for the determination of cardinal directions in the *Kātyāyana-śulbasūtra*. A vertical gnomon is erected on a leveled ground, and a circle is drawn with a cord, whose length is equal to the height of the gnomon, with the centre the foot of the gnomon. At the two points where the tip of the gnomon-shadow touches the circle, pins are placed, and they are joined by a straight line. This line is the east–west line.

The annual and diurnal variations of the length of the gnomon-shadow are recorded in the political work *Arthaśāstra* of Kauṭilya, the Buddhist work *Śārdūlakarṇa-avadāna*, and Jaina works such as the *Sūrya-prajñapti*. These records seem to be based on observations in North India.

The clepsydra is mentioned in the *Vedāṅga-jyotiṣa*, the *Arthaśāstra*, and the *Śārdūlakarṇa-avadāna*. It was like a water jar with a hole at its bottom from which water flowed out in a *nāḍikā* (one-sixtieth of a day).

Towards the end of the *Vedāṅga* astronomy period, certain Greek ideas of astronomy and astrology had some influence in India from the second to the fourth century CE. After that, Hindu astronomy (*Jyotiṣa*) established itself as an independent discipline, and several fundamental texts called *Siddhāntas* were composed. I call this period, from about the fifth to the twelfth centuries CE, the classical Siddhānta period. The main astronomers who described astronomical instruments are Āryabhaṭa (b. CE 476), Varāhamihira (sixth century CE), Brahmagupta (b. CE 598), Lalla (eighth or ninth century CE), Śrīpati (eleventh century CE), Bhāskara II (b. CE 1114), and the anonymous author of the *Sūryasiddhānta*. The *siddhāntas* composed by Brahmagupta, Lalla, Śrīpati, and Bhāskara II contain special chapters on astronomical instruments entitled *Yantra-adhyāya*. The Sanskrit word *yantra* means instrument. No observational data are recorded in the *siddhāntas*, and the extent of actual observations in this period is controversial. Roger Billard maintained that astronomical constants in the *siddhāntas* were determined by actual observations, while David Pingree argued that they were exclusively borrowed from Greek astronomy. In this connection, we should note that the method of determination of astronomical constants by means of observations was correctly explained by Bhāskara II. Let us see the instruments in this period.

The gnomon (*śaṅku*) was continually used in this period. The theory of the gnomon, such as the relationship between the length of gnomon-shadow, the latitude of the observer, and time, was developed in this period, and a special chapter called *Tripraśna-adhyāya* in the *Siddhāntas* was devoted to this subject. Trigonometry, invented in India, was fully utilised for this purpose.

The staff (*yaṣṭi-yantra*) is a simple stick, used to sight an object. There are some variations of the staff, such as V-shaped staffs for determining angular distance with the help of a graduated level circle.

The circle-instrument (*cakra-yantra*) is a graduated circular hoop or board suspended vertically. The sun's altitude or zenith distance is determined, and time is roughly calculated from it. Variations of the circle-instrument are the semi-circle instrument (*dhanur-yantra*) and the quadrant (*turya-golaka*).

A circular board kept horizontally with a central rod is the chair-instrument (*pīṭha-yantra*), and a similar semi-circular board is the bowl-instrument (*kapāla-yantra*). The sun's azimuth is determined by them, and time is roughly calculated from them.

A circular board kept in the equatorial plane is the equator-instrument (*nāḍīvalaya-yantra*). It is a kind of equatorial sundial. The combination of two semi-circular boards, one of which is in the equatorial plane, is the scissors-instrument (*kartarī-yantra*). Its simplified version is the semi-circular board in an equatorial plane with a central rod.

The armillary sphere (*gola-yantra*) was unlike the Greek armillary sphere, which was based on ecliptical coordinates, although it also had an ecliptical hoop. Probably, the celestial coordinates of the junction stars of the lunar mansions were determined by the armillary sphere since the seventh century or so. There was also the celestial globe rotated by flowing water.

The clepsydra (*ghaṭī-yantra*) was widely used until recent times. Unlike the clepsydra of the *Vedāṅga* period, the clepsydra of this period is a bowl with a hole at its bottom floating on water. Water flows into the bowl, and it sinks after a certain time interval. The actual use of the clepsydra of this type was recorded by the Chinese Buddhist traveller Yijing (CE 635–713). The clepsydra actually used can be seen in a museum at Kota (Rajasthan) (see Plate 1). Several astronomers also described water-driven instruments such as the model of fighting sheep.

Plate 1 *Clepsydra at Rao Madho Singh Museum, Kota, Rajasthan, India. Photograph by the author. Used with his permission.*

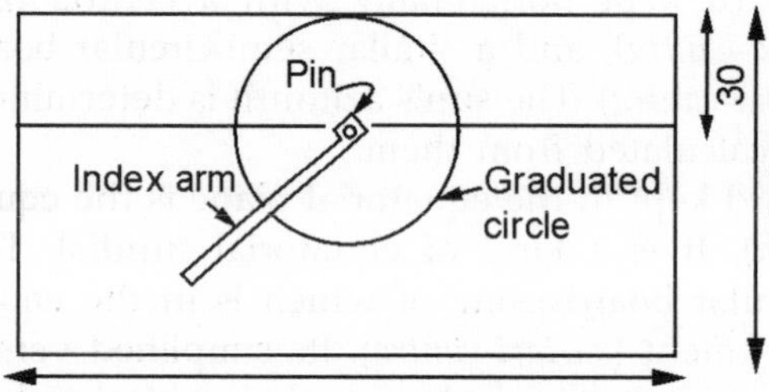

Figure 1 *Phalaka-yantra.*

The *phalaka-yantra* (board-instrument) invented by Bhāskara II is a rectangular board with a pin and an index arm, used to determine time graphically from the Sun's altitude (see Figure 1). This is an ingenious instrument based on the Hindu theory of gnomon.

The astrolabe was introduced into India from the Islamic world at the time of Fīrūz Shāh (r. CE 1351–1388) of the *Tughluq* dynasty. Fīrūz Shāh's court astronomer Mahendra Sūri composed a Sanskrit work on the astrolabe entitled *Yantra-rāja* (King of Instruments, the Sanskrit term for the astrolabe) in CE 1370. This is the earliest Sanskrit work on Islamic astronomy. Use of the astrolabe rapidly spread among some Hindu astronomers, and Padmanābha (CE 1423) and Rāmacandra (CE 1428) described the astrolabe in their works.

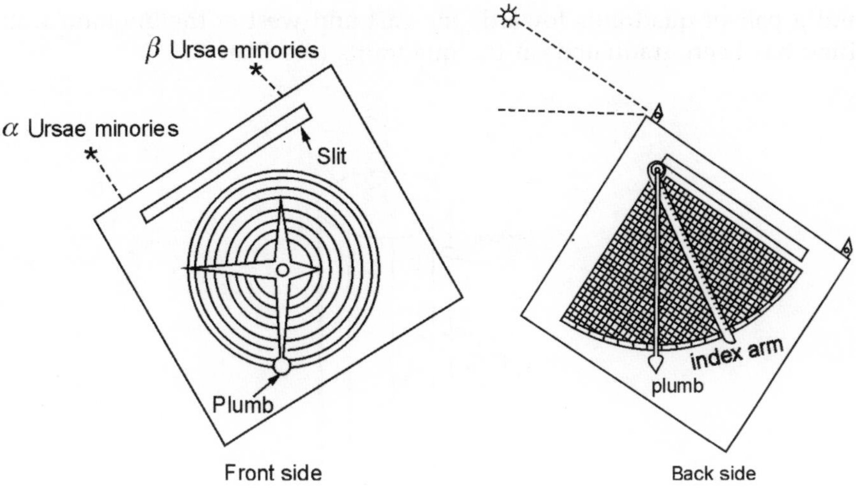

Figure 2 *Dhruva-bhrama-yantra*

Some new instruments were made in the Delhi Sultanate and Mughal periods. Padmanābha invented a kind of nocturnal instrument called *dhruva-bhrama-yantra* (polar-rotation-instrument) (see Figure 2). It was a rectangular board with a slit and a set of pointers with concentric graduated circles. Adjusting the slit to the direction of α and β Ursae Minoris, time and other calculations could be obtained with the help of pointers. Its back side was made as a quadrant with a plumb and an index arm. Thirty parallel lines were drawn inside the quadrant, and trigonometrical calculations were done graphically. After determining the Sun's altitude with the help of the plumb, time was calculated graphically with the help of the index arm.

Later, the quadrant as an independent instrument was described by Cakradhara, and a more exact method to calculate time was explained.

Another new type of instrument in this period was the cylindrical sundial called *kaśā-yantra* (whip-instrument) by Hema (late fifteenth century CE) or *pratoda-yantra* (whip-instrument) by Gaṇeśa (b. CE 1507) (see Figure 3). It is a cylindrical rod having a horizontal gnomon and graduations of time according to the vertical shadow below the gnomon.

The quadrant and the cylindrical sundial exist in the Islamic world also, but the possibility of their influence on these Indian instruments is still to be investigated.

The mahārāja of Jaipur, Sawai Jai Singh (CE 1688–1743) constructed five astronomical observatories at the beginning of the eighteenth century. The observatory in Mathura is not extant, but those in Delhi, Jaipur, Ujjain, and Banaras are. There are several huge instruments based on Hindu and Islamic astronomy. For example, the *samrāṭ-yantra* (emperor-instrument) is a huge sundial which consists of a triangular gnomon wall and a pair of quadrants towards the east and west of the gnomon wall. Time has been graduated on the quadrants (see Plate 2).

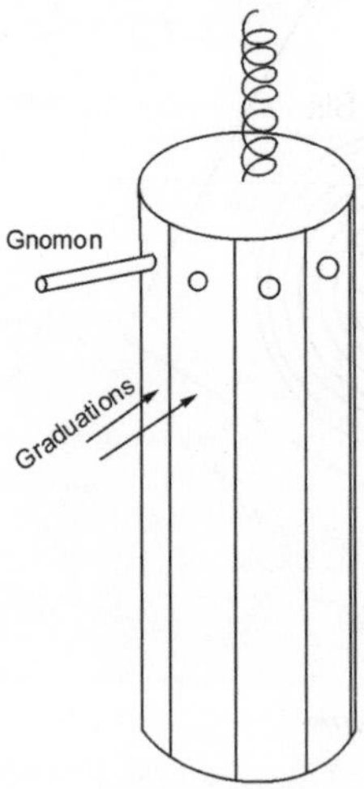

Figure 3 *Pratoda-yantra*

Plate 2 *Samrāṭ-yantra at Jantar Mantar, New Delhi, India, Photograph by the author. Used with his permission.*

By this time, European astronomy had begun to be introduced into India, and Jai Singh had certain information about European astronomy. The earliest European style astronomical observatory in India is a private one of William Petrie, an officer of the British East India Company, which was set up in 1786 at his residence in Madras.

Yukio Ohashi

References

Billard, Roger. *L'astronomie Indienne.* Paris: École française d'extréme-Orient. 1971.

Kaye, G.R. *The Astronomical Observatories of Jai Singh.* Calcutta: 1918. Reprint: New Delhi: Archaeological Survey of India. 1982.

Kochhar, R.K. "Madras Observatory: The Beginning." *Bulletin of the Astronomical Society of India* 13 (3): 162–168, 1985.

Ōhashi, Yukio. "Sanskrit Texts on Astronomical Instruments during the Delhi Sultanate and Mughal Periods." *Studies in History of Medicine and Science* 10–11: 165–181. 1986–1987.

Ōhashi, Yukio. "Development of Astronomical Observation in Vedic and Post-Vedic India." *Indian Journal of History of Science* 28(3): 185–251. 1993.

Ōhashi, Yukio. "Astronomical Instruments in Classical Siddhāntas." *Indian Journal of History of Science* 29(2): 155–313. 1994.

Pingree, David. "History of Mathematical Astronomy in India." In *Dictionary of Scientific Biography*. vol 15. New York: Scribners. 1978. pp. 533–633.

See also: Observatories – Jai Singh – Lunar Mansions – Armillary Sphere – *Śulbasūtras* – Sundials – Gnomon – Clocks and Watches – Āryabhaṭa – Varāhamihira – Brahmagupta – Lalla – Śrīpati – Bhāskara II – Trigonometry – Mahendra Sūri

Astronomy

Astronomy in India, as it was in other ancient civilisations, was interwoven with religion. While the different facets of nature, the shining of the Sun, the waxing and waning of the Moon, and the alternation of the seasons all excited curiosity and evoked wonder, religious practices conformed to astronomical timings following the seasons, equinoxes, solstices, new and full moons, specific times of the day and the like. In the Vedas, the earliest literature of the Hindus, mention of professions such as *Gaṇaka* (calculator) and *nakṣatra-darśa* (star-gazer), and the mention of a branch of knowledge called *nakṣatra-vidyā* (star science) are illustrative of the fascination that the celestial bodies exerted on the Vedic priests.

The Vedas and their vast ancillary literature are primarily works of a religious nature, and not textbooks on astronomy. Still, they inform about the astronomical knowledge, mainly empirical in nature and often mystic in expression, which Vedic Indians possessed and used in their religious life. One finds in the *Ṛgveda* intelligent speculations about the genesis of the universe from non-existence, the configuration of the universe, the spherical self-supporting Earth, and the year of 360 days divided into 12 equal parts of 30 days each with a periodical intercalary month. In the

Aitareya Brāhmaṇa, we read of the moon's monthly elongation and the cause of day and night. Seasonal and yearly sacrificial sessions helped the priests to ascertain the days of the equinoxes and solstices. The shifting of the equinoxes made the Vedic priests correspondingly shift the year backwards, in tune with the accumulated precession, though the rate thereof was not envisaged. The wish to commence sacrifices at the beginning of specific constellations necessitated the identification of the constellations as fitted on the zodiacal frame. They also noticed eclipses, and identified their causes empirically.

The computational components and work rules for times for Vedic rituals are to be found in the *Vedāṅga Jyotiṣa* (Vedic Astronomical Auxiliary), composed by Lagadha. On the basis of the astronomical configurations given in its epoch, the date of this text is ascertained to be in the twelfth century BCE. This work sets out such basic data as time measures, astronomical constants, tables, methodologies, and other matters related to the Vedic ritualistic calendar. It prescribes a five-year luni-solar cycle (*Yuga*) from an epoch when the Sun and the Moon were in conjunction on the zodiac at the beginning of the bright fortnight, at the commencement of the asterism Delphini (*Śraviṣṭhā*) of the (synodic month of) *Maghā,* and of the (solar month) *Tapas,* when the northward course of the sun began. The constants are contrived to be given in whole numbers for easy memorisation. Accordingly, the *Vedāṅga Jyotiṣa* chose a unit of 1830 civil days as its unit, which it divided into five years of 366 days each, the error of the additional 3/4 day in the year being rectified periodically. 1830 days is equal to 62 synodic or lunar months of 29.5 days each, and 60 solar months. The two extra intercalary lunar months are dropped, one at the middle and the other at the end of the cycle, so that the two, the solar and lunar years, commenced together again, at the beginning of each cycle.

During the age that followed, a series of astronomical texts was written, mainly by eighteen astronomers. Passages from some of these texts were quoted in later texts, and five of them were redacted by Varāhamihira (d. 587) in his *Pañcasiddhāntikā.* The texts of this age which are still available are the *Gargasaṃhitā,* and *Parāśarasaṃhitā,* and the Jain texts *Sūryaprajñapti, Candraprajñapti* and *Jambūdvīpa-prajñapti.* The astronomy contained in those is practically the same as that expounded in the *Vedāṅga Jyotiṣa* though with minor differences, including the shifting of the commencement of the year to earlier asterisms due to the precession of the equinoxes.

From about the beginning of the Christian era a number of texts were composed with the generic name *Siddhānta* (established tenet). The scope of these texts was wider, their outlook far-reaching, and their methodology rationalistic. Also, the science began to be studied for its own sake. The stellar zodiac was replaced with the twelve-sign zodiac, and, besides the Sun and the Moon, the planets also began to be reckoned. Their rising and setting, motion in the zodiacal segments, direct and retrograde motion, times of first and last visibilities, the duration of their appearance and disappearance, and mutual occultation began to be computed. These and their synodic motion called *grahacāra* were elaborately recorded. Analyses of these recordings enabled the depiction of empirical formulae for computing their longitudes. While the use of the rule of three (*trai-rāśika*, direct proportion), continued fractions, and indeterminate equations helped computation, the use of trigonometry, both plane and spherical, and geometrical models enabled a realistic understanding of planetary motion, and developed rules, formulae, and tables. This also resulted in fairly accurate prediction of the eclipses and occultation of the celestial bodies. It is to be noted that in Indian astronomical parlance, the word *graha* signifies not only the planets Mars (*Kuja*), Mercury (*Budha*), Jupiter (*Guru*), Venus (*Śukra*) and Saturn (*Śani*), but also the Sun (*Ravi, Sūrya*), Moon (*Candra*), the ascending node (*Rāhu*), and the descending node (*Ketu*).

Yuga in Indian astronomy denotes a relatively large number of years during which a celestial body, starting from a specific point on its orbit, made a certain number of revolutions and returned to the same point at the end of the period. In the *siddhānta* texts the starting point is taken as the first point of Aries (*Meṣa-ādi*) (vernal equinox). Through extended observation over long periods, methods were formulated for accurately ascertaining the motion of the planets, as in the *Ārya-bhaṭīya*. Having obtained in this manner the sidereal periods of all the planets, their lowest common multiple was calculated backwards to provide accommodation to the revolutions of the relevant celestial bodies. Thus the *Vedāṅga Jyotiṣa* used a five-year *yuga* to accommodate only the sun and the moon. Later, the planets were included and the *Romakasiddhānta*, redacted by Varāhamihira, used a *yuga* of 2850 luni-solar years, and the *Saurasiddhānta* used a *yuga* of 108,000 solar years. Āryabhaṭa (b. CE 476) formulated a *mahā-yuga* (grand cycle) of 4,320,000 years which was set to commence at the first point of Aries on Wednesday, at sunrise, at Laṇkā, which is a point on the terrestrial horizon of zero longitude, being the Greenwich meridian of Indian astronomy. He also devised a shorter *yuga*, called *Kali-yuga*, of 432,000 years, which commenced on Friday, February 18, 3102 BCE, where, however, the apogee and node were ahead by 90° and behind by 180°, respectively. Pursuing the same principle, other astronomers have devised *yugas* of different lengths, and still others have suggested zero corrections to the mean position of the planets at the beginning of the *yuga*.

Computing the longitude of a planet when the length of the *yuga* in terms of civil days, the *yuga*-revolutions, and the time at which its longitude is required are given, reduces itself to the application of the rule of three, if the planetary orbits are perfect circles and the planetary motions uniform. Indian astronomers conceive their motion along elliptical orbits according to the epicyclic model or their own eccentric model. In the former, the planets are envisaged as moving along epicycles which move on the circumference of a circle, and in the latter, the planets are supposed to move along a circle whose centre is not at the centre of the Earth, but on the circumference of a circle whose centre is the centre of the earth. Both models give the same result. Several sets of sine tables are also derived and several computational steps called *manda* and *śīghra* are enunciated to give the heliocentric positions in place of the geocentric.

When the true longitude of a planet at a particular point of time is to be found, the *ahargaṇa* (count of days) from the epoch up to the sunrise of the day in question is ascertained. To this is added the time elapsed from sunrise on the relevant day to the required point of time. Since the number of days in the *yuga* and the number of the relevant planetary revolutions (which are constants) are known, the revolutions up to the moment in question are calculated by the rule of three, and the completed revolutions discarded. The remainder would be the position of the mean planet at sunrise of the day in question at zero longitude (i.e. Laṅkā or Ujjain meridian). To get the correct longitude, four corrections are applied to the result obtained above. They are (i) *Deśāntara*, the difference in sunrise due to the difference in terrestrial longitude, (ii) *Cara* (ascensional difference), due to the length of the day at the place, (iii) *Bhujāntara*, the equation of time caused by the eccentricity of the earth's orbit, and (iv) *Udayāntara*, the equation of time caused by the obliquity of the ecliptic (the band of the zodiac through which the sun apparently moves in its yearly course) with the celestial equator.

As a striking natural phenomenon, the eclipse had been taken note of and recorded in several early Indian texts. A solar eclipse was recorded in the *Ṛgveda*, where it describes how, when Svarbhānu (the dark planet *Rāhu* of later legends) hid the sun, sage Atri restored it, first as a black form, then as a silvery one, then as a reddish one, and finally in its original bright form. On account of the popular and religious significance associated with eclipses, their prediction assumed great importance in Indian astronomy. In the *Pañcasiddhāntikā*, details of the computation of the lunar eclipse were given in the *Vāsiṣṭha-Pauliśa* and *Saura Siddhāntas* with geometrical diagrammatic representation, and the solar eclipse in the *Pauliśa, Romaka* and *Saura Siddhāntas*. The treatment in later *siddhāntas*, like those of Āryabhaṭa, Brahmagupta, Śrīpati and Bhāskara II and also the *Sūryasiddhānta*, is accurate. A number of shorter texts were also written solely on the computation of eclipses, aiming at greater perfection.

The Vedas call the intrinsically dark moon *sūrya-raśmi* (sun's light), and thought that it was born anew every day in different configurations. Each of the fifteen lunar days of the bright and the dark fortnights have high religious significance, and different Hindu rites and rituals are fixed on their basis.

From the early *siddhānta* age, computation of the moon's phases is referred to by the term *Candra-śṛṅga-unnati* (elevation of the horns of the Moon) and is computed elaborately. The first work to give this computation is the *Pauliśa Siddhānta*, which devotes an entire chapter to this subject. Most of the later *siddhāntas* also devote one chapter to the subject, adding corrections and evolving newer methods and also giving graphical representations.

Still another topic which finds computational treatment in *siddhānta* texts is the conjunction of planets and stars. Two bodies are said to be in conjunction when their longitudes at any moment are equal. The conjunction of a planet with the Sun is called *astamaya* (setting), with the Moon is *samāgama* (meeting), and with another planet, it is *yuddha* (encounter). The conjunction of a planet with a star is similar, but with the difference that a star is considered a ray of light and has no *bimba* (orb) or motion. The computational methods followed are similar to those in the case of eclipses, with minor modifications. Full chapters are devoted in *siddhānta* texts to computing this phenomenon.

Although naked eye observations and the star-gazer (*nakṣatradarśa*) find frequent mention in Vedic literature, mechanical instruments are of a later origin. The earliest instruments used were the gnomon (*śaṅku*) for finding the cardinal directions, used in the *śulbasūtras*, and the clepsydra (*nāḍī-yantra*) for measuring time. The *Pañcasiddhāntikā*, devotes one long chapter to "Graphical Methods and Astronomical Instruments". While Āryabhaṭa gives only the underlying principles in his *Āryabhaṭīya*, he has a long section of 31 verses on instruments in his *Āryabhaṭasiddhānta*. Almost all later *siddhāntas* have a full chapter on instruments, which include the armillary sphere, rotating wheels, and shadow, water, circle, semi-circle, scissor, needle, cart, tube, umbrella, and plank instruments. Some of these are used for observation; others are for demonstration. After the advent of Muslim astronomy in India, the astrolabe became common and even Hinduised. A number of texts in Sanskrit were also written on the astrolabe, among which the *Yantrarāja* of Mahendra Sūri and *Yantracintāma* of Viśrāma are important. The pinnacle of this activity came with Sawai Jai Singh, ruler of Jaipur, who patronised a group of astronomers, built five huge observatories, in Delhi, Jaipur, Ujjain, Varanasi, and Mathura, and wrote the work *Yantrarāja-racanā* (construction of astrolabes).

Towards the early centuries of the Christian era, texts with the generic name *siddhānta* (tenet), in contrast to the earliest astronomical texts, were composed. These were mathematically based, rationalistic, and adumbrated by geometric models and diagrammatic representations of astronomical phenomena. While Varāhamihira selectively redacted in his *Pañcasiddhāntikā* five of the early *siddhāntas*, his elder contemporary, Āryabhaṭa (b. CE 499) produced a systematic *siddhānta* treatise entitled *Āryabhaṭīya* , in which he speaks also of the diurnal rotation of the Earth. The work is divided into four sections, covering the following subjects: (i) planetary parameters and the sine table (*Gītikā-pāda*), (ii) mathematics (*Gaṇitapāda*), (iii) time reckoning and planetary positions (*Kālakriyā-pāda*), and (iv) astronomical spherics (*Gola-pāda*). This *siddhānta*, which started the Āryabhaṭan school of astronomy, was followed by advanced astronomical treatises written by a great number of astronomers and was very popular in the south of India, where astronomers wrote a number of commentaries and secondary works based on it.

Brahmagupta (b. CE 598) started another school through his voluminous work *Brāhmasphuṭasiddhānta* in twenty chapters. This work shows Brahmagupta as an astute mathematician who made several new enunciations. He also wrote a work by the name *Khaṇḍakhādyaka* in which he revised some of Āryabhaṭa's views. Among the things that he formulated might be mentioned a method for calculating the instantaneous motion of a planet, correct equations for parallax, and certain nuances related to the computation of eclipses. Brahmagupta's works are also significant for their having introduced Indian mathematics-based astronomy to the Arab world through his two works which were translated into Arabic in about CE 800.

Bhāskara I (ca. CE 628), who followed in Āryabhaṭa's tradition, wrote a detailed commentary on *Āryabhatīya* and two original treatises, the *Laghubhāskarīya* and the *Mahābhāskarīya.* Vaṭeśvara (b. CE 880) wrote his erudite work *Vaṭeśvarasiddhānta* in eight chapters, in which he devised precise methods for finding the parallax in longitude (*lambana*) directly, the motion of the equinoxes and the solstices, and the quadrant of the Sun at any given time. He also wrote a work on spherics. Śrīpati, who came later (ca. CE 999), wrote an extensive work, the *Siddhāntaśekhara*, in twenty chapters, introducing several new enunciations, including the Moon's second inequality.

The *Sūryasiddhānta*, the most popular work on astronomy in North India, is attributed to divine authorship but seems to have been composed about CE 800. It adopts the midnight epoch and certain elements from the old *Saurasiddhānta*, but differs from it in other respects. It promulgates its own division of time-cycle (*yuga*) and evinces some acquaintance with *Brāhamasphuṭasiddhānta.*

The *Siddhāntaśiromaṇi* of Bhāskara II (b. CE 1114) comprises four books: *Līlāvatī* dealing with mathematics, *Bījagaṇita* with algebra, *Gaṇitādhyāya* with practical astronomy, and *Golādhyāya* with theoretical astronomy. The author's gloss on the work which he calls *Vāsanā* (fragrance) is not only explanatory but also illustrative and highly instructive. Extremely popular and widely studied in North India, the work has been commented on by generations of scholars.

The *siddhānta* texts composed later, when Muslim astronomy had been introduced into India, bear its influence in the matter of parameters and models though the general set up remains the same. The more important among these are the *Siddhāntaśindhu* (1628) and *Siddhāntarāja* (1639) of Nityānanda, *Siddhāntasārvabhauma* (1646) of Munīśvara, *Siddhāntatattvaviveka* (1658) of Kamalākara and the *Siddhāntasārakaustubha* (1732) of Samrāṭ Jagannātha.

In order to relieve the tedium in working with the very large numbers involved when the *mahā-yuga* or *yuga* is taken as the epoch, a genre of texts called *karaṇas* was evolved which adopted a convenient contemporary date as the epoch. The mean longitudes of the planets at the new epoch were computed using the *siddhāntas* and revised by observation, and the resulting longitudes were used as zero corrections at the epoch for further computations. In order to make computations still easier, planetary mean motions were calculated for blocks of years or of days, and depicted in the form of tables. Each school of astronomy had a number of *karaṇa* texts, produced at different dates and often exhibiting novel shortcuts and methodologies. While the North Indian texts had tables with numerals, South India, particularly Kerala, had its own traditions, and depicted the daily motions of the planets, sines of their equations, and other matters in the form of verses with meaningful words and sentences, employing the facile *kaṭapayādi* notation of numerals. The earliest *karaṇa* texts are the redactions by Varāhamihira in his *Pañcasiddhāntika* of the *Romaka*, *Pauliśa* and *Saura Siddhāntas*, the epoch of all the three being 21st March, 505 CE.

Hindu religious life, which served as the incentive to the development of astronomy in India in its beginnings, has continued to be so even today for the orthodox Indian. The *pañcāṅga* (five-limbed) almanac which is primarily a record of the *tithi* (lunar day), *vāra* (weekday), *nakṣatra* (asterism), *yoga* (sum of the solar and lunar longitudes) and *karaṇa* (half lunar day), can be said to direct and regulate the entire social and religious life of the orthodox Hindu. Though the primary elements of the almanac are identical throughout India, other matters like the sacred days, festivals, personal and community worship, fasts and feasts, and social celebrations, customs, and conventions differ from region to region. These matters are also recorded in the almanacs of the respective regions. In order to bring about some uniformity in the matter, the Government of India appointed a Calendar Reform Committee in 1952 which made several recommendations, but conditions have not changed much.

In astronomy, as in other scientific disciplines, the Indian ethos had been to depict the formulae and procedures, but refrain from giving out the rationale, though much rationalistic work would have gone before formulation. This position is relieved by a few commentators, such as Mallāri in his commentary on the *Grahalāghava* of Gaṇeśa Daivajña and Śaṅkara Vāriyar in his commentaries on the *Līlāvatī* of Bhāskara II and the *Tantrasaṅgraha* of Nīlakaṇṭha Somayājī. It is also interesting that there are texts wholly devoted to setting out rationales, like the *Yuktibhāṣā* of Jyeṣṭhadeva. Even more interesting are texts like *Jyotirmīmāṁsā* (Investigations on Astronomical Theories) by Nīlakaṇṭha Somayājī, which open up a very instructive chapter of Indian astronomy.

K. V. Sarma

References

Āryabhaṭīya of Āryabhaṭa. Ed. and trans. by K.S. Shukla and K.V. Sarma. New Delhi: Indian National Science Academy. 1976.

Billard, Roger. *L'Astronomie Indienne.* Paris: École Francaise d'extréme Orient. 1971.

Bose, D.M. et al. *A Concise History of Science in India.* New Delhi: Indian National Science Academy. 1971.

Dikshit, S.B. *Bhāratīya Jyotiśh śāstra* (History of Indian Astronomy). New Delhi: Director General of Meteorology, 2 pts. 1968, 1981.

Jyotirmīmāṃsā of *Nīlakaṇṭha Somayājī.* Ed. K.V. Sarma. Hoshiarpur: Visveshvaranand Institute. 1971.

Pañcasiddhāntikā of Varāhamihira. Trans and intro. by T.S.K. Sastry and K.V. Sarma. Madras: K.V. Sarma, P.P.S.T Foundation, Adyar. 1993.

Pingree, David. *Jyotiḥśāstra.* Wiesbaden: Otto Harrassowitz. 1981.

Report of the Calendar Reform Committee, Government of India. New Delhi: Council of Scientific and Industrial Research. 1956.

Sarma K.V. *A History of the Kerala School of Hindu Astronomy.* Hoshiarpur: Vishveshvaranand Institute. 1972.

Sastry, T.S.K. *Collected Papers on Jyotisha.* Tirupati: Kendriya Sanskrit Vidyapeetha. 1989.

Sen, S.N. and K.S. Shukla. *History of Astronomy in India.* New Delhi: Indian National Science Academy. 1985.

Subbarayappa, B.V., and K.V. Sarma. *Indian Astronomy: A Source-book.* Bombay: Nehru Centre, Worli. 1985.

Tantrasaṅgraha of Nīlakaṇṭha Somayājī, with Yuktidīpikā and Laghuvivṛti. Ed. K.V. Sarma. Hoshiarpur: Vishveshvaranand Institute. 1977.

See also: Lunar Mansions – Eclipses – Gnomon – Astronomical Instruments – Observatories – Armillary Sphere – Algebra: *Bījagaṇita* – Precession of the Equinoxes

Astronomy in the Indo-Malay Archipelago

> ". . . in recent years anthropology has begun to face up to the implications of the truism that people do not respond directly to their environment but rather to the environment as they conceive of it: e.g. to animals and plants as conceptualised in *their* minds and labelled by *their* language. In other words, understanding human ecology requires understanding human conceptual systems. It is not enough to understand the role of a type of organism in economics and/or ritual. One must also understand how the people involved classify and think about it." (Dentan, 1970)

All societies have their own body of knowledge through which they seek to understand the natural environment and their relationship to it. Thus we may be able to understand a society better by going beyond the categories of Western science and beginning to consider the interrelationship of a society with its environment from the viewpoint of its members. This attempt to understand how members of a society themselves conceive of their environment has come to be known as ethnoecology. Ethnoastronomy, the subject of this article, may be seen as a branch of ethnoecology wherein the interrelationship of human populations with their celestial environment is the focus of interest.

The modern nations of Indonesia and Malaysia, with a combined population of nearly two hundred million, encompass the homelands of well over two hundred distinct ethnic groups whose cultures and languages form part of a common Austronesian heritage, a heritage they share with Polynesians and Micronesians, among others. Inhabiting mountainsides, river valleys, and coastal plains and faced with a rather unpredictable tropical monsoon climate, the peoples of Indonesia and Malaysia have developed a diverse agriculture which includes both inundated rice farming and the shifting cultivation of rice and other food crops. Spread across an archipelago of over 13,000 islands, they have also developed sophisticated systems of navigation. These indigenous agricultural and navigational practices have been found to be informed by an astronomical tradition that is, at once, unique to this cultural area and richly diverse in its local variation. This article describes several of the many techniques of astronomical observation employed by the peoples of Indonesia and Malaysia to help regulate their agricultural cycles and navigate their ships.

The Celestial Landscape

The passage of time is mirrored in all of nature: in the light of day and dark of night, the flowering of plants, the mating and migratory behaviour of animals, the changes in weather, the ebb and flood of the tides, and in the recurring cycles within cycles of the Sun, Moon, planets, and stars. Of these cycles, perhaps the most obvious is the diurnal rising and setting of the Sun, Moon, planets and stars, as well as the synodic or cyclic changes in the phases of the Moon and in the time of day that it rises and sets. More subtle than these might be the annual changes in the sun and stars: the north–south shift in the path of the Sun across the sky (including its rising and setting points and its relative distance above the horizon at noon) and the appearance, disappearance, and reappearance of familiar patterns of stars at various times of night. As an integral part of the natural landscape, these recurring celestial phenomena have long provided farmers and sailors with dependable markers against which agricultural and navigational operations can be timed.

Orientation in space and the art of wayfinding have often relied upon knowledge of these same celestial phenomena. The English term "orient" is derived from the Latin for "rise" and later became associated with the East as the direction in which the Sun appears at dawn. Although the times that individual stars rise and set shifts gradually throughout the year, as viewed from a given latitude, the azimuths at which a star rises and sets vary only slightly over a lifetime, thereby providing a reliable "star compass" by which to determine direction. The Sun, Moon, and planets also rise and set generally east and west, depending upon their individual cycles, affording additional guides by which people are able to orient themselves on land and sea.

Agricultural Time-keeping

Many traditional desert and plains cultures use the shift in the rising and/or setting points of the Sun along the horizon both to mark important dates and seasons and to commemorate the passage of years. These environments are conducive to the use of these horizon-based solar calendars: a permanent location from which to make sightings, a series of permanent distant horizon markers and a clear view are all that is needed for such a calendar. England's Stonehenge and the Big Horn Medicine Wheel of Wyoming are striking examples. Such conditions are not common in Indonesia and Malaysia. Here the landscape may consist of anything from a nearby or distant mountain to, more often, some nearby trees; the horizon is therefore a rather undependable device against which to sight and measure the rising and setting positions of the sun. However, in cultivated areas of the region, even from swiddens located deep within the rainforest, one can usually find a field or homesite from which much of the sky is visible.

There are several types of observations of annually recurring celestial phenomena that can be made where permanent, distant horizon markers are not commonly available. These include cyclic changes in the phases of the Moon, the annual changes in the appearances of familiar groups of stars, and the annual changes in the altitude of the Sun at noon. Variations of these types of observations as practiced by traditional Indonesian and Malaysian farmers will now be presented. For the sake of clarity, they are grouped using Western astronomical categories.

Solar Gnomons

Most often seen on sundials in western cultures, a solar gnomon is simply a vertical pole or other similar device that is used to cast a shadow. The altitude of the Sun above the horizon varies not only through the day, but through the year as well. By measuring the relative length of this shadow each day at local solar noon, one can observe and more or less accurately measure the changing altitude of the Sun above the horizon (or, reciprocally, from the zenith) through the year and thereby determine the approximate date.

Figure 1 *Solar gnomon*

Two distinct types of solar gnomons have been reported. Both measure the altitude of the Sun at local solar noon to determine the date. One type has been attributed to various groups of the Kenyah and may still be in use.

It consists of a precisely measured (= span of maker's outstretched arms + span from tip of thumb to tip of first finger), permanently secured, plumbed, and decorated vertical hardwood pole (*tukar do*) and a neatly worked, flat measuring stick (*aso do*), marked with two sets of notches (see Figure 1).

The first set corresponds to specific parts of the maker's arm and ornaments worn upon it, measured by laying the stick along the radial side of the arm, the butt end against the inside of the armpit. To mark the date, the measuring stick is placed at the base of the vertical pole, butt end against the pole and extending southward. This is done at the time of day that the shadows are shortest, local solar noon. On the day that the pole's noontime shadow is longest (the June solstice), a notch is carved on the other edge of the stick to mark the extent of the shadow made by the pole. This observation indicates that the agricultural season is at hand. From then on, the extent of the noonday shadow is recorded every three days as a record-keeping device. Dates, both favourable and unfavourable, for various operations in rice cultivation such as clearing, burning, and planting are determined by the length of the shadow relative to the marks on the stick that correspond to parts of the arm and to the marks made every three days.

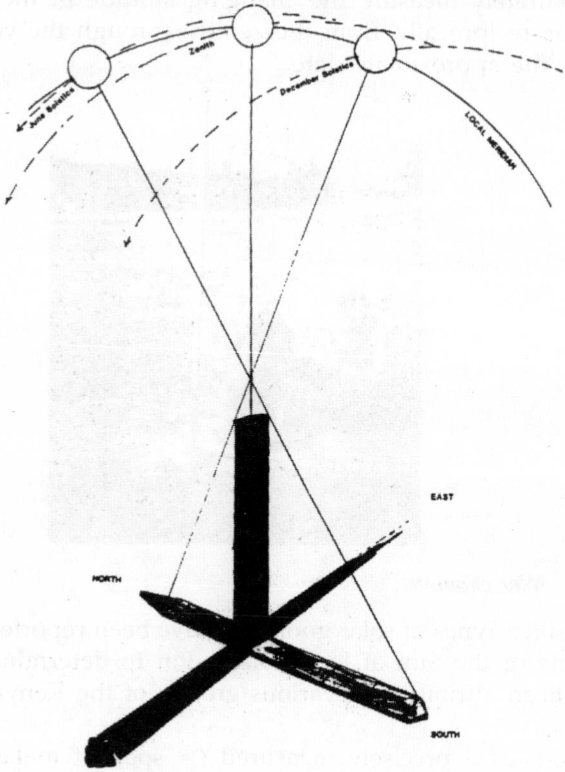

Figure 2 *Javanese bencet*

On Java a highly accurate gnomon, called a *bencet,* was in use from about CE 1600 until 1855 (see Figure 2). A smaller, more portable device than that employed by the Kayan and Kenyah, the *bencet* divides the year into twelve unequal periods, called *mangsa,* two of which begin on the days of the zenith Sun, when the Sun casts no shadow at local solar noon, and another two of which begin on the two solstices, when the sun casts its longest midday shadows. At the latitude of Central Java, 7 degrees south, a unique condition exists which is reflected in the *bencet.* As the illustration shows, when, on the June solstice, the Sun stands on the meridian (that is, at local solar noon) and to the north of the zenith, the shadow length, measured to the south of the base of the vertical pole, is precisely double the length of the shadow, measured to the north, which is cast when the Sun, on the December solstice, stands on the meridian (at noon) south of the zenith. By simply halving the shorter segment and quartering the longer, the Javanese produced a calendar with twelve divisions which are spatially equal but which range in duration from 23 to 43 days. The twelve *mangsa* with their starting dates and numbers of days are shown in Table 1. *Apparitions of Stars at Dawn and Dusk.*

Table 1 *The Pranatamangsa calendar*

Ordinal number	Name(s) of *mangsa*	Duration in days	First day(s)	Civil calendar
nKa-1	Kasa	41	22 June	21 June
Ka-2	Karo or Kalih	23	2 August	1 August
Ka-3	Katelu or Katiga	24	25 August	24 August
Ka-4	Kapator Kasakawan	25	18 September	17 September
Ka-5	Kalima or Gangsal	27	13 October	12 October
Ka-6	Kanem	43	9 November	8 November
Ka-7	Kapitu	43	22 December	21 December
Ka-8	Kawolu	26/27	3 February	2 February
Ka-9	Kasanga	25	1 March	ult. February
Ka-10	Kasepuluh or Kasadasa	24	26 March	25 March
Ka-11	Desta	23	19 April	18 April
Ka-12	Sada	41	12 May	11 May

The second category of observational techniques regularly employed by traditional farmers of the region include all of those which involve apparitions of stars or recognised groups of stars (asterisms) at last gleam at dawn or first gleam at dusk. Because of the Earth's orbital motion about the Sun, it can be observed that each star rises and sets approximately four minutes earlier each night; similarly, each star appears to have moved about 1 degree west, when viewed at the same time each night. As a result, around the time of its conjunction with the Sun, any given star becomes lost in the Sun's glare and is therefore not visible for approximately one month each year. It also means that the altitudes of stars above the horizon vary as a function of both the time of night and the day of the year, such that a certain star or group of stars, when observed at the same time each night (in this case near dusk or dawn) will appear at a given altitude above the eastern or western horizon on one and only one night of the year. Hence the use of any technique or device to measure the altitude of a star or group of stars as it appears at first or last gleam can provide the observer with the approximate date.

In general the term "acronical" refers to any stellar apparition which occurs at first gleam while "cosmical" is used to describe stars at last gleam. A subset of these, "heliacal" (Greek *helios*: Sun) apparitions of stars are those which occur just prior to and shortly after conjunction with the Sun. The heliacal setting of a star or group of stars occurs on the date that the star or stars are last observed before conjunction, just above the western horizon at dusk. Likewise, the heliacal rising of a star or group of stars occurs on the date that the star or stars are first observed, after several weeks' absence, above the eastern horizon at dawn.

Just about three months after its heliacal rise, the star, now about ninety degrees west of the sun, appears on the meridian at first gleam, an event which may be termed a "cosmical culmination" of the star. Two months later the star is nearing opposition with the Sun and undergoes an "acronical rising" in the east at first gleam. In less than one month, after opposition, the "cosmical setting" of the star is seen in the western sky. After two more months the star may be seen on the meridian at first gleam, accomplishing its "acronical culmination". Finally, about three months later, the star undergoes "heliacal setting" as the Sun once again outshines it.

These categories are taken from Western mathematical astronomy and are used here as a way of organising and presenting this material to the western scholar. Actual observations made by local Indonesian and Malaysian farmers may not always be as precise as their assigned astronomical categories might imply. The demand for such precision varies greatly between and within cultures and with local environmental conditions. It would not be unusual to find a local farmer, for example, first noting the heliacal rising of the Pleiades at dawn several days or more after its mathematically calculated reappearance.

The apparitions of stars at first and last gleam have been systematically observed by traditional cultures everywhere. From Indonesia and Malaysia there are references in the literature, too numerous to describe in detail here, to the calendrical use at both dusk and dawn of the stars we know as the Pleiades. Orion and, to a lesser extent, Antares, Scorpius, and Crux. Culminations at both first and last gleam of the Pleiades, Orion, and Sirius are noted in the literature. Interestingly, the observation for calendrical purposes of such culminations seems to be unique to peoples of the Indo-Malay Archipelago.

The first example in this category was practiced by a small Dayak group related to the Kenyah–Kayan complex mentioned earlier. Like their neighbours, they were swidden rice farmers. Unlike their neighbours who tracked the Sun, they depended upon the stars to fix the date of planting. To do so, they nightly poured water into the end of a vertical piece of bamboo in which a line had been inscribed at a certain distance from the open end. The bamboo pole was then tilted until it pointed toward a certain star (unrecorded) at a certain time of night (also unrecorded), causing some of the water to pour out (see Figure 3). It was then made vertical again and the level of the remaining water noted. When the level coincided with the mark, it was time to plant.

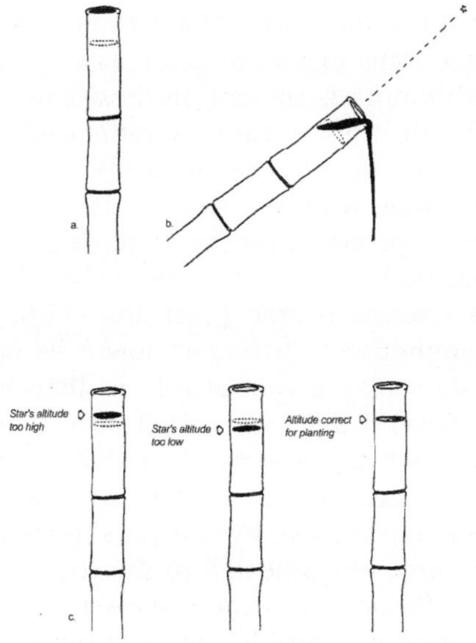

Figure 3 *Bamboo device for measuring the altitude of a star.*

Near Yogyakarta, Central Java, the ritual practitioner raised his hand toward the east in the direction of Orion at dusk, rice seed in his open palm. On the night that kernels of rice rolled off his palm, it was time to sow seed in the nursery. Using a planetarium star projector, the author has fixed the date of this at about 4 January.

Lunar Calendars

Lunar calendars comprise the third category. These calendars are based upon the 29.5 day synodic period, usually measured from new moon to new moon and often subdivided by phase. Because there is not an even number of lunar months in a solar year and because agricultural cycles are, after all, tied to the solar year; simple lunar calendars alone are of little use in farming. However, when it is somehow pegged to the solar year by reference to the apparent annual changes in the positions of the Sun or stars or to other phenomena in nature that regularly recur on an annual basis, a lunar calendar can be of use to the farmer.

Indigenous lunar calendars fall into two general categories: lunar–solar and sidereal–lunar. Examples of the lunar–solar calendar include the Balinese ceremonial calendar, still in use, and the old Javanese *Saka* calendar, used from the eighth through sixteenth centuries. Both are apparently of a common Hindu origin and are primarily lunar; both employ complex mathematical techniques to provide the intercalary days which periodically synchronise the lunar with the solar year.

A second type of lunar calendar, best described as sidereal–lunar, uses the apparitions at dusk and dawn of stars and asterisms as well as the appearance of other signs in nature (such as winds, birds, and flowers) to determine which month is current. In these cases, it is only important to know which month it is for a few months each year (that is, during the agricultural season), thereby obviating the need for codified schemes for realigning the lunar with the solar/stellar year. Such "short" lunar calendars are found spread throughout the region. The Iban calendar provides a good example.

The Iban are a riverine people practicing shifting agriculture in the vicinity of their longhouses, situated in low hills of Sarawak and West Kalimantan. The stars play a central role in Iban mythology and agricultural practices. Several Iban stories tell how their knowledge of the stars was handed down to them by their deities and according to one village headman, "If there were no stars we Iban would be lost, not knowing when to plant; we live by the stars" (Freeman, 1970). The Iban lunar calendar was annually adjusted to the cosmical apparitions of two groups of stars: the Pleiades and the three stars of Orion's belt.

The first observation is probably the most difficult. It is the reappearance of the Pleiades on the eastern horizon just before dawn after two month's absence from the night sky. This heliacal sighting, around June 5th of the civil calendar, informs the observer that the month, taken from new moon to new moon, that is current is the fifth lunar month. It is during this month that two members of the longhouse go into the forest to seek favourable omens so that the land selected will yield a good crop. This may take from two days to a month, but once the omens appear, they return to the longhouse and work clearing the forest begins. If it takes so long for the omens to appear that Orion's belt rises before daybreak (heliacal rising around June 25th), the people "must make every effort to regain lost time or the crop will be poor." This reappearance of

Orion at dawn occurs during the next or sixth lunar month, the time to begin clearing the land.

The remaining observations of the stars are more easily accomplished. They are all cosmical culminations, occurring "overhead" at last gleam, and are seen to be approaching for several weeks. When the Pleiades undergoes its cosmical culmination (September 3rd) and the stars of Orion's belt are about to do so (September 26–30), it is the eighth month and time to burn and plant. For good yields the burn should occur between the time that the two asterisms culminate at first gleam, usually when the two are in balance or equidistant from the meridian (September 16th). Rice seed sown after the star Sirius has completed its cosmical culmination (October 15th) will not mature properly. Planting may carry into the tenth lunar month (October/November), but it must be completed before the Moon is full or the crop will fail. At this point the lunar calendar ends: only months five through ten are numbered and fixed while the remaining months vary according to how quickly the crop matures (e.g. *bulan mantun*, the "weeding month"). The lunar months from November–April are simply not numbered; it is difficult to see the stars during the rainy season and unimportant in any case.

NAVIGATION

As we have just seen, the rice farmers of Indonesia and Malaysia have long noted correlations between celestial and terrestrial cycles and incorporated periodic changes in the sky into their agricultural calendars. Meanwhile, neighbouring seafaring societies have used many of these same phenomena to orient themselves in both space and time for the purpose of navigation. Of these societies, the Bugis of South Sulawesi are perhaps the best known. Maintaining a tradition of seafaring and trade that spans at least four centuries, the Bugis are reputed to have established and periodically dominated strategic trade routes across South-east Asia, stretching north-east from North Sumatra to Cambodia, north to Sulu and Ternate, and east to Aru and Timor. Their maritime prowess notwithstanding, Bugis systems of navigational knowledge and practice have only recently come under study.

Bugis navigators employ a system of dead reckoning which depends upon the knowledge of a variety of features of the natural environment to negotiate the seas in their tall ships. Although these features include land forms, sea marks, currents, tides, wave patterns and shapes, and the habits of birds and fish, navigators rely most heavily upon the prevailing wind directions, guide stars, waves, swells, and, increasingly, the magnetic compass.

The major wind patterns across Island South-east Asia are governed by the monsoons. From approximately May through October, winds from the east and south-east bring generally fair weather and steady breezes; from November through April, the west monsoon brings first calm air, then rain and squalls. For Bugis seafarers these winds are of such fundamental importance that they have named the two monsoons for their respective prevailing directions in the area of the Flores Sea: *bare'* (west) and *timo'* (east).

For the Bugis of the small coral islet of Balobaloang, located midway between Ujung Pandang (formerly Makasar) on Sulawesi and Bima on Sumbawa, trade routes are generally north and south across the Flores Sea. In principle, this allows them to take advantage of both easterly and westerly winds by reaching in either direction, although storms and heavy seas usually confine them to port during the west monsoon.

Bugis navigators have long relied upon the stars and star patterns to set and maintain course. Although most sailors seem to know a few star patterns and their use, the navigators know many more. These star patterns or asterisms are known to rise, stand and/or set above certain islands or ports when viewed from others and thereby pinpoint the direction of one's destination forming a "star compass". For example, in late July and early August *bintoéng balué* (Alpha and Beta Centauri) (Asterism A, Figure 4) is known to make its nightly appearance at dusk in the direction of Bima as viewed from Balobaloang, that is, to the south.

As the night passes and a given asterism is no longer positioned over the point of destination, the navigator's thorough familiarity with the sky allows him to adjust mentally to the new conditions: he can derive through visualisation the points on the horizon at which the stars rise or set relative to wherever they currently appear. When a certain asterism simply is not visible, other associated but unnamed stars may be used to remind the navigator where the original asterism set or is about to rise or they may be used instead of the missing asterism. This, by the way, appears to be analogous to the "star path, the succession of rising or setting guiding stars down which one steers" described by David Lewis for several cultures in Polynesia and Micronesia. The stars identified by Bugis navigators are listed in Table 2 and illustrated in Figures 4 and 5; those which are most relied upon will now be described.

Table 2 *Bugis stars and asterisms familiar to navigators*

Asterism	Bugis name[a]	English gloss	International designation
A	*bintoéng balué*	widow-before-marriage	Alpha & Beta CENTAURI
B	*bintoéng bola képpang*	incomplete house	Alpha-Delta, Mu CRUCIS
B.1	*bembé'*	goat	Coal Sack Nebula in CRUX
C	*bintoéng balé mangngiweng*	shark	SCORPIUS (south)
D	*bintoéng lambarué*	ray fish, skate	SCORPIUS (north)
D.1	(identified w/o name)	lost Pleiad	Alpha SCORPII (Antares)
E	*bintoéng kappala 'é*	ship	Alpha-Eta URSA MAJORIS
F	*bintoéng kappala 'é*	ship	Alpha-Eta UMA; Beta, Gamma UMI
G	*bintoéng balu Mandara'*	Mandar widow	Alpha, Beta URSA MAJORIS
H	*bintoéng timo'*	eastern star	Alpha AQUILAE (Altair)
J	*pajjékoé* (Mak.)[b] or *bintoéng rakkalaé*	plough stars	Alpha-Eta ORIONIS
J.1	*tanra tellué*	sign of three	Delta, Epsilon, Zeta ORI
K	*worong-porongngé* or *bintoéng pitu*	cluster seven stars	M45 in TAURUS (Pleides)
M	*tanra Bajoé*	sign of the Bajau	Large and small Magellanic Clouds
[ ]	*wari-warié* (no gloss)		Venus: morning
[ ]	*bintoéng bawi*	pig star	Venus: evening
	bintoéng nagaé	dragon stars	Milky Way

[a] Note that the Bugis term for 'star(s)' is *bintoéng*; the suffix *é* may be translated as the definite article 'the' in English.

[b] Although *pajjékoé* is a Makasar term, it is most commonly used on Balobaloang to indicate this asterism.

Perhaps the most frequently used asterism among the Bugis of Balobaloang is that *bintoéng balué*, mentioned above. These two bright stars are used to locate Balobaloang from Ujung Pandang and Bima from Ujung Pandang or Balobaloang. With regard to their rise/set points, navigators observe that they appear "in the south" at dusk during the middle of the east monsoon, the peak period for sailing; they further note that they rise southeast and set south-west. Their brightness makes them visible even through clouds. The name *balué* is derived from *balu* which means, "widow from death of the betrothed before marriage" with the affix é forming the definite article. Hence: "the one widowed before marriage". No graphical figure is attributed to this asterism.

Just to the west of Alpha and Beta Centauri is *bintoéng bola képpang* (Crux), visualised as an "incomplete house of which one post is shorter than the other and, therefore, appears to be limping" (Asterism B, Figure

4). Crux is used in conjunction with Alpha and Beta Centauri to navigate along southerly routes; like them, it is known to set southwest. Interestingly, it was emphasised that Crux is also used to help in predicting the weather. This asterism is located in the Milky Way which is known to the Bugis as *bintoéng nagaé* (the dragon), whose head is in the south and whose tail wraps all around the sky. As such, a bright haze of starlight surrounds Crux. On the eastern side of the house, however, there is a small dark patch totally devoid of light which is seen as a *bembé* (goat) (B.1, Figure 4). Between the squall clouds of the rainy season the goat in the sky may be seen standing outside the house trying to get in out of the rain. There are nights, however, when the goat is gone from the protection of the house. Hidden by haze, the missing goat portends a period of calm air and little rain.

In the northern sky the asterism which figures most prominently in Bugis navigation is *bintoéng kappala'é,* (the ship stars) in the western constellation of Ursa Majoris (Figure 5: asterisms E and F = two versions). The "ship" is used when travelling north. In particular, it rises northeast and sets northwest over Kalimantan from Ujung Pandang and Balobaloang. Associations of this group of stars with the hull of a boat or ship appear to be common throughout Indonesia.

Adjoining the ship and likewise used to navigate northward is *balu Mandara'* (widow of the Mandar) (Asterism G, Figure 5). These two stars, Alpha and Beta Ursa Majoris, remind the Bugis of Alpha and Beta Centauri (thus the name *balu*), while "Mandar" recalls their northern seafaring neighbours.

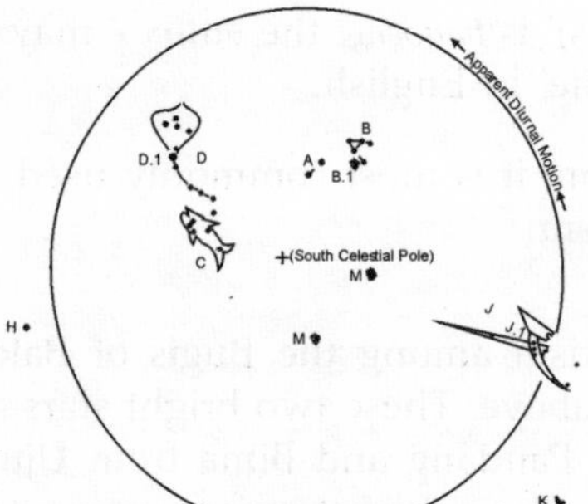

Figure 4 *Bugis asterisms: southern sky.*

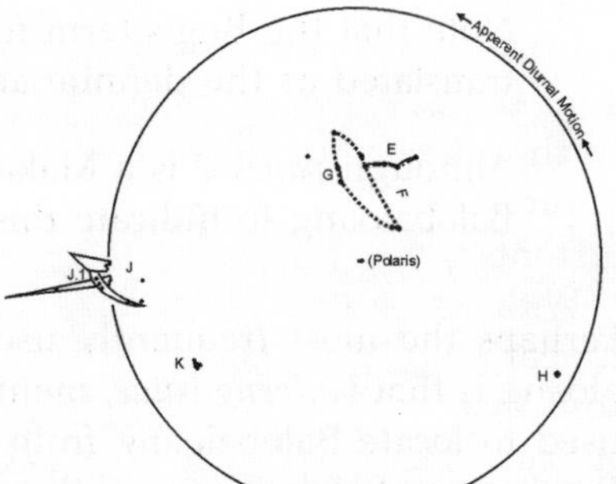

Figure 5 *Bugis asterisms: northern sky*

Several asterisms and a planet are used for sailing east and west: *bintoéng timo'* (eastern star) (Altair; H, Figure 5), *pajjékoé,* Makassar for "the plough" and also known as *bintoéng rakkalaé,* Bugis for "the plough stars" (Orion; Figures 4 and 5), *tanra tellué* (the sign of three) (the belt of Orion; Figures 4 and 5), *wari-warié* [no gloss] or *bintoéng élé'* (morning star; Venus as morning star), and *bintoéng bawi* (pig star; Venus as evening star), so named since it is believed that wild pigs will enter and destroy a garden or orchard when this object shines brightly in the west. Both the pig star and the plough, by the way, speak of an agrarian lifestyle not practiced by Bugis seafarers but culturally shared with their kin who farm the lands of Sulawesi as well as other islands of the archipelago.

Although the stars are useful guides, they are not always visible. Because it is possible to see landfall during the day, navigators appear to plan their voyages so as to maximise its usage. On a voyage from Balobaloang to Bima the captain scheduled departure in mid-afternoon, allowing him to back-sight on Balobaloang and other islands of the atoll and observe the sun as it set in the west until dusk when Alpha and Beta Centauri appeared in the sky. There was, in fact, a period of about thirty minutes where *both* the receding island and the stars could be seen, providing a good opportunity to maintain course as attention was shifted from land forms to the stars.

Except during the height of the west monsoon, it is uncommon to experience extended periods of totally overcast skies. Should clouds conceal the primary guiding asterism, the navigator depends upon his knowledge of other asterisms or unnamed stars to fix his direction. If it is very cloudy, day or night, the navigator turns to wave directions and the magnetic compass to maintain course.

Courses are committed to memory in terms of destinations and their required compass headings under various winds. That is, certain points of the compass are associated with certain destinations from various ports. For example, it is known that to sail from Balobaloang to Bima during the east monsoon, one must head due south, while during the west monsoon one heads south-southwest to south-west, depending on the strength of the wind and current. Like-wise, to reach Ujung Pandang from the island during the east monsoon one travels somewhat east of north, while during the west monsoon a heading to the north-northwest is preferred. This difference takes into account drift from wind and currents, while true directions are also known.

With regard to the study of indigenous astronomical systems, there appear in regional agricultural calendars two types of celestial observations that may be unique to this cultural region. They are (1) observations of "cosmical" and "acronical" culminations – meridian transits at last and first gleam – of groups of stars, and (2) observations of the lunar month for a limited number of months each year, creating discontinuous sidereal-lunar calendars. The use of stars by the Bugis navigators of South Sulawesi likewise appears to represent a system which, along with its transformations used by Oceanic mariners, may be unique to the Austronesian world.

Implicit in the discussion of agricultural calendrical and navigation systems is the understanding that celestial observations do not stand alone. That is, many other environmental markers – changes in wind and weather and the appearances of flora and fauna also inform agricultural and navigational decision-making. It is suggested that by noting more carefully these and other signs in nature to which members of non-Western societies attend, we may gain a deeper appreciation of the true richness of human knowledge across cultures.

Gene Ammarell

REFERENCES

Ammarell, Gene. "Sky Calendars of the Indo-Malay Archipelago: Regional Diversity/Local Knowledge." *Indonesia* 45:84–104. 1988.

Ammarell, Gene. "The Planetarium and the Plough: Interpreting Star Calendars of Java." *Yale Graduate Journal of Anthropology* 3:11–25. 1991.

Ammarell, Gene. *Bugis Navigation*. Doctoral Dissertation. Yale University, 1995.

Casiño, Eric S. "Jama Mapun Ethnoecology: Economic and Symbolic." *Asian Studies* 5:1–3. 1967.

Covarrubias, Miguel. *Island of Bali*. New York: Knopf, 1937.

Dentan, Robert K. "An Appeal to Members of the Society from an Anthropologist." *Malayan Nature* 23:121–122. 1970.

Freeman, J.D. *Report on the Iban*. London School of Economics Monographs on Social Anthropology No. 41. New York: Humanities Press Inc. 1970.

Hose, Charles. "Various Methods of Computing Time among the Races of Borneo." *Journal of the Straits Branch of the Royal Asiatic Society* 42:4–5, 209–210. 1905.

Hose, Charles and W. McDougall. *The Pagan Tribes of Borneo*. London: Macmillan. 1912.

Lewis, David. *We, The Navigators*. Honolulu: University of Hawaii Press, 1972.

Pelras, Christian. "Le Ciel et Les Jours. Constellations et Calendriers Agraires Chez Les Bugis (Celebes, Indonesie)." In *De La Voute Celeste Au Terroir, Du Jardin Au Foyer*. Paris: Éditions de É'Icole des Hautes Études en Sciences Sociales. pp. 19–39. 1987.

van den Bosch, F. "Der Javanische Mangsakalender." *Bijdragen tot de taal-, land- en volkenkunde* 136:251–252. 1980.

See also: Agriculture – Ethnobotany – Navigation – Colonialism and Science – Gnomon

Atomism

Indic atomism is usually traced to Kaṇāda, believed to be the original exponent of *Vaiśeṣika,* one of the earliest of the six well-known Hindu philosophical systems. No precise dates can be offered for the first exposition of atomism, but as the *Vaiśeṣika* system is believed to precede Buddhism a plausible date for either the system or Kaṇāda himself is about the 6th century BCE or earlier. Kaṇāda was known by several other names as well, all of them synonymous with the meaning 'particle-eater', suggesting that he must have been a forceful and visible advocate of what was seen at the time as a novel idea. In some form the idea became part of the later syncretic *Nyāya-Vaiśeṣika* school and of the heterodox philosophies of Jainism and Buddhism as well. (Jain atomism may even have been earlier than *Vaiśeṣika*).

The *Nyāya-Vaiśeṣika* system recognises nine categories of substance, of which the four material ones – earth, water, air and fire – are all considered to be atomic. The atom (called *aṇu* or *paramāṇu*) is very small, indivisible (without parts, indestructible (i.e. eternal – things made up of parts were necessarily transient in this philosophical system)), spherical and supra-sensible. Each of the four substances consists of a different kind of atom, each sharing the attribute of the corresponding substance itself. Thus the earth atom has odour, the water atom has taste, the fire atom colour and the air atom touch.

In keeping with the *Vaiśeṣika* view of causation (called *ārambha-vāda*, meaning a theory of (development of effect from an) initial cause, [state or condition], the atom was supposed to be set in motion at some primordial stage by an unseen force (*adṛṣṭa*). Thus *adṛṣṭa* was the efficient cause of the world, while atoms were the material cause. The initial motion enabled atoms to form dyads (*dvyaṇukaṣ*, which were still infinitesimal and supra-sensible, and not therefore gross matter. The minimum visible unit was thought to be a *tryaṇuka* or triad, or a complex with more atoms, of the size of a mote in a sunbeam. These higher complexes were considered made up of dyads in some way, individual atoms having ceased to play a role after dyad formation. Two like atoms (e.g. of earth) could enter into combination but two unlike atoms (e.g. earth and aiṛ could not. Aggregation into bulk matter was thus explained, but not (it would seem) formation of new compounds among the primary elements. The different properties observed in gross substances were attributed to the different structural arrangement (*vyūha*) of atoms in each substance.

The concept of substance included not only the four elements but also *ākāśa*, time, space, self and mind. In the *Vaiśeṣika* view mind was also atomic.

There were also other schools of atomism in India, in particular among Buddhists and Jains. The Jains also considered atoms indivisible. Atoms were again the prime cause of the four elements, except that sound was the quality of *ākāśa*, which accrued from an aggregation of atoms.

Among Buddhists the *mādhyamika* and *yogācāra* schools were opposed to atomism, but the other two schools, *vaibhāṣika* and *sautrāntika*, admitted the concept. Buddhist atoms were indivisible, invisible, inaudible, untestable and intangible, but they were also momentary, in the sense that they continuously underwent phase changes. This is consistent with the Buddhist philosophical position that reality is instantaneous.

There are some resemblances but many interesting contrasts between Indian and Greek atomistic views. Greek atoms also constituted indivisible building blocks of the physical world. However they had different shapes and sizes but no other qualitative differences; they were in perpetual motion. Democritus thought that the soul was also composed of atoms, whereas the *Vaiśeṣika* considered *ātmā* to be non-material. The Greeks had no concept of atomic complexes like dyads and triads.

Curiously, unlike Greek atomism, Indic atomism endured as a live concept until the 13th century CE, in spite of the vehement opposition of such influential and respected philosophers as Śaṁkara (7th century CE). For example the *Yoga-Vāsiṣṭha* (unknown author, perhaps 6th or 7th century CE, but widely studied even in the 17th century) was strongly atomist, considering consciousness atomic, and even experience as particulate or quantal. It seems as if the general idea of discreteness appeared natural, on both philosophical and logical grounds, to many schools of Indic philosophy (some of them considering even time and space to be discrete). But, throughout those centuries, atomism remained a part of philosophical argument, with no visible direct impact (for example) on the chemistry that was also vigorously practised in India.

B.V. Subbarayappa
Roddam Narasimha

REFERENCES

Bose, D M, S N Sen, B V Subbarayappa. *A Concise History of Science in India.* New Delhi: Indian National Science Academy. 1971.

Hiriyanna, M. *Outlines of Indian Philosophy.* London: George Allen and Unwin. 1932.

Narasimha, R. "A Metaphysics of Living Systems: the Yōga-Vasiṣṭha View." *Journal of Bioscience* 27: 645–650. 2002.

Subbarayappa, B, V. *Indian Perspectives on the Physical World.* New Delhi: Centre for Studies in Civilization. Munshiram Manoharlal Publishers. 2004.

Ātreya

Ātreya was the teacher of Agniveśa, renowned as a writer on medicine. He is popularly known as Punarvasu Ātreya or only Punarvasu. The term

Ātreya implies a person who is either a disciple or a descendant of Atri, the Vedic sage. On the basis of his teachings, Agniveśa and five other disciples, Bhela, Jatukarṇa, Parāśara, Kṣārapāṇi and Hārita, composed separate works. Agniveśa's work is still available today. *Bhela-saṃhitā* is available only in a mutilated form. *Hārīta-saṃhitā*, which is now in print, appears to be the work of a later author who was also known as Hārita. Works of the other three disciples, though quoted profusely by later commentators, are no longer extant.

In the past, there were several other medical authorities with the term Ātreya suffixed to their names. For example, there is Bhikṣu-Ātreya and Kṛṣṇa-Ātreya. Another Ātreya was the head of the Ayurvedic faculty at Taxila University (600 BCE), and his disciple Jīvaka was the physician of Lord Buddha. These should not be confused with Punarvasu Ātreya.

Since Agniveśa's work was reprinted in a period prior to 600 BCE, he and his teacher Ātreya must have flourished much before this time. The *Bhela-saṃhitā* states that Candrabhāgā was the mother of Punarvasu Ātreya, and he was the physician of King Nagnajit of Gāndhāra. This king is mentioned in *Śatapatha-Brāhmaṇa* and *Aitareya-Brāhmaṇa* which were composed prior to 3000 BCE. Therefore, Punarvasu Ātreya must have flourished during or prior to this period.

Though he was the court physician of Gandhāra, evidence in *Caraka-saṃhitā* indicates that he travelled with his disciples in Pāñcāla-kṣetra, the forests of Caitraratha, Pañcagaṅga, Dhaneśāyatana, the northern side of the Himalayas, and Triviṣṭap.

Apart from his teachings, which are codified in *Caraka-saṃhitā*, manuscripts of an independent work called *Ātreya-saṃhitā* are also available. This work has four sections and deals with both the theory and practice of Ayurveda in detail.

Bhagwan Dash

References

Agniveśa. *Carakasaṃhitā*. Bombay: Nirnayasagara Press. 1941.

Bhela. *Bhelasaṃhitā*. Delhi: Central Council for Research in Indian Medicine. 1977.

Mukhopadhyaya, Girindranath. *History of Indian Medicine*. New Delhi: Oriental Books Reprint Corporation. 1974.

Śarmā, Priyavrata. *Āyurveda kā vaijñānika itihāsa*. Varanasi: Caukhambha Oriyantaliya. 1981.

Vāgbhaṭa. *Aṣṭāṅgahṛdaya*. Varanasi: Krishnadas Academy. 1982.

Vṛddhajīvaka. *Kāśyapasaṃhitā*. Varanasi: Chaukhambha Sanskrit Series Office. 1953.

B

Bakhshālī Manuscript

The Bakhshālī Manuscript is the name given to the oldest extant manuscript in Indian mathematics. It is so-called because it was discovered by a peasant in 1881 at a small village called Bakhshālī, about eighty kilometres north-east of Peshawar (now in Pakistan). It is preserved in the Bodleian Library at Oxford University.

The extant portion of the manuscript consists of seventy fragmentary leaves of birch bark. The original size of a leaf is estimated to be about 17 cm wide and 13.5 cm high. The original order of the leaves can only be conjectured on the bases of rather unsound criteria, such as the logical sequence of contents, the order of the leaves in which they reached A.F.R. Hoernle, who did the first research on the manuscript, physical appearance such as the size, shape, degree of damage, and knots, and the partially preserved serial numbers of mathematical rules (9–11, 13–29, and 50–58).

The script is the earlier type of the Śāradā script, which was in use in the north-western part of India, namely in Kashmir and the neighbouring districts, from the eighth to the twelfth centuries. G.R. Kaye, who succeeded Hoernle, has shown that the writing of the manuscript can be classified into at least two styles, one of which covers about one-fifth of the work. There is, however, no definitive reason to think that the present manuscript consists of two different works.

The information contained in the manuscript, the title of which is not known, is a loose compilation of mathematical rules and examples collected from different works. It consists of versified rules, examples, most of which are versified, and prose commentaries on the examples. A rule is followed by an example or examples, and under each one the commentary gives a 'statement', 'computation', and a 'verification' or verifications. The statement is a tabular presentation of the numerical information given in the example, and the computation works out the problem by following, and often citing, the rule step by step.

Thus, the most typical pattern of exposition in the Bakhshālī Manuscript is:

- rule (*sūtra*);

- example (*udāharaṇa*)
  - statement (*nyāsa/sthāpanā*);
  - computation (*karaṇa*);
  - verification(s) (*pratyaya/pratyānayana*).

A decimal place-value notation of numerals with zero (expressed by a dot) is employed in the Bakhshālī Manuscript. The terms for mathematical operations are often abbreviated, especially in tabular presentations of computations. Thus we have: *yu* for *yuta* (increased), *gu* for *guṇa* or *guṇita* (multiplied), *bhā* for *bhājita* (divided) or *bhāgahāra* (divisor or division), *che* for *cheda* (divisor), and *mū* for *mūla* (square root). For subtraction, the Bakhshālī Manuscript puts the symbol, + (similar to the modern symbol for addition), next (right) to the number to be affected. It was originally the initial letter of the word *ṛṇa*, meaning a debt or a negative quantity in the Kuṣāṇa or the Gupta script (employed in the second to the sixth centuries). The same symbol is also used in an old anonymous commentary on Śrīdhara's *Pāṭīgaṇita*, which is uniquely written in the later type of the Śārada script (after the thirteenth century). Most works on mathematics, on the other hand, put a dot above a negative number.

The problems treated in the extant portion of the Bakhshālī work involve five kinds of equations, namely (1) simple equations with one unknown (fifteen types of problems), (2) systems of linear equations with more than one unknown (fourteen types), (3) quadratic equations (two types, both of which involve an arithmetical progression), (4) indeterminate systems of linear equations (three types, including the so-called Hundred Fowls Problem, in which somebody is to buy one hundred fowls for one hundred monetary units of several kinds, and (5) indeterminate equations of the second degree (two types: $\sqrt{x+a}=u$ and $\sqrt{x-b}=v$, where u and v are rational numbers; and $xy=ax+by$).

The rules of the Bakhshālī work may be classified as follows:

(1) fundamental operations, such as addition and subtraction of negative quantities, addition, multiplication, and division of fractions, reduction of measures, and a root-approximation formula,

$$\sqrt{a^2+r} \approx a + \frac{r}{2a} - \frac{(r/2a)^2}{2\,(a+r/2a)};$$

(2) general rules applicable to different kinds of problems: *regula falsi*, rule of inversion, rule of three, proportional distribution, and partial addition and subtraction;

(3) rules for purely numerical problems: simple equations with one unknown, systems of linear equations with more than one unknown, indeterminate equations, and period of an arithmetical progression;

(4) rules for problems of money: equations of properties, wages, earnings, donations, etc., consumption of income and savings, buying and selling, purchase in proportion, purchase of the same number of articles, price of a jewel, prices of living creatures, mutual exchange of commodities, instalments, a sales tax paid both in cash and in kind, and a bill of exchange;

(5) rules for problems of travellers: equations of journeys, meeting of two travellers, and a chariot and horses;

(6) rules for problems of impurities of gold; and

(7) rules for geometrical problems: volume of an irregular solid and proportionate division of a triangle.

All the rules of the first category, namely the fundamental operations, occur only as quotations in the computations of examples. Many of the other rules could belong to either *miśraka-vyavahāra* (the practical problems on mixture) or *śreḍhī* (the practical problems of series) in a book of *pāṭī* (algorithms) such as *Śrīdhara's Pāṭīgaṇita* and *Triśatikā* (eighth century), etc., but they have not been arranged according to the ordinary categories of *vyavahāra* (practical problems).

We apparently owe the present manuscript to four types of persons: the authors of the original rules and examples, the compiler, the commentator, and the scribe. Possibly, however, the commentator was the compiler himself, and 'the son of Chajaka' (his name is unknown), by whom the Bakhshālī Manuscript, or at least part of it, was 'written,' was the commentator, or one of the commentators. The colophon to the section that deals exclusively with the *trairāśika* (rule of three) reads:

> This has been written by the son of Chajaka, a brāhmaṇa and king of mathematicians, for the sake of Hasika, son of Vasiṣṭha, in order that it may be used (also) by his descendents.

Immediately before this statement occurs a fragmentary word *-rtikāvati*, which is probably the same as the country of Mārtikāvata mentioned by Varāhamihira (ca. CE 550) among other localities of north-western India such as Takṣaśilā (Taxila), Gāndhāra, etc. (*Bṛhatsaṃhitā* 16.25). It may be the place where the Bakhshālī work was composed.

A style of exposition similar to that of the Bakhshālī work ('statement,' etc.) is found in Bhāskara I's commentary (CE 629) on the second chapter called *gaṇita* (mathematics) of the *Āryabhaṭīya* (CE 499). Both Bhāskara I's commentary and the Bakhshālī work attach much importance to the verification; it became obsolete in later times. The unusual word *yāva* (*yāvakaraṇa* in Bhāskara I's commentary) meaning the square power, and the apparently contradictory meanings of the word *karaṇī*, the square number and the square root, occur in both works.

Bhāskara I does not use the symbol *yā* (the initial letter of *yāvattāvat* or 'as much as') for unknown numbers in algebraic equations even when it is naturally expected, while he employs the original word *yāvattāvat* itself in the sense of unknown quantities (in his commentary on *Āryabhaṭīya* 2.30). This probably implies that he did not know the symbol. The symbol is, on the other hand, utilised once in the Bakhshālī work in order to reduce the conditions given in an example to a form to which the prescribed rule is easily applicable; after the reduction, the symbol is discarded and the rule is, so to speak, applied mechanically (fol. 54^{v}). This restricted usage of the symbol seems to indicate that the work belongs to a period when the symbol was already invented, but not very popular yet.

There has been quite a bit of dispute over the dates of the manuscript. Hoernle assigned the work to the third or the fourth century CE, Kaye to the twelfth century, Datta to 'the early centuries of the Christian era,' and Ayyangar to the eighth or the ninth century. The above points suggest that the Bakhshālī work (commentary) was composed not much later than Bhāskara I (the seventh century).

Takao Hayashi

References

Ayyangar, A.A.K. "The Bakhshālī Manuscript." *Mathematics Student* 7: 1–16. 1939.

Channabasappa, M.N. "On the Square Root Formula in the Bakhshālī Manuscript." *Indian Journal of History of Science* 2: 112–124. 1976.

Channabasappa, M.N. "The Bakhshālī Square Root Formula and High Speed Computation." *Gaṇita Bhāratī* 1: 25–27. 1979.

Channabasappa, M.N. "Mathematical Terminology Peculiar to the Bakhshālī Manuscript," *Gaṇita Bhāratī* 6: 13–18. 1984.

Datta, B. "The Bakhshālī Mathematics." *Bulletin of the Calcutta Mathematical Society* 21: 1–60. 1929.

Gupta, R.C. "Centenary of Bakhshālī Manuscript's Discovery." *Gaṇita Bhāratī* 3: 103–105. 1981.

Gupta, R.C. "Some Equalijection Problems from the *Bakhshālī Manuscript*." *Indian Journal of History of Science* 21: 51–61. 1986.

Hayashi, T. *The Bakhshālī Manuscript: An Ancient Indian Mathematical Treatise*. Groningen: Egbert Forsten. 1995.

Hoernle, A.F.R. "On the Bakhshālī Manuscript," *Verhandlungen des vii Internationalen Orientalisten Congresses* (Vienna 1886), Arische Section, 1888, pp. 127–147.

Hoernle, A.F.R. "The Bakhshālī Manuscript." *The Indian Antiquary* 17: 33–48 and 275–279. 1888.

Kaye, G.R. "The Bakhshālī Manuscript." *Journal of the Asiatic Society of Bengal* NS 8:349–361. 1912.

Kaye, G.R. *The Bakhshālī Manuscript: A Study in Medieval Mathematics*. Archaeological Survey of India, New Imperial Series 43. Parts 1 and 2: Calcutta, 1927. Part 3: Delhi, 1933. Reprinted, New Delhi: Cosmo Publications. 1981.

Pingree, D. *Census of the Exact Sciences in Sanskrit*. Series A (in progress), vols. 1–4. Philadelphia: American Philosophical Society. 1970–81.

Pingree, D. *Jyotiḥsāstra: Astral and Mathematical Literature*. Wiesbaden: Harrassowitz. 1981.

Sarkar, R. "The Bakhshālī Manuscript." *Ganita Bhāratī* 4: 50–55. 1982.

See also: Zero – Śrīdhara

Baudhāyana

India's most ancient written works are the four *Vedas*, namely *Ṛgveda, Yajur-veda, Sāma-veda,* and the *Atharva-veda*. There are different schools which are represented by various *Saṁhitās* or recensions of the *Vedas*. To assist their proper study, there are six *Vedāṅgas* (limbs or part of the *Veda*), namely *Śikṣā* (phonetics), *Kalpa* (ritualistics), *Vyākaraṇa* (grammar), *Nirukta* (etymology), *Chandas* (prosody and metrics), and *Jyotiṣa* (astronomy, including mathematics and astrology). These auxiliary Vedic works (except the last) are written in *sūtra* or aphoristic style.

The *Kalpa Sūtras* deal with the rules and methods for performing Vedic rituals, sacrifices, and ceremonies, and are divided into three categories: *Śrauta, Gṛhya,* and *Dharma*. The *Śrauta Sūtras* are more specifically concerned with the sacrificial ritual and allied ecclesiastical matters. They often include tracts which give rules concerning the measurements and constructions of *agnis* (fireplaces), *citis* (mounds or altars), and *vedis* (sacrificial grounds). Such tracts are also found as separate works and are called *Śūlba Sūtras* or *Śulbas*. They are the oldest geometrical treatises which represent in coded form the much older and traditional Indian mathematics. The root *śulb* (or *śulv*) means "to measure" or to mete out".

The names of about a dozen *Śulba Sūtras* are known. The oldest of them is the *Baudhāyana Śulba Sūtra*. It belongs to the *Taittirīya Samhitā* of the *Black Yajur-veda* and is the thirtieth *Praśna* or chapter of the *Baudhāyana Śrauta Sūtra*. The title *Baudhāyana Śulba Sūtra* shows that its author or compiler was Baudhāyana, or perhaps more correctly, it belonged to the school of Bodhāyana. Other Vedic works bearing the same name are known. It is more proper to consider these works as belonging to the Baudhāyana school than to regard them as authored by the same person.

Here we are concerned with Baudhāyana the Śulbakāra (that is, the author of the *Baudhāyana Śulba Sūtra*) or the śulbavid (expert in śulba mathematics and constructions). We do not know his biographical details. Georg Bühler believed that he hailed from the Andhra region, but a recent study by Ram Gopal shows that he probably came from northern India. His dates are also uncertain, being any time between 800 and 400 BCE. Taking into acco unt the views of A.B. Keith. W. Caland, David Pingree, and Ram Gopal, he may be placed about 500 BCE or earlier. However, it must be noted that much of the material in *Baudhāyana Śulba Sūtra* is traditional and, thus, still older than its date of compilation and coding.

The *Baudhāyana Śulba Sūtra* is not only the earliest but also the most extensive among the *Śulbas*. The subject matter is presented in a systematic and logical manner. Of course, the language is somewhat archaic, and due to the aphoristic style, the rules are highly condensed. The *Baudhāyana Śulba Sūtra* was commented on by Dvārakānātha Yajva (ca. seventeenth century). His Sanskrit commentary called the *Śulbadīpīkā* was publised more than a century ago by G.F.W. Thibaut (1848–1914) in his edition of the *Baudhāyana Śulba Sūtra*. A recent edition (Varanasi, 1979) of the *Baudhāyana Śulba Sūtra* also contains the above commentary as well as another called *Bodhāyana śulbamīmāṁsā*, which was written by Vyaṅkaṭeśvara (or Veṅkaṭeśvara) Dīkṣita who lived during the Vijaya-nagaram kingdom (ca. 1600).

The text of the *Baudhāyana Śulba Sūtra* is divided into three chapters which comprise a total of 272 passages or 519 aphorisms. The main subject is the measurement, construction, and transformation of various altars and fireplaces. The forms of the three obligatory *agnis* (whose tradition was older than even the *Ṛgveda*) were square, circle, and semicircle, but those of optional *citis* involved all sorts of plane figures, including the above three and also the rectangle, rhombus, triangle, trapezium, pentagon, and some complicated shapes.

Some mathematical topics covered in the *Baudhāyana Śulba Sūtra* are a fine approximation for the square root of two and the so-called Pythagorean theorem. In the latter, the sides of a right-angled triangle obey the rule $a^2 + b^2 = c^2$. This rule is generally called the Pythagorean theorem, although it was known in Babylonia and China, as well as in India, earlier than the time of the Greek philosopher and mathematician Pythagoras.

The *Baudhāyana Śulba Sūtra* also contains formulas for circling a square and squaring a circle, and provides rules for basic simple geometric constructions, such as drawing a perpendicular on a given line or drawing the right bisection of a line. Some other elementary plane figures such as the isosceles triangle, trapezium, and rhombus are also covered for construction.

The *Baudhāyana Śulba Sūtra* deals with the measurements and constructions of a large number of fire-altars (*Kāmya agnis*). These were needed as part of rituals performed to attain certain desired objects according to the religious beliefs of the Vedic people. It was essential that the shape, size, area, and orientation of the relevant altar be according to the prescribed instructions. Otherwise, there was a risk of divine wrath.

The standard forms of some of the optional altars were those which resembled certain birds. The most significant and perhaps the oldest of such altars was the *śyenaciti* (falcon-shaped altar). It was to be constructed when one desired heaven (after death) because "the falcon is the best flyer among the birds." The spatial dimensions, the number of bricks (of prescribed shapes and sizes), the number of layers, etc., are all given. It is interesting to note that the archaeological remains of this most striking and complicated *śyenaciti* reported to be built in Kausambi in the second century BCE, still survive.

Early Indian geometry developed because of the need for accurate altar constructions and transformations which often required quite advanced mathematical knowledge.

R. C. Gupta

References

Bhattacarya, V., ed. *Baudhāyana Śulba Sūtra with Commentaries of Vyaṅkateśvara and Dvārakānātha*. Varanasi: Sampurnanand Sanskrit University. 1979.

Datta, B. *The Science of the Śulba*. Calcutta: University of Calcutta. 1932. Reprinted 1991.

Ganguli, S.K. "On the Indian Discovery of the Irrational at the Time of Śulba Sūtras." *Scripta Mathematica* I: 135–141. 1932.

Gupta, R.C. "Baudhāyana's Value of $\sqrt{2}$." *Mathematics Education* 6(3): 77–79. 1972.

Gupta, R.C. "Vedic Mathematics From the Śulba Sūtras." *Indian Journal of Mathematics Education* 9(2): 1–10. 1989.

Gupta, R.C., "Sundararāja's Improvements of Vedic Circle-Square Conversions." *Indian Journal of History of Science* 28(2): 81–101. 1993.

Kashikar, C.G. "Baudhāyana Śyenaciti: A Study in the Piling Up of Bricks". In *Proceedings of 29th All-India Oriental Conference*. Poona: Bhandarkar Institute. 1980. pp. 191–199.

Kulkarni, R.P. *Geometry According to Śulba Sūtra*. Pune: Vaidika Samsodhana Mandala. 1983.

Prakash, Satya, and R.S. Sharma, eds. *Baudhāyana Śulba Sūtra*. New Delhi: Research Institute of Ancient Scientific Studies. 1968.

Seidenberg, A. "The Origin of Mathematics." *Archive for History of Exact Sciences* 18(4): 301–342. 1978.

Sen, S.N., and A.K. Bag, eds. and trans. *The Śulba Sūtras*. New Delhi: Indian National Science Academy. 1983.

Thibaut, G. *Mathematics in the Making in Ancient India*. Calcutta: Bagchi. 1984.

See also: Śulbasūtras

Bhāskara I

Bhāskara I was the greatest and the earliest exponent of Āryabhaṭa I and his school of astronomy, and explained extensively the all-too-brief aphoristic statements of the *Āryabhaṭīyam* in a celebrated commentary.

Date and Place

There is much evidence to show that Bhāskara I was not a direct pupil of Āryabhaṭa I. Based on statements regarding the time elapsed since the beginning of the present *kalpa*, K.S. Shukla concluded that Bhāskara I wrote his commentary in 629 CE exactly one year after Brahmagupta wrote his famous treatise *Brahmasphuṭa Siddhānta*. The two great stalwarts were thus contemporaries, but they belonged to rival schools and most likely did not know each other.

From the references in the illustrative examples given by him in his work, it appears that Bhāskara's commentary on the *Āryabhaṭīyam* was composed when he was residing at Valabhī in Kathiawar, on the western shore of Cambay about 30 km to the northwest of Bhavanagar. Valabhī was the capital of the Surāṣṭra kingdom and a seat of Buddhist learning in the seventh century CE. Shukla conjectures that Bhāskara might have been an astronomer in the court of King Dhruvabhaṭṭa.

Works

Bhāskara I is known to have composed three works: (i) *Mahā Bhāskarīyam*, (ii) *Laghu Bhāskarīyam* and (iii) *Āryabhaṭīya Bhāṣya*. The first two texts, though independent, are elucidations of *Āryabhaṭa's* astronomy. Bhāskara calls the first *Āryabhaṭa karma nibandha* (referred to as *karma nibandha* later), and says of it that it "has clarity of expression and simple methods (of calculation), and can be comprehended even by those with modest intellect; it was written by Bhāskara after great deliberation." The *Laghu Bhāskarīyam* is an abridgement of this work. It was only later astronomers and commentators who distinguished between the two texts as *Mahā-* and *Laghu-Bhāskarīyam* (respectively the bigger and smaller versions of *karma nibandha*). The third work is a commentary on *Āryabhatīyam* comprising two parts: *Daśagītikā sūtra vyākhyā* and *Āryabhaṭa tantra bhāṣya*.

These three texts constitute the total output of Bhāskara I since his commentator, Śaṅkaranārāyaṇa (869 CE), and later astronomers, mention no other work by him. The chronological order of the texts appears to be: *Mahā Bhāskarīyam, Āryabhaṭīya vyākhyā* and *Laghu Bhāskarīyam*. The first two came to be regarded as ideal textbooks of astronomy up to the end of the sixteenth century.

The works of Bhāskara I have been well known for centuries among astronomers in south India, especially in Kerala. In Kerala, where the Āryabhaṭa school is exclusively followed, Bhāskara I is respected next only to Āryabhaṭa in importance. It is in Kerala that the works of Bhāskara I, with the commentaries of Govindasvāmin, Śaṅkaranāráyaṇa, Parameśvara and Nārāyaṇa – all hailing from Kerala, were obtained. The famous text *Vākya Karaṇa* (or *Vākya Pañcādhyāyī*), on which the *vākya pañcāṅgas* of Tamil Nadu are based, mentions that it is based on *Bhāskarīyam*.

However, for some strange reason, Bhāskara I fell into obscurity in North India after some centuries. There is no mention of Bhāskara I either in Sudhākara Dvivedi's Sanskrit work *Gaṇaka Taraṅgiṇi*, or in S.B. Dikshit's *Bhāratīya Jyotiṣa Śāstra* (in Marathi). However, H.T. Colebrooke rightly guessed (in 1817) the existence of Bhāskara I through a reference by Pṛthūdakasvāmin (860 CE) in his commentary on *Brahmasphuṭa Siddhānta* of Brahmagupta.

Astronomical Results

Conceptual Constructs

M. D. Srinivas (2002) remarks:

> As regards the epistemological status of the planetary models, the Indian astronomical texts present a very clear position that they are conceptual tools which serve the purpose of calculating observationally verified planetary positions. This position is clearly set out in the *Āryabhaṭīyabhāṣya* of Bhāskara I when he starts his exposition of the *manda* and *śīghra* corrections. Bhāskara I says:
>
> 'There are no constraints or limitations imposed on the notions such as the *ucca, nīca, madhyama, paridhi* and so on which are indeed aids to the calculation of the observed motion of the planets. These are only the means for arriving at the desired results. Hence this entire procedure is fictitious, by means of which the observed planetary motion is arrived at. Just as the seekers of ultimate knowledge expound the ultimate truth via untrue means, just as the surgeons practise their surgery etc. on stems and other objects, just as the hair-stylists practise shaving on pots, just as the experts in performance of *yajña* practise

using dry wood, just as the linguists utilise notions such as *prakṛti, pratyaya, vikāra, āgaram, varṇa, lopa, vyatyaya,* etc. to comprehend (well-formed) words, in the same way in our science also the astronomers employ notions such as *madhyama, mandocca, śīghrocca, śīghra-paridhi, jyā, kāsṭha, bhuja, koṭi, karṇa,* etc., in order to comprehend the observed motion of planets. Hence, there is indeed nothing unusual that fictitious means are employed to arrive at the true state of affairs (in all these sciences).'

PRATYABDA ŚODHANA FOR MEAN PLANETS

Bhāskara I gives an interesting and easier method to obtain the mean celestial longitudes of the heavenly bodies and of the special points. In this method (*pratyabda śodhana*), the mean sun is first determined by a short-cut method, and then the mean positions of all other bodies and points (like the Moon, Mars etc., and the apogee and the node of the Moon) from that of the sun already determined.

For example, let MS be the mean position of the Sun and GT be what Bhāskara I calls *graha tanu,* defined by

$$\textit{Graha tanu} = (\textit{Kali}\text{ years} \times 360^\circ) + MS \text{ in degrees.}$$

Then, Bhāskara I gives the formulae:

$$\text{Mean Moon} = \frac{83\ GT^\circ}{225} - \frac{11\ GT''}{50} + 13\ MS,$$

$$\text{Rāhu} = 6\ \textit{Rāśis} - \left[\frac{GT^\circ}{20} + \frac{GT^\circ}{270} + \frac{113\ GT''}{600}\right],$$

where GT° and GT'' denote the *graha tanu* in degrees and seconds of arc respectively. Similar formulae are given for other bodies.

ALGEBRA: INDETERMINATE EQUATION (*KUṬṬAKA*)

An important topic of algebraic interest that engaged the serious attention of Hindu mathematicians was the solution of the first degree indeterminate (i.e. Diophantine) equations (see article on Āryabhaṭa I). These are called *kuṭṭaka,* which literally means "pounding" or "pulverising". The two main types of such equations are called *niragra* (without remainder) and *sāgra* (with remainder) *kuṭṭākāras.* Though the word *kuṭṭaka* refers to the *method* of solution, usually in practice it is also used to mean *equation.*

The way that the problem arises is illustrated below.

Example: At the end of a certain number of days (i.e. *ahargaṇa,* the mean sun is found to be 3 *rāśis,* 12°40′1″. Find the *ahargaṇa* and the whole cycles (i.e., integral number of revolutions) completed by the Sun.

The Sun completes 43,20,000 revolutions in a *mahāyuga* consisting of 1,57,79,17,500 civil days. The number of revolutions for the above specified number of civil days reduces to 576 revolutions in 2,10,389 civil days (by cancelling the common factor 7500).

The mean position of the sun ($3^r 12°40'1''$) becomes 369601/1296000 of a revolution. Then the given problem takes the form of the first degree indeterminate equation :

$$\frac{576}{210389}x = y + \frac{369601}{12969000}$$

where x is the required *ahargaṇa* and y is the number of revolutions completed. We have to find out the values of the unknown integers x and y.

The equation, on simplification, takes the form

$$576x - 60,000 = 2,10,389y.$$

As a solution by Bhāskara, based on Āryabhaṭa's method, we obtain

$$x = 2,10,389n + 35,169 \ ; \ y = 576n + 96,$$

where n is any integer. For $n = 0, 1$ and 2, for example, we get

(i) $x = 35,169, \quad y = 96;$

(ii) $x = 2,45,558, \quad y = 672;$ and

(iii) $x = 4,55,947, \quad y = 1248$

Bhāskara I gives a device to avoid the difficulty when the *śeṣa* (or *kṣepa*), like 60,000 in the above example, is very large.

TRIGONOMETRIC RESULTS

(1) Āryabhaṭa I makes cryptic statements in his *Āryabhaṭīyam* giving some interesting trigonometric results. One of them is to find the sine of an angle greater than 90°. Bhaskara I gives the following results explicitly.

(i) $R \sin(90° + \theta) = R \sin 90° - R \text{ versin } \theta = R \cos\theta$

(ii) $R \sin(180° + \theta) = R \sin 90° - R \text{ versin } 90° - R \sin\theta = -R \sin\theta$

(iii) $R \sin(270° + \theta) = R \sin 90° - R \text{ versin } 90° - R \sin\theta + R \text{ versin } \theta$

$$= -R \sin 90° + R \text{ versin } \theta = -R\cos\theta$$

where R versin $\theta \equiv R(1-\cos\theta)$ is called *utkramajyā* θ. Bhāskara I also clearly explained the method of securing the sine table, obtaining the sine and versine of an intermediate angle by linear interpolation. The *Mahābhāskarīyam* (MB) gives an interesting approximate formula for calculating the sine of an acute angle without using a table. The formula is equivalent to

$$\sin A = \frac{4(180-A)A}{40500-(180-A)A} \tag{1}$$

where A is in degrees (*MB*, VII 17–19: see Bag 1979: 231; Gupta 1967). In Table 1 the values according to (1) are compared with actual values correct to 3 decimal places for angles from 0° to 90° at intervals of 10°.

Table 1 *Bhāskara's sine value*

Angle	sin *A*		Angle	sin *A*	
A	Bhāskara I	Actual	*A*	Bhāskara I	Actual
0°	0.000	0.000	50°	0.765	0.766
10°	0.175	0.174	60°	0.865	0.866
20°	0.343	0.342	70°	0.939	0.940
30°	0.500	0.500	80°	0.985	0.985
40°	0.642	0.643	90°	1.000	1.000

We see that the values from Bhāskara's approximate formula (1) agree with the actual ones to within one unit in the third decimal place, i.e. the maximum error is ± 0.001, which is insignificant in the type of further computations involved.

Note: Although (1) is valid $0 \leq A \leq 90°$, the expression can be used for any angle $\theta (0 \leq \theta \leq 360°)$ by considering the corresponding *acute* angle. E.g.: $\sin(220°) = \sin(180° + 40°) = -\sin 40° = -0.643$ (from Table 1).

Formula (1) has been extensively used by later Indian mathematicians and astronomers. In fact, the popular astronomer Gaṇeśa Daivajña (15th century) composed his famous text *Grahalāghavam* completely dispensing with sine and cosine using formula (1).

In the history of world mathematics Bhāskara I was the first to give such a simple and accurate rational approximation for the sine of an angle.

Geometrical Results

Area of a triangle

Starting from Āryabhaṭa's formula for the area of a triangle,

$$\text{area} = (1/2) \times \text{base} \times \text{altitude}$$

and expressing the altitude in terms of the sides, Bhaskara I gives a formula equivalent to the now well-known result

$$\text{area} = \sqrt{s(s-a)(s-b)(s-c)}$$

where $s = (1/2)(a+b+c)$, the semi-perimeter of the triangle, a, b and c being the three sides. A. K. Bag sketches the proof (Bag 1979: 149).

Some Interesting Examples

To illustrate Bhāskara I's methods three examples from his commentary on the *Āryabhaṭīyam* (*ĀB*, Shukla 1976: 289–334) are given below.

(1) "Correctly state, in accordance with the rules prescribed in *Bhaṭaśāstra* (i.e. *Āryabhaṭīyam*), the cube-root of 8291469824." (Ans: 2024). Ancient Hindu mathematicians clearly explained the method of obtaining the cube root of a number, but their method of cube root extraction is never taught now (*ĀB*, II, 5; Bag 1979: 80–81).

(2) "The two parallel faces of a figure resembling a *paṇava* (a drum-shaped musical instrument) are each 8 units, the central width 2 units, and the length between the faces is 16 units. Give the area of this figure resembling a *paṇava*." – *ĀB*. *Bhā*. II, 9; *Ud*. 4.

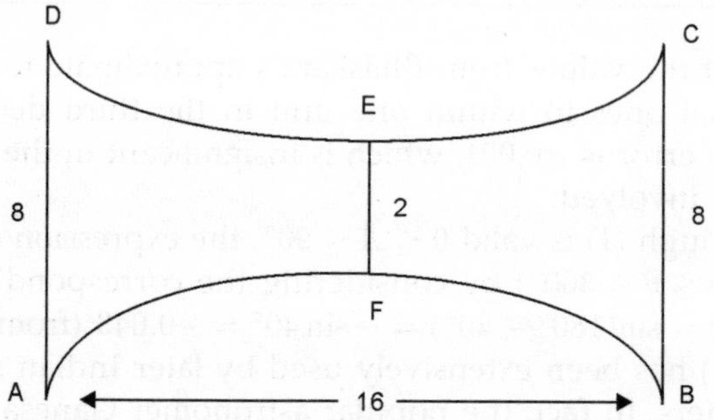

Figure 1 *Area of paṇava-shaped figure*

The figure AFBCED (Figure 1), resembling the musical instrument *paṇava* (*ḍhakkā, ḍamaru* – a double drum), is approximately a combination of two identical trapezia AFED and BFEC. Bhāskara I uses the approximate formula:

$$\text{area} = 1/2\left[\frac{\text{AD}+\text{BC}}{2}+\text{EF}\right]\text{AB}$$

and derives the area as 80 (square units). This example reappears in a later commentary on *Āryabhaṭīyam* of Raghunātharāja of Karnataka. In some *paṇava* shaped instruments the segments DE, EC, AF and FB are straight, in which case Bhāskara's formula is exact.

(3) A hawk is sitting on a pole whose height is 18 cubits. A rat which had gone out of its dwelling at the foot of the pole, to a distance of 81 cubits, while returning to its dwelling, is killed by the cruel hawk on the way. Say how far the rat had gone towards the hole, and determine the horizontal motion of the hawk (the speeds of both rat and hawk being the same) – *ĀB. Bhā.* II, 17; *Ud*, 3.

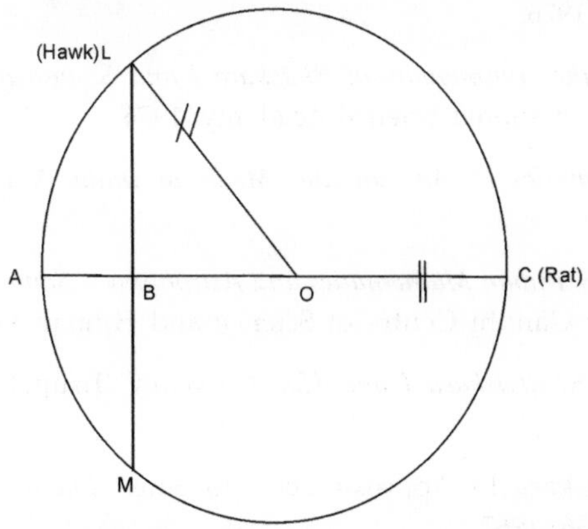

Figure 2 *Hawk-and-rat problem*

Bhāskara I gave two such "hawk-and-rat" problems. He gives the credit for such problems to his predecessors. Bhāskara's method consists of the following:

Draw a circle with centre O (Figure 2). Let ABOC be the horizontal diameter and LBM a vertical chord intersecting the diameter at B. Let BL be the pole, BC the track of the rat, and O the point at which the rat is caught. The hawk is perched at the top L of the pole and B is the hole, the abode of the rat. The rat has traversed 81 cubits to C. The hawk, sitting at L, sees the rat moving from C towards B and swoops down to attack it. The hawk catches the rat at O traversing a distance LO. As hawk and rat travel at the same speed, LO = CO. Drawing a circle with centre O and radius OL, we have

$$LB.\ BM = AB.\ BC \text{ or } LB^2 = AB\ .\ 81.$$

With LB = BM = 18 and BC = 81 we get AB = 4 cubits. Therefore, BO = (BC – AB)/2 = $38\frac{1}{2}$ cubits and CO = (BC – AB)/2 = $42\frac{1}{2}$ cubits.

The same or similar problems are repeated by later mathematicians; for example by Bhāskara II, who replaces the hawk-and-rat combination by a peacock-and-serpent pair.

S. Balachandra Rao

REFERENCES

Āryabhata I and his Astronomy. Tirupati: Rashtriya Sanskrit Vidyapeetha. 2003.

Āryabhaṭīyam, Ed. and Trans. K. S. Shukla et al. New Delhi: Indian National Science Academy. 1976.

Āryabhāṭīyam with the commentary of Bhāskara I and Someśvara. Ed. K. S. Shukla, New Delhi: Indian National Science Academy. 1976.

Bag, A. K. *Mathematics in Ancient and Medieval India*. Varanasi: Chaukambha Orientalia. 1979.

Balachandra Rao, S. *Indian Mathematics and Astronomy – Some Landmarks*. Revised 3rd Ed. Bangalore: Gandhi Centre of Science and Human Values. 2004.

Balachandra Rao, S. *Bhāskara I and His Astronomy*. Tirupati: Rashtriya Sanskrit Vidyapeetha. 2003.

Gupta, R. C. "Bhāskara I's Approximation to Sine." *Indian Journal of History of Science* 2(2): 121–136. 1967.

Mahā-Bhāskarīyam of Bhāskara I with the Bhāṣya of Govindsvāmin and super-commentary Siddhānta-dīpikā of Parameśvara. Cr. Ed. T. S. Kupanna Sastri, Madras: Government Oriental Manuscripts Library. 1957.

Srinivas, M. D. "Geometrical Picture of Planetary Motion According to Nīlakaṇtha," In *500 years of Tantrasaṅgraha: A Landmark in the History of Astronomy*. Ed. M.S. Sriram, K. Ramasubramanian, and Mandyam Doddamane Srinivasa. Shimla: Indian Institute of Advanced Study. 2002.

Bhāskara II

Undoubtedly, the greatest name in the history of ancient and medieval Indian astronomy and mathematics is that of Bhāskarācārya (b. CE 1114). His *Līlāvatī* is the most popular book of traditional Indian mathematics. He is usually designated as Bhāskara II in order to differentiate him from his earlier namesake who flourished in the early part of the seventh century.

According to Bhāskara's own statement towards the end of his *Golādhyāya,* he was born in Śaka CE 1036 or CE 1114. He also adds that he came from Vijjaḍaviḍa near the Sahya mountain. This place is usually identified with the modern Bijapur in Mysore. S.B. Dikshit is of the opinion that Bhāskara's original home was Pāṭaṇa (in Khandesh), where a relevant inscription was discovered by Bhau Daji in 1865. According to the inscription, Manoratha, Maheśvara, Lakṣmīdhara, and Caṅgadeva were the names of the grandfather, father, son, and grandson, respectively, of Bhāskara. Caṅgadeva was the chief astronomer in the court of King Siṅghaṇa and had established in CE 1207 a *maṭha* (residential institution) for the study of the works of his grandfather. Bhāskara's father, Maheśvara, was also his teacher.

Bhāskara's *Līlāvatī* (The Beautiful) is a standard work of Hindu mathematics. It belongs to the class of works called *pāṭī* or *pāṭīgaṇita*; that is, elementary mathematics covering arithmetic, algebra, geometry, and mensuration. Its popularity is shown by the fact that it is still used as a textbook in the Sanskrit medium institutions throughout India. It provides the basic mathematics necessary for the study of almost all practical problems, including astronomy. The subject matter is presented through rules and examples in the form of about 270 verses which can be easily remembered. There is a story that the author named the work to console his daughter Līlāvatī, who could not be married due to some unfortunate circumstances; but the truth of the story cannot be ascertained. Bhāskara addressed the problems to a charming female Līlāvatī, who, according to some scholars was his wife (and not his daughter).

The great popularity of *Līlāvatī* is illustrated by the large number of commentaries written on it since it was composed about CE 1150. Some of the Sanskrit commentators were: Parameśvara (about 1430), Gaṇeśa (1545), Munīśvara (about 1635), and Rāmakṛṣṇa (1687). Only a few of these have been published. Gaṇeśa's gives a good exposition of the text with a demonstration of the rules. However, the best traditional commentary is the *Kriyākramakarī* (ca. 1534), which is a joint work of Śaṅkara Vāriyar and Mahiṣamaṅgala Nārāyaṇa (who completed it after the demise of Śaṅkara).

There are a number of commentaries and versions in regional Indian languages. Quite a few modern scholars have edited, commented on, and translated *Līlāvatī*. At least three Persian translations are known, the earliest being by Abū al-Fayḍ Fayḍī (CE 1587). The English translation by H.T. Colebrooke (London, 1817) was based on the original Sanskrit text and commentaries.

Bhāskara's *Bījagaṇita* (Algebra) is a standard treatise on Hindu algebra. It served as a textbook for Sanskrit medium courses in higher mathematics. In it the author included an exposition of the subject based on earlier works. Among the sources named were the algebraic works of Śrīdhara and Padmanābha. Besides operations with various types of numbers (positive, negative, zero, and surds), it deals with algebraic, simultaneous, and indeterminate equations. There is a separate chapter on the Indian cyclic method called *cakravāla*. He attributes the method to earlier teachers but does not specify any name. Due to the difficult nature of some of the topics, the *Bījagaṇita* was not as popular as the *Līlāvatī*.

Bhāskara's *Siddhānta-śiromaṇi* (CE 1150) is an equally standard text-book on Hindu astronomy. It has two sections: *Grahaṇita* (Planetary Mathematics) and the *Golādhyāya* (Spherics). Often these two sections appear as independent works. There is a lucid commentary on the whole work by the author himself. It is called *Mitākṣarā* or *Vāsanābhaṣya*. Other commentators include Lakṣmīdāsa, Nṛsiṁha, Munīśvara (1638/1645), and Rāmakṛṣṇa. The fourteenth chapter of the *Golādhyāya* is the *Jyotpatti*, which may be regarded as a small tract on Hindu trigonometry.

Usually it is customary to regard *Līlāvatī, Bījagaṇita, Grahaṇita,* and the *Golādhyāya* as the four parts of the *Siddhānta Śiromaṇi* to make it a comprehensive treatise of Bhāskarācārya's Hindu mathematical sciences. His two other works are the *Karaṇa-kutūhala* (whose epoch is 1183), a handbook of astronomy, and a commentary on Lalla's *Śiṣyadhīvṛddhida-tantra* (eighth century CE). Some other works have also been attributed to him, but his authorship of *Bījopanaya* is questionable.

Bhāskara introduced a simple concept of arithmetical infinity through what he calls a *khahara,* which is defined by a positive quantity divided by zero, e.g. 3/0. His arithmetical and algebraic works are full of recreational problems to provide interesting pedagogical examples.

Perhaps the most important part of Bhāskara's *Algebra* is his exposition of the Indian cyclic method. We now know that the method was already known to Jayadeva (eleventh century CE or earlier). A modern expert, the late C.-O. Selenius, praised it by remarking that "no European performance in the whole field of algebra at a time much later than Bhāskara II, nay nearly up to our times, equalled the marvellous complexity and ingenuity of cakravāla method." Fermat proposed the equation $61x^2 + 1 = y^2$ in 1657 to Frénicle as a challenge problem. However, by applying the above method, Bhāskara had already solved the problem five centuries earlier. Bhāskara's solution (which he got in just a few lines) in its smallest integers was x = 226, 153, 980, y = 1, 766, 319, 049.

The feat was possible not only due to the technique but also because of a well-developed symbolic notation. Colebrook remarks

> Had an earlier translation of Hindu mathematical treatises been made and given to the public, especially to the early mathematicians in Europe, the progress of mathematics would have been much more rapid, since algebraic symbolism would have reached its perfection long before the days of Descartes, Pascal, and Newton.

Another gem from Bhāskara's *Algebra* is a very short proof of the so-called Pythagorean theorem.

The geometrical portion of *Līlāvatī* covers mensuration regarding triangles, quadrilaterals, circles, and spheres. A special rule gives the numerical lengths of the sides of regular polygons (from three to nine sides) in a circle of radius 60,000. The last chapter entitled *Aṅka-pāśa* is devoted to combinatorics.

Bhāskara's *Jyopatti* contains many trigonometrical novelties which appear in India first in this tract.

Although an equivalent of the differential calculus formula

$$\Delta \sin \theta = \cos \theta \Delta \theta$$

already appeared in Muñjāla's *Laghumānasa* (CE 932), Bhāskara II gave its geometrical demonstration. He knew that when a variable attains an extremum, its differential vanishes. He is credited even with a knowledge of the mean value theorem and the Rolle's theorem of differential calculus. A crude method of infinitesimal integration is implied in his derivation of the formula for the surface of a sphere. This he gave in the *Vāsanābhāṣya* on the *Golādhyāya* (chapter III).

Bhāskara's *Siddhānta-śiromaṇi* is one of the most celebrated works of Hindu astronomy. It is a comprehensive work of Brahma-pakṣa. He praised Brahmagupta, who belonged to the same school, but his own astronomical work became more famous.

Although based on the works of predecessors, rather than on any fresh astronomical observations, the *Siddhānta-śiromaṇi* served as an excellent textbook. Systematic presentation of the subject matter, lucidity of style, and simple rationales of the formulas made it quite popular. It also contained some improved methods and new examples. For instance, he gave a very ingenious method of finding the altitude of the Sun in any desired direction in his *Golādhyāya* (III, 46). His professional expertise and all-round knowledge made him a truly great and revered *ācārya* (professor) of Hindu astronomy and mathematics for generations.

R. C. Gupta

References

Colebrooke, H.T. *Algebra with Arithmetic and Mensuration from the Sanscrit of Brahmegupta and Bháscara*. London: Murray, 1817. Reprinted Wiesbaden: Martin Sandig. 1973.

Datta, B. and A.N. Singh. "Use of Calculus in Hindu Mathematics." *Indian Journal of History of Sciences* 19(2): 95–104. 1984.

Gupta, R.C. "Bhāskara II's Derivation for the Surface of a Sphere." *Mathematics Education* 7(2): 49–52, 1973.

Gupta, R.C. "Addition and Subtraction Theorems for the Sine and their Use in Computing Tabular Sines." *Mathematics Education* 8(3B): 43–46. 1974.

Gupta, R.C. "The *Līlāvatī* Rule for Computing Sides of Regular Polygons." *Mathematics Education* 9(2B): 25–29. 1975.

Jha, A., ed. *Bīja Gaṇita of Bhāskarācārya*. Banaras: Chowkhamba. 1949.

Kunoff, Sharon. "A Curious Counting/Summation Formula from the Ancient Hindus." In *Proceedings of the Sixteenth Annual Meeting of the CSPM*. Edited by F. Abeles et al. Toronto: CSHPM. 1991. pp. 101–107.

Sastri, Bapu Deva. ed. *The Siddhānta-śiromaṇi by Bhāskarācārya with His own Vasanābhāṣya*. Benares: Chowkhamba. 1929.

Selenius, Clas-Olof. "Rationale of the Chakravāla Process of Jayadeva and Bhāskara II." *Historia Mathematica* 2(2): 167–184. 1975.

Srinivasiengar, C.N. *The History of Ancient Indian Mathematics*. Calcutta: World Press. 1967.

See also: Algebra – Mathematics – Geometry – Parameśvara – Munīśvara – Trigonometry – Combinatorics – Muñjāla

Al-Bīrūnī

Abū Rayḥān Muḥammad ibn Aḥmad al-Bīrūnī was born on Thursday, 3rd of Dhū al-Hijjah, 362 H (4th September, CE 973) at "Madīnah Khwārizm". His exact birthplace is still a matter of dispute. It is conjectured that he was born in the outskirts (*bīrūn*) of Kāth, at al-Jurjāniyah, Khwārizm or at a place called Bīrūn, as implied by his nickname al-Bīrūnī. The only clue given by al-Bīrūnī was that he was born in a city in Khwārizm. The name Abū Rayḥān (perfume or herb) was given to him because of his love for sweet fragrances. Al-Bīrūnī died in 443 H (CE 1051).

Al-Bīrūnī was a devout Muslim, yet there was no conclusive evidence of his adhering to any particular *madhḥab* (denomination) throughout his life. His native language was the Khwarizmian dialect. He knew Persian but preferred Arabic because he believed that Arabic was more suitable for academic pursuit. He received some of his early education under the tutelage of the astro-mathematician Abū Naṣral-ʿIrāq and Abd al-Samad from Khwārizm in addition to his formal education at the *madrasah*, an institution where Islamic sciences are studied.

Al-Bīrūnī's first patron was the Sāmānid Sultan Abū Sālih Manṣūr II who reigned in Bukhara until the city was invaded by the Ghaznavid Sultan Maḥmūd in 389 H (CE 999). Later al-Bīrūnī went to Jurjān to the court of Abu'l Haṣṣan Qābūs ibn Washmjīr Shamṣ al-Ma ʿalī (r. CE 998–1012), under whose patronage he wrote *alāthār al-bāqiya min al-Qurūn al-khāliya* (Chronology) which was completed in 390 H (CE 1000). Al-Bīrūnī found the Sultan indiscriminate and harsh.

His next sojourn was in Khwārizm and Jurjāniyah, under the service of the Sāmānid Prince Abu'l Abbas al-Maʾmūn ibn Muḥammad II. He was the best patron and al-Bīrūnī received the respect he very much deserved. It was during this period that al-Bīrūnī met the physician al-Jurjāni. His *Taḥdīd Nihāyāt al-amākin li-taṣḥiḥ masāfāt alMasākin* (Determination of the Coordinates of Cities) was completed in CE 1025, and his *Kitāb fī taḥqīq ma li'l Hind* (Book on India) was finally published in 421 H (CE 1030). Al-Bīrūnī's other book, *Kitāb altafhīm li-awāʾil sināʿat al-tanjīm* (The Book of Instruction in the Art of Astrology) which was dedicated to Rayḥānah, daughter of al-Hassan, was written in Ghaznah, CE 1029. Al-Bīrūnī's *magnum opus* was an astronomical encyclopaedia, *al-Qānūn al-Masūdī fī al-hay'ah wa 'ltanjīm* (Canon Masudicus) which comprises eleven treatises divided into 143 chapters. It was completed in 427 H (CE 1035). Apart from emphasising the importance of astronomy, he gives accurate latitudes and longitudes and also geodetic measurements. His *Kitāb al-jamāhir fī maʿrifat al-jawāhir* (Mineralogy) was completed less than a decade later (435 H/ CE 1043). Al-Bīrūnī's *Kitāb al-ṣaydanah fiʿl-ṭibb* (Materia Medica, or Pharmacology) was completed only in the form of a rough draft before he died. The list of books mentioned thus far is not exhaustive; altogether he wrote about 146 treatises.

Al-Bīrūnī was both a scientist and a philosopher, but he never used the word "science" in the sense the word is understood today—that knowledge which is popularly thought to be exact, objective, veritable, deductive, and systematic. He used the Arabic word *ʿilm,* which means knowledge.

Solving scientific problems, which to al-Bīrūnī is analogous to "untying knots", (*Kitāb fī ifrād al-maqāl fī umr al-ẓlāl* [The Exhaustive Treatise On Shadows]) is the main activity of scientists. That science to al-Bīrūnī was a problem-solving activity and a scientific problem was a problem circumscribed by the Holy *Quʾrān* and *Sunnah* can be discerned by examining, in particular, the Introductions of his major books. *India,* for example, was written by al-Bīrūnī primarily because in his opinion, "while Muslims had been able to produce fairly objective works on such religions as Judaism and Christianity, they had been unable to do so with regard to Hinduism". In *Taḥdīd,* he clearly states another aspect of a scientific problem. He says that "geography is very essential for a Muslim for knowing the right direction of Mecca (*qibla*)".

Al-Bīrūnī was certainly a very prolific, multidimensional scholar. He did serious work in almost all branches of science in his time and his 146 treatises range from 10 to 700 pages each.

Abdul Latif Samian

References

Primary sources

al-Āthār al-bāqiya min al-gurūn al-khāliya. English trans. by Edward Sachau, *The Chronology of Ancient Nations*. London: Minerva GMBH. 1879.

al-Qānūn al-Masᶜūdī (Canon Macudicus). Hyderabad: Osmania Oriental Publications Bureau. 1954–1956.

Kitāb al-jamāhir fi marifat al-jawāhirr. Ed. by F. Krendow. Hyderabad-Dn: Osmania Oriental Publications Bureau. 1936.

Kitāb al-ṣaydanah fiᶜl-ṭibb. Ed. and trans. by Hakid Mohammed Said et al. Pakistan: Hamdard Academy. 1973.

Kitāb al-tafhīm li-awāᶜil ṣinā ᶜat al-tanjīm. Tehran: Jalal Humāᶜl, 1940. English trans. by R. Ramsay Wright, *The Book of Instruction in the Art of Astrology*. London: Luzac. 1934.

Kitāb fi ifrād al-maqāl fi amr al-ẓilāl. Hyderabad-Dn: Osmania Oriental Publications Bureau, 1948. Ed. and trans. as *The Exhaustive Treatise on Shadows*. E.S. Kennedy. Syria: University of Aleppo. 1976.

Kitāb fi taḥqīq ma liᶜl Hind. English trans. by Edward Sachau, *Alberuni's India*. London: Trubner. 1888.

Taḥdīd nihāyat al-amākin li-taṣḥiḥ. English trans. by Jamil Ali, *The Determination of the Coordinates of Cities, al-Biruni's Tahdid al-Amakin*. Beirut: American University of Beirut. 1967.

Secondary sources

Boilot, D.J. "L'oeuvre d' al-Bīrūnī: essai bibliographique." *Mélanges de l' Institut dominicain d'études orientales* 2:161–256, 1955.

Kennedy, E.S. "al-Bīrūnī". In *Dictionary of Scientific Biography,* vol. II. New York. Charles Scribner's Sons, 1970.

Khan, Ahmad Saeed. *A Bibliography of The Works Of Abū'l Rāihān Al-Bīrūnī*. New Delhi: Indian National Science Academy. 1982.

Nasr, Seyyed Hossein and William C. Chittick. *al-Bīrūnī: An Annotated Bibliography of Islamic Science.* Tehran: Imperial Iranian Academy of Philosophy, 1975.

Nasr, Seyyed Hossein. *An Introduction to Islamic Cosmological Doctrines*. Revised edition. Albany: State University of New York Press. 1993.

Said, Hakim Mohammed and Ansar Zahid Khan. *Al-Bīrūnī—His Times, Life and Works*. Karachi: Hamdard Foundation. 1979.

Al-Bīrūnī: Geographical Contributions

The geographical knowledge of the Muslims, in part derived from the Greeks and others, and contemporaneously developed and advanced by themselves, had reached a very high level of development by the tenth century. It is in this development that the work of al- Bīrūnī is significant. Al-Bīrūnī presented a critical summary of the total geographical knowledge up to his own time. He made some remarkable theoretical advances in general, physical, and human geography. Al-Bīrūnī did not confine himself to a simple description of the subject matter with which he was concerned. He compared it with relevant materials and evidence, and evaluated it critically, offering alternative solutions.

George Sarton identifies al-Bīrūnī as one of the great leaders of this period because of his relative freedom from prejudice and his intellectual curiosity. Although his interests ranged from mathematics, astronomy, physics and the history of science to moral philosophy, comparative religion and civilization, al-Bīrūnī became interested in geography at a young age. He is considered to be the greatest geographer of his time.

In the area of physical geography, he discussed physical laws in analyzing meteorology and climatology. He wrote of the process of streams development and landscape evolution. He introduced geomorphological enquiry to elucidate a history of landscape. He developed the mathematical side of geography, making geodetic measurements and determining with remarkable precision the coordinates of a number of places. Some of his noteworthy contributions to geography include: a theory of landform building processes (erosion, transportation, and deposition); proofs that light travels faster than sound; explanations of the force of gravity; determination of the sun's declination and zenithal movement; and discussion of whether the Earth rotates on its axis. He described various concepts of the limits for which he seems to have had recourse to contemporary sources not available to earlier geographers. He made original contributions to the regional geography of India.

In the study of physical phenomena, including landforms, weather, and geology, al-Bīrūnī adopted the methods of the physical sciences and drew conclusions with scientific precision. Long before Bernhard Varenius (CE 1622–50), al-Bīrūnī developed a schema for physical geography: (a) terrestrial conditions, describing the shape and size of the Earth; (b) cosmic concepts, dealing with the measurement of the circumference of the Earth and the establishment of the exact location of places; (c) classification of natural phenomena either in accordance with their nature or with their position in time and space. He studied phenomena in time (chronological science) and also tried to study them in space. In his view, geography was an empirical science.

Based on available knowledge concerning the surface of the Earth, he deduced and described the shape and forms of land surface. Al-Bīrūnī examined questions concerning the Earth's shape, size, and movement.

He explained running water as the most effective agent by which the surface of the land is sculpted. He further asserted that as the rivers of the plains of India approached the sea they gradually lost their velocity and their power of transportation, while the deposition process along their beds increased proportionately. Al-Bīrūnī considered the changes in the course of a river a universal phenomenon. He also recognised the influence of the sun upon the tide and suggested that heavenly bodies exert a gravitational effect on the tides.

Al-Bīrūnī recognised that the heat of the atmosphere and the Earth's surface is derived from the sun through the transfer of energy by rays, and that it varies with the length of time that the Earth is exposed to the rays. He recognised the wind's force and velocity and argued that the wind, in all its phases, it determined by certain causes.

Al-Bīrūnī noticed the peculiarities of the Indian monsoon, observed the time of its breaking, and described its westward and northward movements and the unequal distribution of rain in different areas of India.

Finally, he added that the habitable world does not reach the north on account of the cold, except in certain places where it penetrates into the north in the shape, as it were, of tongues and bays. In the south it reaches as far as the coast of the ocean, which in the east and west is connected with what he calls the comprehending ocean (*India*, Vol. I, p. 196).

In short, al-Bīrūnī recognised geography as an empirical science, and he dealt with the terrestrial globe as a whole. He stressed its nature and properties. He also tried to investigate the causes of global phenomena and described them as they exist.

Akhtar H. Siddiqi

REFERENCES

Ahmad, S. Maqbul. "Djughrafia." (Geography) In *The Encyclopedia of Islam*, Vol. II. Leiden: E.J. Brill, 1965, pp. 575–87.

Barani, S. Hasan. "Muslim Researches in Geodesy." In *Al-Bīrūnī Commemorative Volume.* Calcutta: Iran Society, 1951, pp. 1–52.

Kazmi, Hassen Askari. "Al-Bīrūnī's Longitudes and their Conversion into Modern Values." *Islamic Culture* 49: 165–76, 1974.

Kazmi, Hassen Askari. "Al-Bīrūnī on the Shifting of the Bed Amu Darya." *Islamic Culture* 50: 201–11, 1975.

Kramer, J.H. "Al-Bīrūnī's Determination of Geographical Longitude by Measuring the Distance." In *Al-Bīrūnī Commemorative Volume.* Calcutta: Iran Society, 1951, pp. 177–93.

Memon, M.M. "Al-Bīrūnī and his Contribution to Medieval Muslim Geography." *Islamic Culture* 33: 213–18, 1959.

Sachau, Edward C. *The Chronology of Ancient Nations*. London: W.H. Allen and Co., 1879.

Sachau, Edward C. *Alberuni's India*, Vols I and II. London: Kegan Paul, Trench, Trubner and Co., 1910.

Sarton, George. *Introduction to the History of Science*, Vol. I. Baltimore: Williams and Wilkins, 1927, pp. 693–737.

Sayili, Aydm. "Al-Bīrūnī and the History of Science." In *Al-Bīrūnī Commemorative Volume*, ed. M.M. Said. Karachi: Times Press, 1979, pp. 706–12.

See also: Geography in Ancient India – Maps and Mapmaking in India

Brahmagupta

"Brahmagupta holds a remarkable place in the history of Eastern civilisation. It was he who taught the Arabs astronomy before they became acquainted with Ptolemy" (Sachau, 1971). Bhāskara II described Brahmagupta as *Gaṇakacakracūdāmani* (Jewel among the circle of mathematicians).

Brahmagupta was born in CE 598 according to his own statement:"... when 550 years of the Śaka era had elapsed, Brahmagupta, son of Jisṇu, at the age of 30, composed the *Brāhmasphuṭasiddhānta* for the pleasure of good mathematicians and astronomers". Thus he was 30 years old in Śaka 550 or CE 628 when he wrote the *Brāhmasphuṭasiddhānta.* That he was still active in old age is clear from the title epoch of CE 665 used in another of his works called *Khaṇḍa-khādyaka*. Pṛthūdaka Svāmin, an ancient commentator on Brahmagupta, calls him Bhillamālācārya, which shows that he came from Bhillamāla. This place has been identified with the modern village Bhinmal near Mount Abu close to the Rajasthan–Gujarat border.

We have no knowledge of Brahmagupta's teachers, or of his education, but we know he studied the five traditional *Siddhāntas* on Indian astronomy. His sources also included the works of Āryabhaṭa I, Lāṭadeva, Pradyumna, Varāhamihira, Simha, Śrīṣeṇa, Vijayanandin, and Viṣṇucandra. He was, however, quite critical of most of these authors.

The *Brāhmasphuṭasiddhānta* is Brahmagupta's most important work. It is a standard treatise on ancient Indian astronomy, containing twenty-four chapters and a total of 1008 verses in *āryā* meter. The *Brāhmasphuṭasiddhānta* claimed to be an improvement over the ancient work of the Brahmapakṣa, which did not yield accurate results. Brahmagupta used a great deal of originality in his revision. He examined and criticised the views of his predecessors, especially Āryabhaṭa I, and devoted two chapters to mathematics. There have been many commentators on this work. The earliest known was Balabhadra (eighth century CE), but his commentary is not extant.

Chapter 7 is on *Gaṇita* (Mathematics). It deals with elementary arithmetic, algebra, and geometry. The subject is presented in twenty-eight topics of logistics (arithmetical operations) and determinations, including problems related to mixtures, plane figures, shadows, series, piles, and excavations. He wrote in a concise and understandable style, whether dealing with simple mathematics or complex geometry. In the treatment of surds, Brahamagupta is remarkably modern in outlook. The *Brāhmasphuṭasiddhānta* includes formulas for the rationalisation of the denominator, as well as a marvellous piece of pure mathematics in the rule for the extraction of the square root of a surd. Still more remarkable algebraic contributions are contained in a chapter entitled *Kuṭṭaka*, which is a traditional name for indeterminate analysis of the first degree. The second order indeterminate equation

$$Nx^2 + c = y^2 \qquad (1)$$

is called *varga-prakṛti* (square nature). An important step towards the integral solutions of such equations is what is called Brahmagupta's *Lemma* in the history of mathematics. In modern symbology the Lemma is as follows:

> If (α, β) is a solution of (1) with $c = k$, and (α', β') is its solution with $c = k'$, then $(\alpha\beta' \pm \alpha'\beta, \beta\beta' \pm N\alpha\alpha')$ will be its solution with $c = kk'$.

This lemma not only helps in finding any number of solutions from just one solution, but it also helps in solving the most popular case of $c = 1$, provided we know a solution for $c = -1$, or ± 2 or ± 4. Euler rediscovered it in Europe in 1764.

In geometry, Brahamagupta's achievements were equally praiseworthy. He wrote a fine symmetric formula for the area of a cyclic quadrilateral, which appeared for the first time in the history of mathematics. Even more important are his expressions for the diagonals of a cyclic quadrilateral.

Brahmagupta's name has been immortalised by yet another achievement. A "Brahmaguptan quadrilateral" is a cyclic quadrilateral whose sides and diagonals are integral (or rational) and whose diagonals intersect orthogonally. He gave a simple rule for forming such figures in the *Brāhmasphuṭasiddhānta* (Chapter 7, Verse 38): If a, b, c and α, β, γ are the sides (integral or rational) of two right-angled triangles (c and γ being hypotenuses), then $a\gamma, b\gamma, c\alpha$ and $c\beta$ are the required sides of a Brahamaguptan quadrilateral.

Prior to Brahmagupta, the usual method for computing the functional value intermediary between tabulated values was that of linear interpolation, which was based on the rule of proportions. He was the first to give second order interpolation formulas for equal as well as unequal tabulated argumental intervals. Mathematically, his rule is equivalent to the modern Newton–Stirling interpolation formula up to the second order.

The *Khaṇḍa-khādyaka* is a practical manual of Indian astronomy of the *Karaṇa* category. The author claims that it gives results useful in everyday life, birth, marriage, etc. quickly and simply, and is written for the benefit of students. The work consists of two parts called the *Pūrva* and the *Uttara*. The former comprises the first nine chapters and expounds the midnight system. The latter six chapters provide corrections and additions. This work has been studied by a great number of commentators, from Lalla in the eighth century to Āmarāja in the twelfth. It was translated into Arabic first by al-Fazārī (eighth century) and then by al-Bīrūnī.

Brahmagupta's genius made use of mathematics (traditional as well as that which he developed) in providing better astronomical methods. He used the theory of quadratic equations to solve problems in astronomy. He knew the sine and cosine rule of trigonometry for both plane and spherical triangles. He supplied standard tables of Sines and Versed Sines.

The historian of science George Sarton called him "one of the greatest scientists of his race and the greatest of his time".

R. C. Gupta

References

Primary sources

Sharma, R.S. et al., eds. *The Brāhma-sphuṭa-siddhānta* (with Hindi translation). 4 volumes. New Delhi: Indian Institute of Astronomical and Sanskrit Research. 1966.

Chatterjee, Bina, ed. and trans. *Khaṇḍakhādyaka with the commentary of Bhaṭṭotpala*. Calcutta: World Press. 1970.

Sengupta, P.C., trans. *The Khaṇḍakhādyaka of Brahmagupta*. Calcutta: Calcutta University, 1934.

Secondary sources

Colebrooke, H.T. *Algebra with Arithmetic and Mensuration from the Sanscrit of Brahmegupta and Bháscara*. London: Murray. 1817.

Gupta, R.C. "Second Order Interpolation in Indian Mathematics." *Indian Journal of History of Science* 4: 86–98. 1969.

Gupta, R.C. "Brahamagupta's Rule for the Volume of Frustum-like Solids." *Mathematics Education* 6(4B): 117–120. 1972.

Gupta, R.C. "Brahmagupta's Formulas for the Area and Diagonals of a Cyclic Quadrilateral." *Mathematics Education* 8(2B): 33–36. 1974.

Kak, Subhash. "The Brahmagupta Algorithm for Square-Rooting." *Gaṇita Bhāratī* 11: 27–29. 1989.

Kusuba, Takanori. "Brahmagupta's Sūtras on Tri- and Quadrilaterals." *Historia Scientarum* 21: 43–55. 1981.

Pottage, John. "The Mensuration of Quadrilaterals and the Generation of Pythagorean Triads etc." *Archive for History of Exact Sciences* 12: 299–354.

Sachau, Edward, trans. *Alberuni's India*. New York: Norton. 1971.

Sarton, George. *Introduction to the History of Science*. Baltimore: Williams and Wilkins. 1947.

Venkutschaliyenger, K. "The Development of Mathematics in Ancient India: The Role of Brahmagupta." In *Scientific Heritage of India*. Ed. B.V. Subbarayappa and S.R.N. Murthy. Bangalore: Mythic Society. 1988. pp. 36–47.

See also: Geometry – Arithmetic: *Pāṭīgaṇita* – Algebra: *Bījagaṇita* – Astronomy.

Bricks

Bricks in the Vedic Age

The earliest textual references to bricks are contained in the *Śatapatha Brāhmaṇā* (S.B.) which gives us the etymology of the word *iṣṭakā* for brick, "*tat yat istat samabhavan tasmāt istakah; tasmāt angina istaka pacanti*. (that using which *iṣṭi* (a sacrifice) is performed is *iṣṭakā* (brick); the *iṣṭaka* are baked in the fire). As these bricks were invented in the course of *isti* or sacrifice, they are called *istaka* (S.B. Vi. W. 1.1 8.9; vi 2.3.2). Bhagwan Singh (1995: 285, f.n.5) refers to the *Yajurveda*, which mentions a large variety and names of bricks and concludes that the art of bricklaying and craft of manufacturing bricks of various shapes were well developed in the Vedic period. The *Taittarīya Samhitā* (T.S.) mentions circular bricks (*vikrani*) (T.S.v. 3.7), conical bricks (*coda*) (T.S.v. 4.3), gold headed bricks (*vimbhrt*) (T.S.V.2.9 ; 5.3 etc), pot shaped bricks (*kumbhestake*) (T.S.v.6.1) and others with markings. According to Sarkar, the brick mantras refer to the manner in which Angiras placed the bricks firmly or invented them (Bhagwan Singh 1995: 289, f.n.6). In fact the use of bricks in sacrificial as well as secular structures by the Harappans (Indus Valley Civilisation) was recorded in Vedic literature. For instance, they mention *kupa* (well), *puṣkaraṇi* (public bath), *durga durna* (fortification), *surmi* (covered drain) and *kakud* (water storage chamber). The water ducts and public bath of Mohenjodaro, the water chamber of Dholavira and the covered drains of Lothal have all been well studied in the Indus Civilisation.

Bricks as Building Materials

The earliest users of mud bricks or sun-dried bricks were the Neolithic people who had settled at Mehrgarh (Jarrige, 1984) by 6000 BCE in Baluchistan. They knew agriculture and gradually used pottery also. Subsequently the Harappans, as the inhabitants of the Indus Valley Civilisation are known, used both kiln-fired bricks and mud bricks, the former mostly for drains, baths, dockyards and sacrificial altars. Three phases of this civilisation have been recognised in the Indus (Pakistan) and Saraswati (Haryana, Punjab, Rajasthan, and Western Uttar Pradesh) valleys and in Gujarat. The three phases are Pre-Harappa, Harappa (Mature) and Late Harappa, corresponding to the pre-urban, urban and de-urbanised phases of Harappan settlements, dating from 3400 to 1500 BCE (Rao, 1991).

The Neolithic people of Mehrgarh built cell-like buildings with complex plans using mud bricks mixed with chaff. These structures were divided into narrow rectangular compartments and very small square cells arranged symmetrically on either side of the corridor. Perhaps they were prototypes of the famous granaries of Harappa and Mohenjodaro, used by the Mehrgarh people for storing grains by 4000 BCE. They laid bricks in alternate layers of headers and stretchers corresponding to the later "English bond" in masonry. In the early Harappan site of Banawali in Haryana, house walls were built in single mud bricks. Burnt bricks were used for drains and both were in the standard ratio of 1:2:3 (thickness, breadth, length).

Brick Size

According to V.B. Mainkar (1984) of the Indian Standards Institution, the bricks of the Indus Valley Civilization from Baluchistan to the Godavari River were made in dimensions which were integral multiples of large gradations of the ivory scale of the Harappan port of Lothal in Gujarat, namely 25.56 mm. The smallest division on the ivory scale is 1.704 mm. The burnt brick wall of the dockyard above ground at Lothal (2300 BCE) measures 1.04 m which is equal to 40 large gradations on the Lothal scale (Figure 1). In the pre (Early) Harappa phase (3000 BCE) of Kalibangan in Rajasthan, the bricks measure $10 \times 20 \times 30$ cm. At Banawali the brick sizes are $12 \times 24 \times 36$ cm and $13 \times 26 \times 39$ cm. Mud bricks made of fine alluvial clay without any *kankar* (lime) nodules used in the Indus settlements of the third millennium BCE are found to be stronger than those made of black clay containing *kankar* used in the second millennium BCE. Burnt radial bricks were used in the construction of wells and at the turnings of drains (Figure 2).

Figure 1 *Ivory scale with decimal gradation marks, Lothal (Gujarat). (Courtesy of the Archaeological Survey of India).*

Figure 2 *Radial bricks used in the construction of a well, Lothal. (Courtesy of the Archaeological Survey of India).*

Indus Valley bricks come in two sizes. The normal size of Harappa (Mature) mud bricks in the Indus Valley is $10 \times 20 \times 40$ cm, keeping the ratio 1:2:4. The smaller ones are $7.5 \times 15 \times 30$ cm. This ratio of one length to two breadths helped the masons to use complete bricks as headers or stretchers. Because of the extensive use of mud bricks for building anti-flood structures such as the protective walls enclosing the citadel as in Harappa, or the entire township as in Kalibangan and Lothal, it was not possible or economical to produce millions of kiln-fired bricks. It is only in the case of the most important structures, namely the dock at Lothal, that millions of kiln-fired bricks were used. This is because tidal water entered the basin of the dock where sea sailing ships were berthed. The prosperity of Lothal depended on overseas trade which necessitated maintaining its carefully built dock in good condition to withstand erosion by floods and tidal water. The builders even used overburnt bricks to withstand the salt effect. When floods eroded the massive mud brick protective walls of Harappa and the platforms on which houses were built in Lothal, burnt bricks were also used. Burnt brick revetments had to be provided to the affected mud brick platforms at both places.

The enormity of cooperative effort put in by the Harappans in saving their cities from floods can be gauged from the 4 m-high mud brick podium of the warehouse of Lothal (Figure 3), covering an area of 1930 sq.m. Originally 74 cubical blocks of mud bricks stood here under a timber superstructure where cargo was handled and sealed. Lothal withstood major floods in 2200 BCE and again in 2000 BCE but not the deluge in 1900 BCE which destroyed the dock and warehouse. The residents who returned after the floods could not afford to produce new burnt bricks and therefore started using old damaged bricks in baths. In most houses soakage jars replaced burnt brick. If the most common size of the mud bricks is compared with that of burnt bricks, the difference is within the limits of permissible fire-shrinkage, namely 8 to 15 per cent. The difference also suggests the use of common moulds for both bricks. Brick firing was done on modern principles.

Figure 3 *Mud brick podium of the warehouse built 4 m high as an anti-flood device, Lothal. (Courtesy of the Archaeological Survey of India).*

THE GREAT BATH AT MOHENJODARO

Within the citadel is a public bath of burnt bricks built in a courtyard (Figure 4). The bath is 11.89 × 7.0 m; the depth is 2.44 m. The brick-paved courtyard is surrounded by verandahs, at the back of which are ranged rooms on three sides. A double-ringed well in one of the rooms supplied water to the bath. The floor of the bath is reached by a flight of steps constructed by laying brick-on edge in gypsum mortar which is also used for the walls of the bath. As a further precaution, a damp-proof course of bitumen, 2 cm thick, was introduced between the facing bricks of the basin and intermediate wall of baked bricks, which itself was retained by a mud brick packing and the outermost baked brick wall. In the treads of the brick steps leading to the floor of the bath, timber was set in bitumen or asphalt. An outlet of the bath led to a large drain with a corbelled arch. A staircase from a room of the verandah led to the terrace of an upper storey. The longer side of the Great Bath is equal to a length of 40 bricks, the smaller side to a length of 24 bricks and the depth equal to the length of 8 bricks, the ratio being 5:3:1.

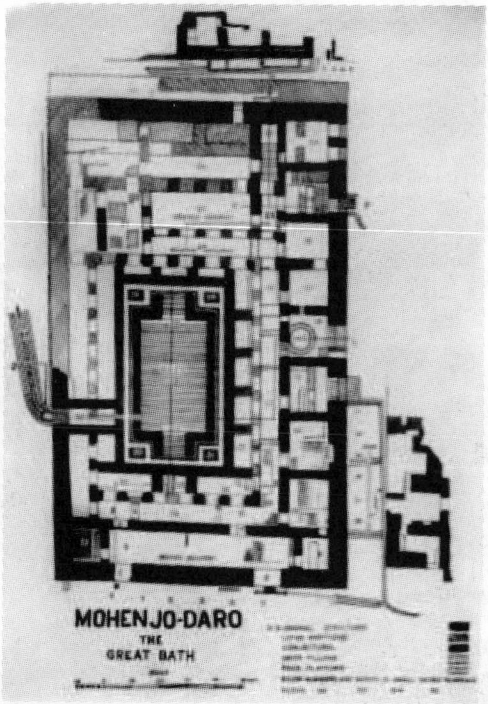

Figure 4 *The Great Bath of burnt bricks, Mohenjodaro. (Courtesy of the Archaeological Survey of India).*

PRIVATE BATHS

Lothal had private baths larger than those in Mohenjodaro. A row of twelve houses with brick-paved baths was connected to the public sewer through a runnel. In the Acropolis at Lothal each runnel had an inspection sump to clear solid waste manually so that only liquid waste entered the sewer (Figure 5). The bricks were rubbed down to obtain fine joints to make the masonry watertight. The floor of polished bricks was plastered with lime and wainscoted or skirted with bricks laid on edge or flat. The houses in the Lower Town had large baths measuring 2.5 × 1.8 m.

Figure 5 *Brick-paved baths connected through runnels and sump to an underground sewer, Lothal. (Courtesy of the Archaeological Survey of India).*

ROLE OF BRICKS IN SANITARY ENGINEERING

All major Harappan towns had well-planned sanitary arrangements to keep the entire town clean by providing surface and underground drains, cesspools and water chutes. The internal width of the main sewers to which private drains were connected varied from 1 to 0.6 m. The drop in the terminal part of the sewer was 1.02 m over a length of 9 m, thereby providing a steeper gradient for self-clearance. The curve at the turn of the drain was smooth enough to minimise friction. A unique feature of the main sewer in the Acropolis was the provision of drops from 0.5 m to 1.5 m to reduce the velocity of liquid waste, but its depth increased at the terminal to 1.5 m before discharging into the cesspool. The grooves in the apron wall of the terminal were meant to insert a wooden screen to hold back solid waste, if any. Finally another sewer discharged the waste into the dock where it was washed away at high tide. Another feature of the main sewer is that its floor was paved with polished brick slopes on either margin, keeping the central one-third part flat to provide two smaller marginal channels. The purpose of this measure was to change the angle of flow and thereby reduce its velocity. Second, the marginal drains carried normal sewage while the whole drain carried both sewage and storm water as in modern sewers (Figure 6).

Figure 6 *Underground sewer designed to carry both sewage and storm water, with stepped floor to reduce velocity of the stream, Lothal. (Courtesy of the Archaeological Survey of India).*

THE DOCK AT LOTHAL

The largest structure built in the Indus Civilization using millions of burnt bricks was the tidal dock at the port town of Lothal for sluicing ships sailing from the Gulf of Cambay into the basin in high tide. The importance of the tidal dock in the economy of Lothal is evident from the use of burnt bricks which were too precious to be used even for the dwellings in the Acropolis. The basin of this dock measures 210 m north-to-south and 35 m east-to-west with an inlet channel 12 m wide in the northern (shorter) arm. The Lothal dock had a unique lock-gate system, as early as 2300 BCE. There was provision for a wooden door to close the outlet in its southern arm, which could be lifted at high tide and closed at low tide to retain about 1.5 m water depth. The original height of the embankment of baked brick walls was 4.15 m but after two great floods it got reduced to 3.3 m with 42 courses of bricks. At the foundation level the width of the wall was 1.75 m but with two offsets it was reduced to 1.2 m.

Figure 7 *Brick wall of the Dockyard in plumb, Lothal. (Courtesy of the Archaeological Survey of India).*

The inner face of the wall was in plumb (Figure 7) to enable boats to touch the edge of the wharf for easy handling of cargo. From the dock five perforated stone anchors were recovered, one of which was triangular. L.S. Leshnik (1968) suspected that the basin was a water storage tank and the stone anchors might be counterweights for lift irrigation. H.D. Sankalia, who subsequently studied the dock, wrote, "Leshnik's outlook is initially vitiated and prejudiced and therefore he refuses to regard Lothal as a port though he concedes its proximity to the sea. The use of small stones as counterweights of shaduf in lift irrigation is confined to narrow mouthed wells" (Sankalia 1974: 374). They are not used in tanks. Oza (1960, see Rao 1991:143), Director of Ports in Ahmedabad, has concluded that an artificial enclosure (at Lothal) was constructed for shipping and the comfortable working of cargo and safety of boats. N.K. Panikkkar (1971), Director of the National Institute of Oceanography (NIO), who examined the dock, observed that, "since the Lothal dock was a purely tidal one, the Lothal engineers must have possessed adequate knowledge of tidal effects, including their amplitude, erosion and thrust. Using this knowledge, they were capable of receiving ships at high tide and ensuring that ships at low tide could float. This is perhaps the earliest example of tidal phenomena being put to a highly practical purpose both in the selection of a site having the highest tidal amplitude and in adopting a method of operation for entry and exit of ships" (Panikkar and Srinivasan 1971). I R.Nigam, a scientist at NIO who collected samples from the floor of the dock in 1986, says, "The analysis shows the profuse presence of the microorganisms *foraminifera* indicating that tidal sea water used to enter the basin" (Nigam 1988: 20–22).

MORTAR

Generally mud mortar was used as a binding material, reserving the use of lime mortar for drains, baths, sinks and water chutes. In the dock, mud mixed with lime was used. Gypsum was used as a binding material in Mohenjodaro.

MASON'S TOOLS

Masons used terracotta plumb bobs, shell and ivory scales and compasses. V.B. Mainkar (1984) has described at length the integrated Indus scale in decimal gradation used for linear measurement. Two hollow cylindrical shell objects, one from Dholavira with six slits on each of the two margins of the ring, and another ring from Lothal, with four slits on each margin, seem to have served the purpose of a compass similar to the one used in modern plane table surveying. The lines joining opposite slits form a 30″ angle in the Dholavira instrument and 45° in the Lothal compass. For fixing alignment of streets and lanes the Harappans must have used the compass.

Masonry

The walls of baked bricks are in plumb on both faces in Lothal, while they are slightly battered in Mohenjodaro. The walls had hardly any footing as they stood on solid platforms. But the embankment walls of the dock had one or more offsets. In addition to the trabeated arch in the spillway of the Lothal dock, a vaulted arch seems to have been attempted in a kiln at Lothal. The public drains of Mohenjodaro had corbelled roofs, while there are flat roofs for drains in Lothal for which large size bricks were used. The Indus people used burnt bricks for constructing kilns, manholes and dye vats (Figure 8).

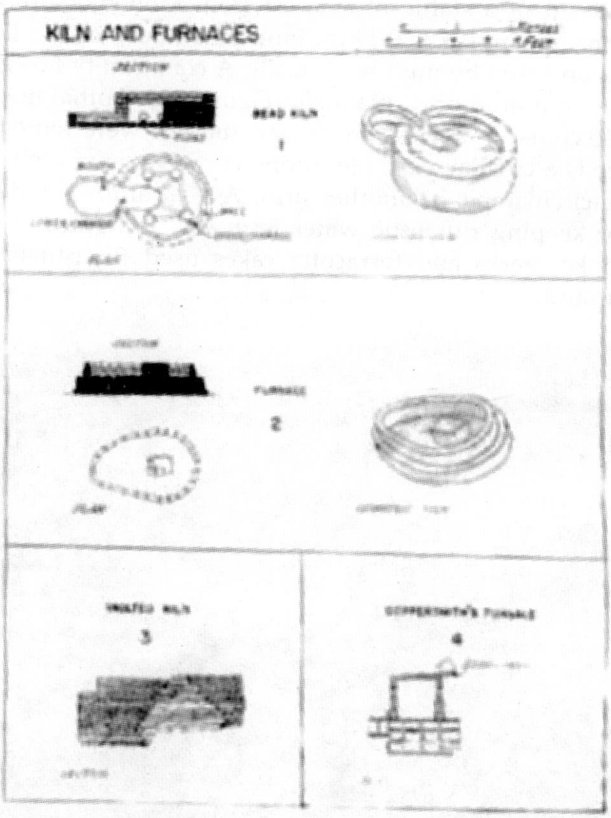

Figure 8 *Brick masonry of kilns, manholes and dye vats of the Indus Valley Civilisation. (Courtesy of the Archaeological Survey of India).*

Sacrificial Brick Altars

On one of the mud brick platforms in the citadel at Kalibangan, archaeologists uncovered seven oblong fire altars sunk into the ground. In the centre of the fire altar stood a cylindrical or faceted stele. On another platform a fire altar and a rectangular pit lined with kiln-baked bricks were discovered. There were also a well and bath-platforms for ritual ablution during and after the sacrifice, which were identified as animal sacrifices from the charred bones. The sacrificial altar at Lothal is a rectangular enclosure of burnt bricks built on a platform in a house. The gold pendant found in the altar, along with bovine bones, must have been used as a *rukma* (adornment) by the priest or sacrificer as shown on the forehead of a statue of a priest from Mohenjodaro. One of the fire altars, measuring 0.8 × 0.25 m, built in a house in the Lower Town of Lothal, is enclosed by mud brick walls. A community fire altar built of burnt bricks on a mud brick platform (Figure 9) at Lothal measures 0.3 x 0.8 m. Three courses of bricks are *in situ*; the rest were removed by later users. There is a circular posthole in one corner, a square pit in one arm and a semi-circular pit in another arm. A beautifully painted jar used perhaps for keeping ritualistic water is *in situ*. A terracotta ladle (*sruk*) bearing smoke marks and terracotta cakes used for ritualistic offering were also found.

Figure 9 *The Fire Altar of burnt bricks built on a mud brick platform for community worship, Lothal. (Courtesy of the Archaeological Survey of India).*

Early Historic Inscribed Bricks

Four massive sacrificial altars of burnt bricks were unveiled by T.N. Ramachandran (1954) of the Archaeological Survey of India in the village Jagatgram, 3 km from Kalsi in the Dehradun district of Uttar Pradesh. The importance of one of these vast sacrificial altars lies in its being recognised as a *śyenaciti* (hawk shaped altar). The inscriptions in the Brahmi script of the third century are repeated on several bricks used to construct the altar. The language of the inscription is Sanskrit. The inscription of Asoka at Kalsi is also situated nearby. The brick inscriptions say, "King Silavarman who was yogeswara and belonged to the Vrsagana (gotra) performed the Asvamedha sacrifice to which this brick belongs". On the bricks of another nearby altar the inscription refers to the fourth Asvamedha sacrifice performed by him, the lord of Lustrum. The *Brsagana gotra* is mentioned by the grammarian Paṇini (fourth century BCE). Such sacrifices were performed for the purification of the nation or people. The Greek kings performed them once every five years (Ramachandran, 1954)

Pathway

At Banawali there is a one metre-wide pathway of bricks on edge, running on the inner side of the wall of the citadel. In Kalibangan a house floor was paved with beautifully stamped brick tiles. The bathrooms of Lothal were paved with burnt bricks which were carefully rubbed to a polish.

Buddhist Monuments

The earliest Buddhist monument, namely the Dharmarajika Stupa at Sarnath near Varanasi (Banaras in Uttar Pradesh), was built of burnt bricks by King Aśoka in the third century BCE. It consists of a hemispherical dome with a low terrace at the base. Here stood Asoka's monolithic pillar with a lion capital which is now the emblem of the Indian Republic. Hundreds of *stūpas,* some containing relics and associated with *chaitya* halls and Viharas (monasteries) were built of bricks. The brick edifices at Guntapalli in Andhra Pradesh, dating to the second century CE, are noteworthy. Among the earliest Hindu temples built of bricks, the one at Bhitargaon in the Kanpur district of Uttar Pradesh is unique. It is a massive pyramidal edifice of diminishing tiers profusely decorated with beautiful terracotta figures in niches depicting gods, demigods, human, animal and bird figures. The Indo-Saracenic monuments such as masjids, tombs and idgahs were built of small bricks with a veneer of stone. For arches, vaults and domes, bricks were extensively used in the sixteenth century.

S. R. Rao

REFERENCES

Bhagwan, Singh. *The Vedic Harappans*. Delhi: Aditya Prakashan. 1995.

Jarriage, J.F. "Towns and Villages of Hill and Plain." In *Frontiers of the Indus Civilization*. Eds. B.B. Lal and S.P. Gupta. New Delhi: Books and Books. 1984. pp. 289–300.

Leshnik, L.S. "The Harappan 'Port' at Lothal: Another View." *American Anthropologist* 70: 911–922. 1971.

Mainkar, V.B. "Metrology of the Indus Civilization." In *Frontiers of the Indus Civilization*. Eds. B.B. Lal and S.P. Gupta. New Delhi: Books and Books. 1984. pp. 141–152.

Marshall, (Sir) John. *Mohenjo-daro and the Indus Civilization. 3 vols. London:* Arthus Probsthain. 1931.

Nigam, R. "Was Lothal Basin a Dock or Water Storage Tank." *In Marine Archaeology of Indian Ocean Countries*. Ed. S.R. Rao. Goa: National Institute of Oceanography (NIO). 1988, pp. 20–21.

Panikkar, N.K., and T.N. Srinivasan. "Concept of Tides in Ancient India." *Indian Journal of Marine Sciences* 6(1). 1971.

Ramachandran, T.N. *Indian Archaeology* (1953–54): *A Review*. New Delhi: Archaeological Survey of India. 1954.

Rao, S.R. *Dawn and Devolution of the Indus Civilization*. New Delhi: Aditya Prakashan. 1991.

Sankalia, H.D. *Prehistory and Protohistory of India and Pakistan*. Poona: Deccan College, Postgraduate and Research Institute. 1974.

Sarkar, S.C. *Some Aspects of the Earliest Social History of India.* London: H. Milford, Oxford University Press. 1928.

Sanskrit Texts

Aṣṭādhyāyī of Pannṇini. Ed. S.C. Vasu. Delhi: Motilal Banarasidas. 1962.

Taittarīya Samhitā. Ed. A. Mahadeva Sastry and A Rangachary. Delhi: Motilal Banarasidas. 2004.

The Hymns of the Ṛgveda. Trans. R. T. H. Griffith. Delhi: Motilal Banarasidas. 1973.

C

Calculus

Calculus, both integral and differential, was known *to a limited extent* and used for specific purposes in India during early and medieval times, and is attested to by the works of Āryabhaṭa (b. 476), Brahmagupta (b. 598), Muñjāla (b. 932), Bhāskarācārya (b. 1114), Nārayāṇa Paṇḍita (1356), Nīlakaṇṭha Somayājī (b. 1444), and Jyeṣṭhadeva (ca. 1500–1610). In the same manner as Hindu geometry grew in response to the religious needs of designing sacrificial altars, it would seem that calculus too evolved when there was the need to ascertain favourable and unfavourable times for religious rites and rituals on the basis of the moments of the eclipses, conjunction of planets, and conjunction of planets and stars. Calculus did not evince further growth in India.

In order to determine the accurate motion of a planet at a particular moment (*tātkālika-gati*), Bhāskarācārya divides the day into a very large number of intervals of time units called *truṭi*, equal to 1/33750 of a second, and compares the successive positions of the planet, the motion during that very small unit of time being considered constant. He also suggests that the differential coefficient vanishes when the interval is diminished to the absolute minimum. He illustrates this by three examples.

Hindu ideas on infinitesimal integral calculus are shown in the methods employed and formulae arrived at for the calculation of the area of a circle, and the surface area and volume of a sphere, all of which are virtually the same as these derived by modern mathematics.

Two methods are adopted for finding the surface area of a sphere. One of them is to draw, from a point on the sphere taken as the pivot, a very large number of circles, parallel to each other and of small and equal interstices, and to slice the sphere at these circles. The sum of the areas of the strips peeled off each slice, the longest of them being at the equator, and the smallest, equal to zero, at the pivot and the bottom, would be equal to the surface area of the sphere. It is explained that when the breadth of the strips tends to zero, the total surface area could be obtained by the formula $4\pi r^2$, where r is the radius of the sphere.

In the same manner, the volume of a sphere is equal to the sum of the numerous cones that would be formed with their bases on the surface of the sphere and their apexes being at the centre of the sphere. When the number of cones is infinite, their bases could be taken as flat, and the volume of the cones could be calculated by the usual formula. From this postulate, the formula for the volume of the sphere is identified as $\frac{4}{3}\pi r^3$. The several methods enunciated for the determination of the value of the circumference of a circle, involving the irrational quantity, π, as exposited in several works from Kerala, like the *Yuktibhāsā* of *Jyeṣṭhadeva*, also concern integral calculus.

K. V. Sarma

References

Āryabhaṭīya of Āryabhata, Critical Edition. Ed. K.S. Shukla and K.V. Sarma. New Delhi: Indian National Science Academy. 1976.

Balagangadharan, K. "Mathematical Analysis in Medieval Kerala." In *Scientific Heritage of India: Mathematics*. Ed. K.G. Poulose. Tripunithura: Government Sanskrit College. 1991. pp. 29–42.

Datta, Bibhutibhusan, and Awadesh Narayan Singh. "Bhaskara and the Differential Calculus." In *The History of Ancient Indian Mathematics*. Ed. C.N. Srinivasiangar. Calcutta: The World Press. 1967. pp. 91–93.

Datta, Bibhutibhusan, and Awadesh Narayan Singh. "Infinitesimal Calculus." In *Mathematics in Ancient and Medieval India*. Ed. A.K. Bag. Varanasi-Delhi: Chaukhamba Orientalia. 1979. pp. 286–98.

Datta, Bibhutibhusan, and Awadesh Narayan Singh. "Use of Calculus in Hindu Mathematics." *Indian Journal of History of Science* 19(2): 95–104. 1984.

Sengupta, P.C. "Infinitesimal Calculus in Indian Mathematics." *Journal of the Department of Letters, Calcutta University* 22:1–17. 1932.

Siddhāntaśiromaṇi of Bhāskarācārya. Ed. Muralī Dhara Chaturvedi. Varanasi, 1981, *Spaṣṭādhikāra*. 36–38: pp. 119–21: *Buvanakośa*. 58–61: pp. 363–67.

Yuktibhāṣā (of Jyeṣṭhadeva) (in Malayalam). Ed. Rama Varma (Maru) Thampuran and A.R. Akhileswara Iyer. Trissivaperur: Mangalodayam Limited. 1948. Trans. K.V. Sarma (Ms).

See also: Āryabhaṭa – Brahmagupta – Bhāskarā – Nārayāna Paṇdita

Calendars

Calendrical science developed in India in three distinct phases, and each phase influenced the succeeding one. The first phase produced the Vedic calendar, covering a period from an unknown antiquity to the Mauryan emperor Aśoka (ca. 300 BCE); the second had the Greco-Indian calendar for a short period between the post-Aśokan and post-Kuṣāṇa period (ca. CE 300). And the third produced the Siddhāntic calendar from the Gupta period (CE 319) which is still used in India today. All three phases are characterised by formulations of fitting a lunar year into a sidereal solar one by suitably intercalating lunations.

References to calendrical elements like lunation, intercalation, year, and solstice occur even in the earliest part of Vedic literature. A separate calendrical literature, deemed a part of the Vedas, was developed in a small text, *Vedāṅga Jyotiṣa*, but this text has not been fully deciphered yet. There are some astronomical statements in the text, and also in the earliest part of Vedic literature, which do not hold true in the latitude belt and in the time period when the Vedic Aryans are believed to have settled in India.

It is a separate course of study as to how and when the Vedic Indians obtained this calendar, but they adopted it without verifying it.

This calendar is based on the following imperfect parameters:

5 sidereal years = 1830 days.

62 lunations = 1830 days.

67 sidereal lunar months = 1830 days.

The star *beta Delphini* indicates the winter solstice (which was indeed the case around 1500 BCE) which is apparently a fixed point. The lunar orbit is divided into twenty-seven equal arcs from *beta Delphini* each equal to 13°20′, called *nakṣatra,* the first arc named *Dhaniṣṭhā*. The arcs are named after a prominent star of the division. The theory is that if, in five years, two lunations are intercalated, then the lunar year will remain tied with the solar one. This period is called a 5-yearly cycle or *yuga*.

In practice, a cycle begins when the Sun and the Moon are in conjunction at *beta Delphini* (i.e. from the winter solstice). The first, second and fourth year each contain twelve new moon ending lunations. One lunation is intercalated at the middle of the third year and another at the end of the fifth. The second cycle then again begins from a similar new moon at winter solstice.

This imperfect calendar was destined to collapse, as the succeeding cycles will begin from new moons shifted by some 4.5 days from the winter solstice at *beta Delphini* per cycle, accumulating to one lunation in six or seven cycles. However, the calendar did not collapse, and so the accumulated lunation was extracalated, but we are not told in the text how this was done. Perhaps the rule is hidden in an obscure part of the text, and so we have to guess it.

A lunation is divided into thirty parts called *tithis*. Accordingly, 30 *tithis* = 1830/62 days. The excess 30/62 of a *tithi* over a day is called an omitted *tithi*. A month's days are designated by the *tithi* of the day and not by ordinal numbers. The Jainas used this cycle in their calendar with a marginal change: that months were full-moon-ending and a cycle began from a full moon at summer solstice in the middle of the *Āśleṣā* division (180° away from *beta Delphini*).

Kautilya wrote a book entitled *Arthaśāstra* which is believed to reflect the social and administrative conventions of the Mauryan empire. This book, on the subject of measuring time, only reproduced the Jain school of the 5-yearly cycle. We are thus assured of its use until then.

The second phase of the Indian calendar began by the first century BCE after Śaka penetration into north-western India.

The Śakas, a central Asian people, conquered Bactria in 123 BCE and established an era with epoch at 123 BCE, now called the Old Śaka Era. This epoch has variously been fixed at 88 BCE (Konow) and 110 BCE (Herzfeld), but, as M.N. Saha has shown, 123 BC fits into all the circumstantial evidence.

The Śakas used the Greek calendar based on the Metonic cycle in their homeland. This was a nineteen year cycle at whose beginning and end the Sun and Moon are in the same relative position to each other. When they reached India they adopted Indian culture and Aryanised or Indianised the Greek calendar, producing a Greco-Indian calendar. It replaced the classical 5-yearly cycle and it circulated in India up to ca. CE 200. The Kuṣāṇas, another Śaka tribe, penetrated further inside India, adopting Indian culture and using this calendar. Kings of the Śaka and Kuṣāṇa dynasties left behind some inscriptions bearing dates. These always refer to an era, as in era of King Kaṇiṣka. Further, they contain a mysterious expression – *etaye purvam* – whose meaning has not been deciphered.

The Indian form of the calendar is that the months are full moon ending; the 19-year cycle begins from a full moon at the autumnal equinox in the month Kārtika; months are assigned Sanskrit or Sanskritised Greek names; and, in special cases, the day is designated by the moon's *nakṣatra* on that day. The Greek form is that days are designated by ordinal numbers and not by *tithis*, and intercalation is made at the middle of the year in Chaitra.

All these features are fully reflected in the inscriptions of the Kuṣāṇa kings. For example, present researches have shown that Kaṇiṣka established his era in CE 78 after deleting the hundredth place in the old Śaka era for that year.

Now, 123 + 78 = 201 = 1, deleting the hundredth place.

Hence, CE 78 = 1 Kaṇiṣka era (current) and

0 Kaṇiṣka era (elapsed).

We interpret *etaye purvam* as elapsed year, and accordingly the Julian equivalent of Kaṇiṣka era in these inscriptions is:

CE 78 = 0 Kaṇiṣka era.

The day of the month can also be identified with the *tithi* with tolerable accuracy. We have seen that the Moon gains 12° over the Sun per *tithi*. In case of new-moon-ending months, we should expect, on 20th Āṣāḍha:

Moon–Sun = $20 \times 12° = 240°$ (near about)

or Sun = Moon − 240°.

If the Moon is near *Uttara Phalgunī* (tropical longitude in CE 90 = 144°46′), we should have: Sun = 144°46′ − 240° = 265° (say), so that the Sun's position comes near the winter solstice, which cannot happen in Āṣāḍha. However, if the month is full-moon-ending, then the Moon – Sun = $5 \times 12° = 60°$, or Sun = Moon-60° = 144°46′ − 60° = 85° (say), i.e. near the summer solstice, which indeed happens in Āṣāḍhā.

If we compute the mean tropical longitude of the sun (L) and mean elongation of the moon in days (D), we get:

Date	L	D
June 21, CE 90	87°.16	4.81

These luni-solar positions are in complete agreement with the inscriptional dates. We may perhaps identify these dates as Julian equivalents of the inscriptional dates. Using this scale we can decipher all the other dates in the inscriptions of the Kuṣāṇa kings and use the same method to decipher the dates of the old Saka Era.

Table 1 *In a period of 4,320,000 years*

	Āryabhaṭa	*Sūryasiddānta*
Sidereal revolution of the Sun	4,320,000	4,320,000
Sidereal revolution of the Moon	57, 753, 336	57, 753, 336
Sidereal revolution of Jupiter	364, 224	364, 220
Total civil days	1, 577, 917, 500	1, 577, 917, 828
Intercalary lunations	1, 593, 336	1, 593, 336
Omitted *tithis*	25,082,580	25,082,252

This second phase was short-lived, but had far-reaching effects on the later Indian calendar. In fact, this was perhaps the most productive period in Indian calendrical history.

This Greco-Indian calendar was discontinued after the fall of the Kuṣāṇa empire.

In the third phase, the Indian calendar was thoroughly revised. New calendrical elements were introduced, new techniques in computational works were formulated, and the application of cycles for intercalation was abandoned. The new calendar that emerged is called the Siddhāntic calendar. This calendar was perfected by later Indian astronomers and is still used all over India with some marginal changes in some regions. In this phase the solar zodiac with twelve signs was introduced. The *nakṣatra* division was recast so that the first arc-division, renamed *Aśvinī*, began from the first sign *Meṣa* (Aries). This common point was the vernal equinox. By CE 550, the vernal equinox shifted to *zeta Piscium*, and this star (or a point very close to it) has since been the beginning of the Indian zodiac.

The new technique for computations was that an epoch, *Kaliyuga*, was formulated such that all the luminaries, like the Sun, Moon, and planets, were assumed to be in conjunction at the initial point of the zodiac at this epoch. Sidereal motions of the luminaries were so assigned that each of them made an integral number of revolutions in a period of 4,320,000 years (see Table 1).

Astronomical parameters in this phase were not uniform in different schools. We cite below some calendrical parameters from the Āryabhaṭian school (CE 476) and the Sūryasiddhānta school of the tenth century.

For any day in question, elapsed days from *Kali* are computed from the above figures, and from the rates of sidereal motion of the luminaries their positions are found in the zodiac. Bentley computed the epoch of *Kali* at 18th February, 3102 BCE midnight at Ujjain (longitude 75°45′ east). All astronomers have generally accepted this date.

Āryabhaṭa formulated corrections of mean motions of the luminaries, but such formulations are not found in earlier works. Our presumption is that only mean motions were considered in the pre-Āryabhaṭian period.

In this revised calendar, the luni-solar year begins from the new moon just preceding the Sun's entry to the initial point in the spring month *Caitra*. Now, twelve lunations fall short of 10.8 days from a sidereal year, and this accumulates to one lunation in two or three years. Whenever such an extra lunation forms, it is intercalated in the year; there is no role of any cycle here. This luni-solar year, with days designated by the *tithi* of the day, is used all over India for religious purposes.

A solar year is also devised beginning from the Sun's entry into the first sign *Meṣa*, and is divided into twelve solar months, each month being the period the Sun stays in a sign. Days of solar months are designated by ordinal numbers. The first month is *Vaiśākha*. Datings are recorded by luni-solar year in some regions, and by solar year elsewhere. The era generally used is the Kaṇiṣka era renamed the Śaka era.

In a year containing twelve lunations, the twelve new moons will, most generally, be distributed over the twelve signs, i.e. each sign will contain a new moon. However, in an intercalary year when thirteen lunations are distributed over twelve signs, two new moons must occur in one sign, and in extreme cases, one sign may not contain any new moon at all. Such situations are natural phenomena and are no problem for astronomers.

Out of the thirteen lunations occurring in an intercalated year, any one may be earmarked as the extra month. However, metaphysics has taken a toll here. As one sign will contain two new moons, so two lunations, one from each new moon, will occur in that sign. One lunation out of these two has to be selected as the intercalary one. Volumes of religious literature have been written, perhaps more voluminous than their astronomical counterparts, on which lunation is the proper one and which is the intercalary one. Similarly, when a sign becomes void of a new moon, religious literature is controversial as to which lunation is to be assigned to that sign. Almost every region of India has its own convention on these points.

A.K. Chakravarty

REFERENCES

Chakravarty, A.K. *Origin and Development of Indian Calendrical Science*. Calcutta: Indian Studies. 1975.

Chatterjee, S.K. and Chakravarty, A.K. "Indian Calendar from Post Vedic Period to 1900 AD." In *History of Astronomy in India*. New Delhi: Indian National Science Academy. 1985. pp. 252–307.

Chattopadhyaya, Debiprasad. *History of Science and Technology in Ancient India*, vol. 1. *The Beginnings*. Calcutta: Firma KLM Private Ltd. 1966.

Konow, Sten. *Corpus Incriptionum Indicarum* vol. II, part 1. Calcutta: Government of India Central Publication Branch. 1929.

Report of the Calendar Reform Committee. New Delhi: Government of India, Council of Scientific and Industrial Research. 1955.

Report of the State Almanac Committee. Alipur: Government of West Bengal. 1963.

Shamasastry, R., ed. *Vedāṅga jyotiṣa*. Mysore: Government Press. 1936.

Tilak, B.G. *Vedic Chronology and Vedāṅga Jyotiṣa*. Poona: Tilak Brothers. Gaikwar Wada. 1925.

See also: Astronomy – Lunar Mansions – Astrology

Candraśekhara Sāmanta

Candraśekhara Sāmanta (1835–1904) was a self-made astronomer who had the distinction of revising the traditional calendar of Orissa during the nineteenth century. He was a scion of a junior branch of the chiefs of the small estate of Khandapara and bore the title *Sāmanta* (Feudatory) on that account. His full traditional name was Sāmanta Śrī Candraśekhara Singh Harichandan Mohāpatra, and he was called locally Pathani Sāmanta. Young Candraśekhara had little modern education, but learned Sanskrit and astrology from his uncle, which he further developed by an intensive study of two of the authoritative texts of the times, the *Sūryasiddhānta*, and the *Siddhāntaśiromaṇi* of Bhāskarācārya. Exhibiting an uncanny interest in watching the skies, he became aware that the times of the rising and setting of the Sun and Moon and of the other celestial bodies were at variance with the times indicated in local almanacs arrived at by computation based on traditional texts. Often, when almanacs differed in their indication of times, and even of dates (*tithis*), there was confusion in the matter of fixing the days for domestic and social festivals and of sacred worship at temples like that of Lord Jagannātha at Puri.

Deeply religious and equipped with the fundamentals of traditional astronomy, Candraśekhara took it upon himself to remedy the prevailing state of affairs, and from the age of 23 he commenced watching the transit, conjunction, rising and setting of the celestial bodies, and recorded his observations consistently and systematically. As aids to accuracy in observation, he designed, all by himself, several astronomical instruments which had been described in the *Sūryasiddhānta* and the *Siddhāntaśiromaṇi* and their commentaries. Among the instruments were an armillary sphere and a vertical wheel, which in modern terms served the purpose of the transit and alta-azimuth instruments. He also used the clepsydra (water clock) for measuring sidereal time. An instrument which he designed and used constantly for celestial observation in place of a telescope was a T-square frame which he called *mānayantra* (measuring instrument) with the main limb twenty-four digits and the crosspiece marked with notches and having holes at distances equal to the tangents of the angles formed at the free end of the main piece. He also designed an automatically revolving wheel (*svayam-vāhaka*), with spokes partly filled with mercury, which he also used to measure time. He effectively used the gnomon, which when fitted with a small mirror could be used at night for measuring time and angular distances.

Candraśekhara's observations, experiments, and recordings, on the basis of which he coined corrections to earlier enunciations and innovated new methodologies and practices, lasted thirty-four years. This enabled him to introduce reforms to all aspects of astronomical computation, which he set out in an extensive work *Siddhāntadarpaṇa*, in 2500 verses. This is an astronomical manual in five chapters, devoted, respectively, to

Mean planets (*Madhyama-adhikāra*), True planets (*Sphuṭa-adhikāra*), Problems of Direction, Time and Place (*Tripraśna adhikāra*), Spherics (*Gola-adhikāra*), and Time and Appendices (*Kāla adhikāra* and *Pariśiṣṭa*). The main contributions of Candraśekhara include corrections to the sidereal periods of the star planets and to the main inclinations of the planetary orbits to the ecliptic (the band of the zodiac through which the sun apparently moves in its yearly course), identification, and evaluation of evection, variation, and annual equation among the moon's inequalities, and horizontal parallax of the sun and the moon. Recognition and honours were late in coming when the Government of India conferred on him the title of *Mahāmahopādhyāya* (Scholar of scholars), the highest title for a Sanskritist.

K. V. Sarma

REFERENCES

Bandopadhyay, Amalendu. "Astronomical Works of Samanta Chandrasekhar." *Journal of the Asiatic Society* 30: 7–12. 1989.

Baral, Haris Chandra. "Chandra Sekhar Samanta." *Indian Review* 23: 459–61. 1922.

Ray, Joges Chandra. "Centenary of Chandra-sekhara and a Reformed Hindu Almanac." *Modern Review* 1936. pp. 56–60.

Sarma, S.R. "Perpetual Motion Machines and their Design in Ancient India." *Physics* 29(3): 665–76. 1992.

Sengupta, P.C. "Hindu Luni-solar Astronomy." *Bulletin of the Calcutta Mathematical Society* 24: 1–18. 1932.

Siddhāntadarpaṇa by Candraśekhara Sinha. Ed. Joges Chandra Ray. Calcutta: Indian Depository. 1897.

See also: Armillary Sphere – *Sūryasiddhānta*

Caraka

The oldest of the Ayurvedic classical works still in use by Ayurvedic physicians is the *Carakasaṃhitā*, and its celebrated editor was Caraka. Caraka's contribution was so significant that the work bears his name, rather than that of Agniveśa, the original author.

Caraka was a popular name in ancient India from the Vedic period up to the time of Kaniṣka (CE 1), but the editor of the *Carakasaṃhitā* belongs to a period prior to 600 BCE, the pre-Buddhist period. We know this because the prose style of this work resembles the Brāhmaṇas and Upaniṣads, because there are no references to Buddha or Buddhist philosophy, because there are references to the Vedas and Vedic gods, and because there is no Puranic mythology in the text.

The *Carakasaṃhitā* consists of 120 chapters which are divided into eight sections. Unfortunately, the original complete work is no longer extant. Seventeen chapters of the sixth section, and the entire seventh and eighth sections were later supplemented by another scholar-physician called Dṛḍhabala (CE 12).

Out of the eight specialised branches of Ayurveda, Caraka's work primarily deals with six: (1) internal medicine, (2) toxicology, (3) psychiatry, (4) pediatrics, (5) rejuvenation therapy, and (6) sexology, including the use of aphrodisiacs. The other two, surgery and the treatment of eye, ear, nose, and throat diseases, are scantily described. The important topics dealt with in this text are anatomy, physiology, etiology (disease causation), pathology, therapeutics, form and time of treatment, conduct of the physicians, medicaments and appliances, and rules for diet and drugs. In addition, it contains a detailed classification and nomenclature for diseases, as well as information on embryology, obstetrics, personal hygiene, sanitation, and the training of physicians. It also describes the cosmological, biological, physico-chemical, metaphysical, ethical, and philosophical ideas prevalent in ancient India.

Bhagwan Dash

REFERENCES

Agniveśa. *Carakasaṃhitā*. Bombay: Nirnayasagara Press. 1941.

Dash, Bhagwan. "Caraka." In *Cultural Leaders of India: Scientists*. Ed. V. Raghavan. New Delhi: Publication Division, Ministry of Information and Broadcasting, Government of India. 1976. pp. 24–43.

Ray, Priyadaranjan, and Hirendra Nath Gupta. *Caraka saṃhitā: A Scientific Synposis*. New Delhi: National Institute of Sciences of India. 1965.

Śarmā, Priyavrata. *Caraka-cintana*. Varanasi: Chowkhamba Vidyabhawan. 1970.

Śarmā, R.K., and Bhagwan Dash. *Carakasaṃhitā*: *Text with English Translation and Critical Exposition Based on Cakrapāṇidatta's Āyurveda dīpikā*. Varanasi: Choukhambha Sanskrit Series Office. 1976.

City Planning

As the birthplace of Indian culture, the towns of the Indus Valley represent an important and rich source of information concerning urban development in the Indian subcontinent. The Indus River is one of the largest and most important rivers in south central Asia. From its source innn the Himalayan Mountains to its terminus in the Indian Ocean it traverses a course of over two thousand miles. For millennia it has been an essential route for travel, trade, and communication and has been the source of much of India's agricultural production. The valley which surrounds the Indus River has witnessed the birth, growth, and death of many cities.

Excavations of Indus towns have demonstrated the most ancient town planning in the world. The grid pattern (straight streets intersecting other straight streets at right angles) is among the most common and universal types of town planning. The discovery and excavation of the most famous of these sites, Mohenjodaro, by Sir John Marshall (1931) and Ernst Mackey (1938) supported the contention of Dan Stanislovski (1946) that the development of the grid pattern in western societies – Greece, Rome, and Europe in the Middle Ages – can be linked directly to the development of town planning in the Indus valley.

A recent excavation on the Western plains of the Indus – the Rahman Dheri site – reveals a town plan from the early Indus period. This site comprises an area of roughly 22 hectares and could have been home to ten to fourteen thousand people. Radio carbon-dating of artefacts places development somewhere in the first half of the fourth millennium BCE. Rahman Dheri is built in the classic grid pattern. It is rectangular in shape surrounded by an immense wall and bisected by a major traffic artery which runs roughly south-east to north-west. Perpendicular to this road exists a pattern of regularly occurring laneways which appear to create individual dwelling lots.

The successor to the plan of Rhaman Dheri is thought to be the Surkotada site in Kutch which dates to ca. 2500–2000 BCE. Though smaller in scale than its predecessor the Surkotada site is built along the same grid pattern as Rahman Dheri. This settlement was enclosed by a stone rubble and brick fortifications. This produced two separate areas each of roughly sixty square meters.

Similar to Surkotada, only on a larger scale, the Kalibangan site was completely surrounded by an enormous rampart. This site, about 200 kilometres to the south-east of Hirappa on the banks of the (now dry) Ghaggar river, was composed of two mounds. The smaller mound, named the citadel, was to the west and was roughly 240 metres by 120 metres. The larger mound, named the lower city, was to the east and measured 360 metres by 240 metres. The lower city demonstrates the grid plan divided into blocks of which the east–west side was roughly 40 metres in width. The width of lanes and streets in the lower city ranges from 1.8 to 7.2 m. Interestingly each lane or street is some multiple of 1.8 metres. This site is thought to date to the early Indus period, roughly 3000 BCE.

The evolution of Indus town planning reached its zenith at Mohenjodaro. The town is geographically located on a floodplain of the Indus river; the city occupies about one square kilometre – over five times larger than Kalabangan. The basic grid of Mohenjodaro is about 180 metres square, subdivided into 16 sections of about 40 square metres. Mohenjodaro existed from ca. 2500 to 2000 BCE. During this time it bore the brunt of often severe flooding, and as a result the original grid plan was often modified and transformed. The city itself was originally planned in the same way as Kalabangan: an oblong walled city divided into a citadel and a downtown area divided by an open space. Similar to its predecessor communities, the town plan of Mohenjodaro indicates the same parallel grid street structure that has come to characterise urban planning in this region at this time.

As this brief history of the evolution of town planning in the Indus river valley demonstrates the grid pattern was commonplace. The importance of the Indus rivenr valley communities to the historical development of culture and civilisaton in the Indian subcontinent and the East is well known. Stanisloski has suggested that the grid pattern form of town planning has its roots in the Indus valley. Parallel developments in Nepal, Sri Lanka (formerly Ceylon), Burma, Korea, Vietnam, and China also indicate how widespread the influence of the Indus river valleys communities were. It is suggested that this form of town planning was taken up by the Greeks, Romans, and other Western Europeans and eventually became a standard form of urban organisation in Europe and the New World.

Excavations at and along the Indus river valley reveal cities strikingly similar in topography to contemporary Western European and North American cities. While some may suggest that the grid pattern evolved in a happenstance fashion without benefit of central planning or administration, substantial evidence exists to support the opposite point of view. First, the grid pattern can be seen in a variety of towns all along the Indus river which existed and flourished over a several thousand year period. Second, the similarity between these towns and the general division between a lower city area suggests the development of town planning conventions as early as 2500 BCE. Third, there is a remarkable mathematical symmetry in Kalabangan where lanes and streets (which range from 1.8 to 7.2 m in width) all occur in multiples of 1.8 m.

The contribution of this historic period to the development of Western cities should not be underestimated. Much of what we take to be enlightened urban planning was considered over four thousand years ago by planners in south-eastern Asia.

Paul Gregory

Colonialism and Science

Colonisation was not merely a political phenomenon; it had far-reaching economic and cultural ramifications. It was an exercise in power, control, and domination. Scientific and technological changes greatly facilitated this progress. Techno-science and colonialism are closely linked and to some extent share a cause and effect relationship. In recent years a good deal of work has been done on the nature, course, and consequence of this relationship in different geographical and culture areas. Some scholars see in it utilitarian and developmental images; many others find it utterly exploitative, while some prefer to opt for a middle path and emphasise both the regenerative and retrogressive aspects of the science and colonisation nexus. So the debate continues, and several works have appeared with case studies on Africa, Latin America, and Asia. India, being a prime example of classic colonisation, has also received considerable attention.

The fact that India has a very long techno-scientific tradition and a rich cultural heritage is fairly well known. It was, however, during the seventeenth to eighteenth centuries, the post-Renaissance epoch (that of Descartes and Newton) that Europe began to outdistance India in scientific and material advancement. The rise of modern science in Europe profoundly disturbed the balance of scientific development among traditional societies. It is also possible that the various sciences and technologies were on a decline in India around 1790. There was definitely no "conscious" spirit of technological innovation and scientific enquiry to match the spirit of Europe. The result was colonisation.

The advancing European trading companies became deeply involved in political and military rivalries, culminating in the establishment of the British paramountcy over the Indian subcontinent. A new empire was in the making. The colonisers were out to collect the maximum possible information about India, its people and resources. They reported what was best in India's technological traditions, what was best in India's natural resources, and what could be most advantageous for their employers. The English East India Company, for its part, was quick to realise that the whole physical basis of its governance was dependent upon the geographical, geological, and botanical knowledge of the areas being conquered. The colonisers fully recognised the role and importance of science in empire building.

The most interesting feature of this early phase of colonial science lies in its highly individualistic character. State followed the trade, and certain individuals on the spot would largely determine what was advantageous for both. These colonial scientists would try their hands at several fields simultaneously, and were in fact botanist, geologist, zoologist, physicist, chemist, geographer, and educator — all rolled into one. As data gatherers they had no peers, but for analysis and recognition, they had to depend upon the metropolitan scientific culture whose offshoots they were and from which they drew sustenance. The colonial government quickly patronised geographical, geological, and botanical surveys; after all these were of direct and substantial economic and military advantage. Medical or zoological research did not hold such promises. Research in physics and chemistry was simply out of the question, for there were no laboratories, equipment, or specialised training. The reigning spirit remained that of exploration. Systematisation or analysis of its results had to wait for some time, but then, even disjointed and often haphazard studies served some purpose.

Other positive achievements were the establishment of scientific bodies and museums. Pre-British India never had anything like a scientific society, not to say a journal, which could provide some sort of a platform for scientific workers. William Jones was the first to realise this and founded the Asiatic Society in Calcutta in 1784. This society soon became the focal point of all scientific activities in India. It was followed by the Madras Literary and Scientific Society (1805), the Agricultural and Horticultural Society of India (1817), Calcutta Medical and Physical Society (1823), and the Bombay Branch of the Royal Asiatic Society (1829). Trigonometrical and topographical surveys were organised under the Great Trigonometrical Survey of India (1818), and a Geological Survey was established in 1851. Scientific research thus for a long time remained an exclusive governmental exercise, and this largely determined the nature and scope of scientific research in India. Colonial science primarily implied "natural history" and its star (if not sole) attraction was the exploitation of natural resources. It was basically plantation research with emphasis on experimental farms, the introduction of new varieties, and the various problems of cash crops. Next came surveys in geology and meteorology. Another major area of concern was health. The survival of the army, the planters, and other colonisers was at stake. The importance of medical research was always recognised, but the quantum of emphasis varied from time to time. In any case, however, research was not to be a curiosity-oriented affair. Financial considerations were invariably there. The colonial administrators consistently held that scientists in India should leave pure science to Britain and apply themselves only to the applications of science. They would goad the various organisations to work along only economically beneficial lines. Colonial researchers often found themselves unable to distinguish between basic and applied research. This was particularly true of the geologists and the botanists. Their problem was how to discover "the profitable mean course" in which scientific research, having a general bearing, would at the same time solve the local problems of immediate economic value. The dilemma was fairly acute.

In the field of education, science was unfortunately never given a high priority. In 1835 Thomas Macaulay not only succeeded in making a foreign language, English, the medium of instruction, but his personal distaste for science led to a curriculum which was purely literary. A few medical and engineering colleges were opened, but they were meant largely to supply assistant surgeons, hospital assistants, overseers, etc. What India got was some sort of a hybrid emerging out of a careless fusion between literary and technical education. What is more, adoption of English as the sole medium of instruction in science rather hampered its percolation to the lower classes. Colonial education widened the social gulf and accentuated the age-old divide. Even in government institutions, growth was kept under a self-regulatory check. The Tokyo Engineering College was established in 1873, much later than the Engineering College at Roorkee, and by 1903 it had a staff of 24 professors, 24 assistant professors, and 22 lecturers. The Massachusetts Institute of Technology was established in 1865 and by 1908 it had 306 teachers. And Roorkee, even after 100 years of its existence (i.e. in 1947), had only three professors, six assistant professors, and twelve lecturers.

Colonialism involved not only exploration and classification but also coding and decoding cultures. Its cultural projects showed deeper penetration and greater resilience than its economic forays. With the help of schools, universities, textbooks, museums, exhibitions, newspapers, etc. local discourse was influenced and colonised. Modernity was presented as a colonial import and not something intrinsic to humanity's rational nature. Colonial rule, with its sharp tools, dissected and bared differences and was not inclined to synthesise. Colonialism usually stalls the possibilities of exchange and prefers one-way traffic. One may talk of transfer — transfer of knowledge, systems, or technologies — but it was a transfer restricted or guided to achieve certain determined objectives. As education and awareness grew, several Indians participated in the official scientific associations and institutions, but very often they searched for a distinct identity and established institutions, scholarships, and facilities of their own.

In the first half of the nineteenth century, Bal Gangadhar Shastri and Hari Keshavji Pathare in Bombay, Master Ramchandra in Delhi, Shamhaji Bapu and Onkar Bhatt Joshi in Central Provinces, and Aukhoy Kumar Dutt in Calcutta worked for the popularisation of modern science in Indian languages. In 1864 Syed Ahmed founded the Aligarh Scientific Society and called for the introduction of technology to industrial and agricultural production. Four years later Syed Imdad Ali founded the Bihar Scientific Society. These societies did not live long. In 1876 Mahendra Lal Sarkar established the Indian Association for Cultivation of Science. This was completely under Indian management and without any government aid or patronage. Sarkar's scheme was very ambitious. It aimed not only at original investigations but at science popularisation as well. It gradually developed into an important centre for research in optics, acoustics, scattering of light, magnetism, etc. In Bombay, Jamshedji Tata drew up a similar scheme for higher scientific education and research. This was opposed by the then Governer-General, Lord Curzon. Yet it finally led to the establishment of the Indian Institute of Science at Bangalore in 1909. There was thus greater awareness by the turn of the century.

In the first quarter of the twentieth century, those who put India on the scientific map of the world were J.C. Bose, who studied the molecular phenomenon produced by electricity on living and non-living substances, Ramanujan, a mathematical genius, and P. C. Ray who analysed a number of rare Indian minerals to discover in them some of the missing elements in Mendeleef's Periodic Table. C.V. Raman's research on the scattering of light later won him the Nobel Prize in 1930 and gave the name to Raman spectroscopy. Meghnad Saha pioneered the field of astrophysics, while S.N. Bose's collaboration with Einstein led to what is known as the Bose–Einstein equation. These were great sparks, individual and sometimes lonely, yet imbued with both scientific and national spirit. They thought over what role science and technology would play in building modern India, and they dreamt of freedom. The colonial government was aware of its limitations and discomfiture, and gradually permitted greater indigenisation of its scientific institutions and cadre. In the wake of the First World War the government realised that India must become more self-reliant scientifically and industrially. It appointed an industrial commission in 1916 to examine steps that might be taken to lessen India's scientific and industrial dependence on Britain. However, few of the Commission's recommendations were actually implemented. Discontent continued to grow. India's national leaders appreciated the importance of science and technology in national reconstruction and worked closely with the scientific talent of the time. A National Planning Committee was formed in 1937 for this purpose, in which several leading Indian scientists and technologists participated. In 1942, the Council for Scientific and Industrial Research was established. The end of colonial science was pretty near. With the A.V. Hill Report in 1944 on Scientific Research in India, the curtain dropped.

The foregoing analysis illustrates that in a colonial situation field sciences may have been developed through imported scientists as an economic necessity, but little fundamental research was possible. A few colonial scientists made important contributions that no doubt enriched science in general, but their activities hardly succeeded in introducing science to the Indian people or in ameliorating their condition. Colonial science did, on the whole, support and help sustain exploitation and underdevelopment.

Deepak Kumar

REFERENCES

Adas, Michael. *Machines as the Measure of Men: Science, Technology and Ideologies of Western Dominance*. Ithaca: Cornell University Press, 1989.

Alvarez, Claude. *Homo Faber: Technology and Culture in India, China and the West, 1500 to Present Day*. The Hague: Nijhoff. 1980.

Arnold, David. ed. *Imperial Medicine and Indigenous Societies*. Manchester: Manchester University Press. 1988.

Bala, Poonam. *Imperialism and Medicine in Bengal*. New Delhi: Sage Publications. 1991.

Brockway, Lucile, H. *Science and Colonial Expansion: The Role of the British Botanic Gardens*. New York: Academy Press. 1978.

Chakravarty, Suhash. *The Raj Syndrome: A Study in the Imperial Perceptions*. New Delhi: Chanakya Publications. 1991.

Cipolla, Carlo, M. *Guns and Sails in the Early Phase of European Expansion. 1400–1700*. London: Collins. 1965.

Dharampal. *Indian Science and Technology in the Eighteenth Century: Some Commentary European Accounts*. Delhi: Impex India. 1971.

Ellsworth, Edward, W. *Science and Social Science Research in British India, 1788–1880*. Westport, Connecticut: Greenwood Press. 1991.

Gaeffke, Peter and Utz, David A., eds. *Science and Technology in South Asia*. Philadelphia: Department of South Asia Regional Studies, University of Pennsylvania. 1985.

Goonatilake, Susantha. *Crippled Minds: An Exploration into Colonial Culture*. New Delhi: Vikas. 1982.

Grove, Richard. *Green Imperialism*. Delhi: Oxford University Press. 1995.

Headrick, Daniel, R. *The Tools of Empire: Technology and European Imperialism in the Nineteenth Century*. New York: Oxford University Press. 1981.

Hoodbhoy, Pervez. *Islam and Science*. London: Zed Books. 1991.

Hutchins, Francis, G. *The Illusion of Permanence: British Imperialism in India*. Princeton: Princeton University Press. 1967.

Kumar, Deepak, ed. *Science and Empire: Essays in Indian Context*. Delhi: Anamika Prakashan. 1991.

Kumar, Deepak. *Science and the Raj 1857–1905*. Delhi: Oxford University Press. 1995.

Kumar, Krishna. *Political Agenda of Education: A Study of Colonialist and Nationalist Ideas*. New Delhi: Sage Publications. 1991.

Leslie, Charles, ed. *Asian Medical Systems*. Berkeley: University of California Press. 1977.

MacKenzie, John M., ed. *Imperialism and the Natural World*. Manchester, Manchester University Press. 1990.

MacLeod, Roy and Kumar, Deepak, eds. *Technology and the Raj*. New Delhi: Sage Publications. 1995.

MacLeod, Roy and Lewis, M., eds. *Disease, Medicine, and Empire*. London: Routledge. 1988.

McClellan, J.E. *Colonialism and Science: Saint Domingue in the Old Regime*, Baltimore, Johns Hopkins University Press. 1992.

Meade, Teresa, and Walker, Mark, eds. *Science, Medicine, and Cultural Imperialism*. New York: St. Martin's. 1991.

Mendelssohn, K. *Science and Western Domination*. London: Thames and Hudson. 1976.

Nandy, Ashis. *Alternative Sciences*. New Delhi: Allied. 1980.

Nandy, Ashis. *The Intimate Enemy: Loss and Recovery of Self Under Colonialism*. Delhi: Oxford University Press. 1983.

Petitjean, Patrick et al. eds. *Science and Empires*. Dordrecht: Kluwer. 1992.

Pyenson, Lewis. *Cultural Imperialism and Exact Sciences: German Expansion Overseas, 1900–1930*. New York: P. Lang. 1985.

Pyenson, Lewis. *The Empire of Reason: Exact Science in Indonesia, 1850–1950*. Leiden: Brill.

Reingold, Nathan and Rothenberg, Marc, eds. *Scientific Colonialism*. Washington: Smithsonian Institution Press. 1986.

Sangwan, Satpal. *Science, Technology and Colonialization: An Indian Experience, 1757–1857*. Delhi: Anamika Prakashan. 1991.

Zimmerman, F. *The Jungle and the Aroma of Meats: An Ecological Theme in the Hindu Medicine*. Berkeley: University of California Press. 1987.

See also: Western Dominance – Ramanujan

Combinatorics in Indian Mathematics

Having prescribed the rule

$$C_n^r = \prod_{k=1}^{r} \frac{n-k+1}{k}$$

for the number of combinations of r things taken at a time from n things, Bhāskara II (CE 1150) remarked: "This [rule] has been handed down [to us] as a general [method], being employed [for their own purposes] by the experts [of specific fields of study], namely, for the tabular presentation of possible meters in metrics, for the number of ways of opening ventilating holes, etc., and the diagram called Partial Meru in arts and crafts, and for the varieties of tastes in medicine. [But], for fear of prolixity, they are not explained here" [*Līlāvatī* 113–114]. It is impossible to tell exactly when and by whom this rule was formulated, but from ancient times Indian peoples have had a keen interest in arranging things in order in various aspects of human life, and in theorising in various areas of study.

The *Bhagavatī*, one of the twelve canonical books of the Jainas, counts by enumeration the number of all the possible cases when one person, two persons, three persons, etc. are distributed in the seven nether worlds.

Most books of Sanskrit and Prakrit prosody, such as the *Chandaḥsūtra* of Piṅgala (ca. CE 200?), *Vṛttajātisamuccaya* of Virahāṅka (between the sixth and the eighth centuries), *Jayadevacchandaḥ* of Jayadeva (before CE 900), *Chando 'nuśāsana* of Jayakīrti (ca. CE 1000), *Vṛttaratnākara* of Kedāra (before CE 1000), *Chando'nuśāsana* of Hemacandra (ca. CE 1150), etc., devote one of their chapters (usually the last one) to "six kinds of ascertainment" (*ṣaṭpratyaya*), namely:

(1) spread (*prastāra*), to spread a list of all the possible variations of a given type of meter consisting of short and long syllables according to a certain method;

(2) lost (*naṣṭa*), to find out a lost variation when its serial number in the list is given;

(3) mentioned (*uddiṣṭa*), to calculate the serial number in the list when a particular variation is mentioned;

(4) short- (and-long-) calculation (*laghukriyā/galakriyā*), to calculate the number of the variations in the list that have a given number of short or long syllables (for this purpose, a diagram called the Mount Meru-like spread (*meruprastāra*), equivalent to the so-called Pascal's triangle, is constructed);

(5) number (*saṃkhyā*), to calculate the number of all the variations in the list; and

(6) way *(adhvan/adhvayoga)*, to calculate the space of writing materials required for writing down the list.

A chapter on prosody in the *Nāṭyaśāstra* of Bharata (before CE 600), a dramaturgical work, and one in the *Agnipurāṇa* (ca. CE 800), a work on sacred traditions, also contains a section on these topics.

Four out of the six kinds of ascertainment, namely, spread, lost, mentioned, and number, were applied to the melody (combinations of seven notes) and the rhythm (combinations of a half, one, two, and three beats) in Indian music [Śārṅgadeva's *Saṅgītaratnākara*, ca. CE 1250], for which a diagram called Partial Meru (*khaṇḍameru*) was employed. This was also the case with the combinations of five sub-categories taken severally from five categories of "carelessness" (*pramāda*) when they contain different numbers of sub-categories, in Jaina philosophy [Nemicandra's *Gommaṭasāra*, ca. CE 980].

Medical treatises such as the *Carakasaṃhitā* (ca. CE 100) and *Suśrutasaṃhitā* (ca. CE 200) treated combinations of the six basic tastes (*rasa*), and of the three humours (*doṣa*) of the body. Varāhamihira dealt with the problem of combinations of nine and sixteen basic perfumes in the *Bṛhatsaṃhitā* (ca. CE 550).

Indian mathematicians prior to Bhāskara, too, are known to have been interested in these problems of combinatorics. Brahmagupta devoted one whole chapter consisting of nineteen (or twenty) stanzas of his astronomical work, *Brāhmasphuṭasiddhānta* (CE 628), for combinatorics related to Sanskrit prosody. The problem of tastes was taken up by the mathematician Śrīdhara in his *Pāṭīgaṇita* (ca. CE 750). Mahāvīra treated the six kinds of ascertainment of Sanskrit prosody as well as tastes in his mathematical treatise, *Gaṇitasārasaṃgraha* (ca. CE 850).

In addition to the traditional rule for combination mentioned above, Bhāskara gave six rules for permutations concerning sequences of numerical figures in a chapter called "nets of numbers" (*aṅkapāśa*) of his famous *Līlāvatī*. However, it is Nārāyaṇa who for the first time treated various problems of permutation and combination, including the system of "spread-lost-mentioned", of different areas systematically from the viewpoint of mathematics with the help of variously defined sequences of numerals. Chapter 13 entitled "nets of numbers" of his mathematical treatise, *Gaṇitakaumudī* (CE 1356), consists of about one hundred stanzas for rules of combinatorics and forty-five for examples.

Takao Hayashi

References

Alsdorf, L. "Die Pratyayas. Ein Beitrag zur indischen Mathematik." *Zeitschrift für Indologie und Iranistik* 9: 97–153. 1933. Translated into English with notes by S.R. Sarma: "The Pratyayas: Indian Contribution to Combinatorics." *Indian Journal of History of Science* 26: 17–61, 1991.

Bag, A.K. "Binomial Theorem in Ancient India." *Indian Journal of History of Science* 1: 68–74. 1966.

Chakravarti, G. "Growth and Development of Permutations and Combinations in India." *Bulletin of the Calcutta Mathematical Society* 24 (2): 79–88. 1932.

Das, L.R. "La théorie des permutations et des combinations dans le gaṇita hindou." *Periodico di matimatiche* 12: 133–140. 1932.

Datta, B. "Mathematics of Nemicandra." *The Jaina Antiquary* 1(2): 25–44. 1935.

Datta, B. and Singh, A.N. "Use of Permutations and Combinations in India." Revised by K.S. Shukla. *Indian Journal of History of Science* 27: 231–249. 1992.

Gupta, R.C. "Varāhamihira's Calculation of nC_r and Pascal's Triangle." *Ganita Bhāratī* 14: 45–49. 1992.

Hayashi, T. "Permutations. Combinations, and Enumerations in Ancient India" (in Japanese). *Kagakusi Kenkyu*, Ser. II 18(131): 158–171. 1979.

Kusuba, T. "Combinatorics and Magic Squares in India. A Study of Nārāyana Paṇḍita's *Gaṇitakaumudī*, Chaps 13–14." Ph.D. Dissertation, Brown University. 1993.

Singh, P. "Nārāyaṇa's Treatment of Net of Numbers." *Gaṇita Bhāratī* 3: 16–31, 1981.

Singh, P. "Contributions of Some Jain Ācāryas to Combinatorics." *Vaishali Institute Research Bulletin* 4: 90–110. 1983.

Singh, P. "The so-called Fibonacci Numbers in Ancient and Medieval India." *Historia Mathematica* 12: 229–244. 1985.

See also: Bhāskara – Brahmagupta – Śrīdhara – Varāhamihira – Mahāvīra – Nārāyaṇa

D

Decimal Notation

Decimal notation is a system which imparts to nine figures (digits) an absolute numerical value and also a positional value which latter increases their value ten times by being shifted by one place to the left. Thus, the digits: 1, 2, 3, 4, 5, 6, 7, 8, and 9, coupled with the figure "0" which stands for zero or *śūnya* (nothing, empty), while expressing just their individual values when standing alone, can express also any quantity of any magnitude by their repeated use in the same number, and shifting of places, as needed. The importance of this contrivance is apparent from the words of the great French mathematician P.S. Laplace, when he says: "The idea of expressing all quantities by nine figures whereby both an absolute value and one by position is imparted to them is so simple that this very simplicity is the reason for our not being sufficiently aware how much admiration it deserves" (Srinivasiengar, 1967). Halstead observes: "The importance of the creation of the zero mark can never be exaggerated. This giving to airy nothing, not merely a local habitation and a name, a picture, a symbol, but helpful power, is the characteristic of the Hindu race from whence it sprang. It is like coining the *Nirvāṇa* into dynamos. No single mathematical creation has been more potent for the general on-go of intelligence and power".

In Indian tradition, the need for enumeration and decimal notation stemmed from the adoration of gods and for ritualistic purposes. From Vedic times, the Hindus used the decimal notation for numeration. The *Ṛgveda* (ca. 2000 BCE) groups gods into three (1.105.5); there are three dawns (8.41.3); there were seven rays of the Sun-god (1.105.9); there were seven sages (4.42.8), and seven seas (8.40.5). There were 180 Marut-gods, or three times sixty (8.96.8); the God Śyāvā gave as gifts cows numbering 210 or three times seventy (8.19.37). There were 21 followers of Indra, or three times seven (1.133.6), and the number of horses prayed for was thrice seven times seventy or $3 \times 7 \times 70$(8.46.26). In Vedic literature, besides the primary numbers, one to nine, expressed by the terms, *eka, dvi, tri, catur, pañca, ṣaṭ, sapta, aṣṭa* and *nava,* the decuple terms from ten to ninety, expressed by *daśa, viṃśati, triṃśat, catvāriṃśat, pañcāśat, ṣaṣṭi, saptati, aśīti* and *navati* are found. These are then sequentially multiplied by ten, taking terms from 100 to 10 to the power of 12, the terms being *śata, sahasra, ayuta, niyuta, prayuta, koṭi, arbuda, nyarbuda, samudra, madhya, anta,* and *parārdha* . In the matter of the arrangement of decuples in compound number-names, the practice generally followed in Vedic literature was to put the term of higher denomination first, except in the case of the two lowest denominations, where the reverse method was followed. See, for example, *sapta śatāni viṃśatiḥ* (seven hundreds and twenty, *Ṛgveda* 1.164.11), *sahasrāṇi śata daśa* (thousands hundred, and ten, *Ṛgveda* 2.1.8) and *ṣaṣṭi, sahasra navatim nava* (sixty thousands, ninety and nine, 60,099 *Ṛgveda* 1.53.9).

With respect to written symbols for numbers, since no palaeographic records of the Vedic age have been preserved, little can be said. However, a few Vedic passages occur where written numerical symbols are mentioned. In the *Ṛgveda* (10.62.7) certain cows with the mark of "8" *(aṣṭa-karṇī)* are referred to, and *Yajurveda-Kāṭhaka Saṃhitā* makes mention of pieces of gold with the mark "8" imprinted on them *(aṣṭa-pruddhiraṇyam, aṣṭā-mṛdamhiraṇyam,* 13.10). Inscriptions and manuscripts, of later ages, all over India, use numerical symbols profusely. The tendency had been, from early ages, to spell out the numbers or make use of things permanently associated with a number to represent that number. For instance, eyes, hands, etc. were used for 2; Moon, sky, etc. were used for 1, seasons for 6, and week for 7. Another method was to attribute specific numerical values for the letters of the alphabet and use those letters to indicate the specified numbers, a method which was mentioned by the Sanskrit grammarian Pāṇini of the fourth century BCE. These methods were very popular in the classical age in India, especially with mathematicians and astronomers.

K. V. Sarma

References

Datta, Bibhutibhusan, and Avadhesh Narayan Singh. *History of Hindu Mathematics: A Source Book*. Bombay: Asia Publishing House. 1962.

Gupta, R.N. "Decimal Denominational Terms in Ancient and Medieval India." *Ganita Bharati* 5(1–4): 8–15. 1986.

Halstead, G.B. *On the Foundation and Technic of Arithmetic*. Chicago. Open Court. 1912.

Ray, Priyadranjan, and S.N. Sen. *The Cultural Heritage of India. vol. VI: Science and Technology*. Calcutta: Ramakrishna Mission Institute of Culture. 1986.

Ṛgveda: The Hymns of the Ṛgveda. Trans. R.Th. Griffith. Delhi: Motilal Banarasidass, 1986.

Sharma, Mukesh Dutt. "Indian Invention of Decimal System and Number Zero." *Vedic Path* 44(1): 32–37. 1982.

Srinivasienagar, C.N. *The History of Ancient Indian Mathematics.* Calcutta: World Press. 1967.

See also: Mathematics – Sexagesimal System – Zero

Deśāntara

In Indian astronomy, the *Deśāntara* of a place is its terrestrial longitude, i.e. the "distance of the place" from a universally accepted zero meridian. In modern times, the meridian at Greenwich in England is accepted as the zero meridian, and the longitude is expressed in terms of the angle subtended, at the pole, by the Greenwich meridian and the meridian of the place in question. Indian astronomy had, from early times, taken as the zero meridian the meridian passing through the ancient city of Ujjain in Central India, cutting the equator at an imaginary city called Laṅkā and passing through the south and north poles. Again, in order to facilitate the conversion of the local time to that of the zero meridian and vice versa, the *Deśāntara* was expressed in terms of time-measures like *nāḍī* (or *ghaṭī*), equal to 24 minutes, as converted from the corresponding degrees. Since the earth completes one eastward rotation of 360° in 24 hours, it is 15° an hour or 1 degree in 4 minutes. In terms of Indian measures, since 60 *nāḍīs* are equal to 24 hours or two and half *nāḍīs* make one hour, the rotation of 15° corresponds to a period of two and a half *nāḍīs*. Since 1 *nāḍī= 60 vināḍīs*, and 1 *vināḍi= 6 prāṇas*, the rotation will be 1 degree in 10 *vināḍīs*, or 1 minute in 10 *prāṇas*. The *deśāntara* which is expressed in terms of time-measure is done through either *nāḍīs, vināḍīs* or *prāṇas.*

Since the planetary positions derived by Indian astronomical computation are all related to the mean sunrise at the zero meridian, viz., the Ujjain meridian, to arrive at the positions at local places, a longitude correction or *deśāntara-saṃskāra* is called for. It is calculated in time-measures, as above, and is subtracted if the place in question is east of Ujjain and added if it is west of Ujjain.

The *deśāntara-saṃskāra* is expressed also in terms of the distance, i.e. in *yojanas,* of the desired place from the Ujjain meridian in the same latitude, at the rate of 55 *yojanas* for 10 *vināḍīs* or 1 minute.

K. V. Sarma

See also: Astronomy

Devācārya

Devācārya, son of Gojanma and author of the astronomical manual *Karaṇaratna* (lit. Gem of a Manual), hailed from Kerala in South India. The epoch of *Karaṇaratna,* i.e. the date from which planetary computations were instructed to be commenced in that work, is the first day of the year 611 in the Śaka era, which corresponds to February 26 of CE 689. This places Devācārya in the latter half of the seventh century. We know he came from Kerala because he used the *Kaṭapayādi* system of letter numerals and the *Śakābdasaṃskāra,* which is a correction applied to the mean longitudes of the planets from the Śaka year 444 or CE 521, and a unique method of computing the solar eclipse, all of which are peculiar to Kerala, and because his work is popular in that part of the land.

Devācārya uses the elements of the Āryabhaṭan school of astronomy as the basis of his work, as he himself states towards the commencement of *Karaṇaratna.* His work is based both on the *Āryabhaṭīya* and on the second work of Āryabhaṭa, the *Āryabhaṭa-siddhānta,* as abridged in the *Khaṇḍakhādyaka* of Brahmagupta. The influence of the *Sūryasiddhānta* and of Varāhamihira are also apparent. It is also noteworthy that Devācārya himself innovated several thitherto unknown methodologies and techniques.

In eight chapters, the *Karaṇaratna* encompasses almost all the generally accepted aspects of Hindu astronomical manuals. Chapter I of the work is concerned with the computation of the longitudes of the Sun and Moon, and also the five basic elements of the Hindu calendar *(pañcāṅga).* Computation and graphical representation of the lunar and solar eclipses are the subjects of Chapters II and III. Chapter IV deals with problems related to the gnomonic shadow and Chapter V with the calculation of the time of the moonrise and allied matters. In Chapter VI, heliacal rising of the Moon and elevation of the Moon's horns are dealt with. The last two chapters are concerned with the derivation of the longitudes of the planets, planetary motion, and planetary conjunctions.

Several peculiarities characterise Devācārya's work. Among these are the computation of the Sun, Moon, Moon's apogee, and Moon's ascending node using the "omitted" lunar days, the *Śakābda* correction, and the application of a third visibility correction for the Moon. However, what is most significant in the work is the recognition of the precession of the equinoxes and the rule that he gives for its determination on any date. Devācārya's measure of the rate of precession is 47 seconds per annum, its modern value being 50 seconds. Devācārya's importance lies in the fact that his work formed a record of the astronomical practices and methodologies for the quick derivation of astronomical data that prevailed in India during the seventh century CE.

K. V. Sarma

REFERENCES

The Karaṇa-ratna of Devācārya. Ed. Kripa Shankar Shukla. Lucknow: Department of Mathematics and Astronomy, Lucknow University. 1979.

Pingree, David. *Census of the Exact Sciences in Sanskrit*. Philadelphia: American Philosophical Soc. Ser. A, vol. 3, 1976. p. 121.

See also: Āryabhaṭa – *Sūryasiddhānta* – Varāhamihira – Precession of the Equinoxes

Dyes

The human urge to paint the body with symbolic, warlike, or identifying colours may have been among the earliest impulses which led to the discovery of colour-yielding clays and plants. The dyeing of human clothing followed. Encounters, both warlike and peaceful, between groups then led to the identification of certain colours with specific regions or groups of producers, and exchanges began. This specialisation and trade eliminated many of the poorer dyes of prehistoric times, which may have numbered many hundreds, and by the time recorded history took note of dyes only a relative few still saw widespread use.

Dye exchanges at first were very local. The major ones in ancient times were usually confined to the great areas of early culture and urbanisation such as China, northern India, and the eastern Mediterranean. Later, trade in dyestuffs became more long distance, and, with European intrusions into Asia and invasion of the Americas, transoceanic and worldwide. This huge trade in natural dyes was destroyed and again reduced to a local level by the invention of coal tar or aniline dyes and other chemical dyes, in the mid-nineteenth century.

The production and exchange of colourants was always associated with certain other industries. Cloth manufacture and weaving were certainly the main ones. Others arose out of the limitations of some natural dyes. Many were not "fast"; that is they tended to fade or discolour when exposed to light, frequent washing, or wear and tear. Accordingly, much effort was historically expended in the search for mordants (dye fixers). Many of these were readily available, such as blood, dung, or urine. Various acids, alkalis, and wetting agents were widely used. Alum, an astringent, was probably the most common, and large cargoes were

or kutch, a brown dye, has been manufactured in India for over two thousand years and comes from the leaves and twigs of various acacia and mimosa trees.

Humans have elaborated dyes from the bodies of insects and animals for millennia. The most expensive, prestigious dye of ancient times was Tyrian purple. It was manufactured in Crete by 1600 BCE, but is usually associated with the Phoenicians. Tyre became the great market centre for this dye until captured by the Arabs in CE 638. Phoenician traders had spread its use all over the Mediterranean. It was so rare and costly that in many areas this fast, blue to purple dye was restricted to the clothing of high ecclesiastics, the aristocracy, or simply the ruling family. This colourant is extracted from a gland found in several shellfish, most notably *Murex trunculus*. A few similar dyes were used in the Americas before and after the European invasions, especially in Nicoya (Costa Rica) and on the Peruvian coast, but the American sources never produced enough dye to be of importance beyond local markets.

Insects have also been significant to the natural dye industry. Kermes is the oldest and most widespread of these dyestuffs. The insects harvested, *Coccus arborum* and *Coccus ilicis*, live on the holm oak (*Quercus ilex*), the shrub oak (*Quercus coccifera*) and a few other trees. Kermes is Armenian for "little worm" and is a scarlet dye. It is mentioned in the Old Testament and in the writings of ancient Greece, but its origins are probably Asiatic and it was much used in India.

The European invasions of South and South-east Asia and of the Americas greatly changed dye usage by bringing new and better colourants into the international markets. Indigo, which produces a range of blues, became for about two centuries the most important of all dyestuffs. It is a vegetable dye of considerable fastness, and has been known in parts of Asia for over four thousand years. *Indigofera tinctoria*, of which only the leaves bear dye, is of the order leguminosae, and belongs to the pea family. It was found in India, South-east Asia, Africa and America. It has been discovered in both Egyptian and Inca tombs, and its continent of origin is obscure.

Dutch ships carried indigo throughout the Indian Ocean, and then, in the seventeenth century, to Europe, where, despite a struggle, it eventually displaced woad. Bengal supplied large quantities in the late eighteenth century to the British textile industry. When American indigo, mostly produced in Central America, the Carolinas, and Georgia, began to flood the market in the eighteenth century, the woad industry collapsed in the face of this cheaper and better dyestuff.

Cochineal, a scarcer and more expensive dye, had a similar history. It is made from an American insect, *Coccus cacti*, which feeds on a cactus (*Nopalia* or *Opuntia cochinellifera*), and was used in Mesoamerica, especially in the Oaxaca region, long before the Europeans arrived. While its production is elaborate and costly, the result is a superior scarlet or crimson dye, and it soon replaced kermes in Asian and European markets and dyeworks. After bullion it was the most expensive item carried by the Spanish treasure fleets.

The brazilwood industry grew after the products of the American continents began to enter world commerce. Vast new stands were found in Brazil – hence the name – and to this was added logwood or campeachy wood, a large, tropical American tree *(Haematoxylon campeacheanum)* found especially in Campeche, Tabasco, and Belize. It yields black or blue dyes, plus edible seeds called allspice. By the seventeenth century it was in use in Africa and Europe.

One crop from America moved to Asia. Annato (*Bixa orellana*) was used by Mesoamerican peoples largely as a food additive and colourant, but when taken to South-east Asia and India this dye, which is a poor, fugitive yellow to red, was used for cloth, especially monks' robes. It is so culturally accepted in these regions today that many writers describe it as indigenous.

The new dominance achieved by Asian and above all American dyes such as indigo and cochineal, both of which were spread by European expansion, lasted less than two centuries. They, in their turn, were overwhelmed by the new coal tar or aniline dyes invented in the mid-nineteenth century. Many of the natural dyes, however, remain in use in local and peasant economies.

Murdo J. Macleod

References

Cannon, John. *Dye Plants and Dyeing*. Portland, Oregon: Timber Press. 1994.

Donkin, R.A. *Spanish Red: An Ethnographic Study of Cochineal and the Opuntia Cactus*. Philadelphia: American Philosophical Society. 1977.

Leggett, William Ferguson. *Ancient and Medieval Dyes*. Brooklyn: Chemical Publishing Co. 1944.

Robinson, Stuart. *A History of Dyed Textiles*. Cambridge, Massachusetts: MIT Press. 1969.

Zanoni, Thomas A., and Eileen K. Schofield. *Dyes From Plants: An Annotated List of References*. New York: The New York Botanical Garden. 1983.

See also: Textiles

E

East and West

As traditionally used in the West, the terms East and West imply that the two are somehow of equal importance. While that might be arguable in the nineteenth and twentieth centuries it was certainly not true during the long reaches of human history prior to the nineteenth century. By 500 BCE the globe supported four major centres of civilisation: the Chinese, the Indian, the Near Eastern, and the Western, considering Greek culture antecedent to what eventually became the West. Of the four the West was probably the least impressive in terms of territory, military power, wealth, and perhaps even traditional culture. Certainly this was the case after the fall of the Roman Empire in the fifth century CE. From that time until about CE 1500 the West probably should be regarded as a frontier region compared to the other centres of civilisation.

From roughly 500 BCE to CE 1500 a cultural balance was obtained between the four major centres of civilisation. During these millennia each centre continued to develop its peculiar style of civilised life, and each continued to spread its culture and often its control to peoples and lands on the periphery. While the inhabitants of each centre of civilisation were aware of the other centres, sometimes traded with them, and occasionally borrowed from them, the contacts were sufficiently thin so that no one centre threatened – commercially, militarily, politically, or culturally – the existence of the others. During this long period, that is through much of civilised human history, there was no question of Western superiority or hegemony. No visitor from Mars would likely have predicted that the West would eventually dominate the globe.

Obviously the Greeks and Romans knew quite a bit about the Near-Eastern world, especially about the Persian empire, that of Alexander the Great, and the successor states formed after its collapse. About India and China they knew much less, and what they knew was much less accurate. Although Herodotus (ca. 484–425 BCE) reported some things about India, most of the information available to the Greeks came from the writers who described Alexander's campaigns in the Indus Valley (326–234 BCE) and from Megasthenes. They described India as fabulously rich, the source of much gold and precious stones. It was hot; the Sun stood directly over-head at midday and cast shadows toward the south in summer and toward the north in winter. They described huge rivers, monsoons, tame peacocks and pheasants, polygamy, and the practice of *suttee* (widow burning). However, they also reported fantastic things such as gold-digging ants, cannibals, dog-headed people, and people with feet so large that they served as sun shades when sitting. The Romans knew that India was the source of spices and that China, which they called Serica and well as Sinae, was the source of silk. There are even possible traces of Asian influence in some Roman silver and ivory work and perhaps even some influence of Buddhism on Neo-Platonism and Manichaeanism.

From about the fourth century, even before the fall of the Roman empire, until the return of Marco Polo from China in the late thirteenth century, Europe or the West added little factual information to its understanding of India and China. After the fall of the Roman empire there was no direct trade between Europe and Asia, and thus there were no opportunities to test the stories by observation. The rise of Islam in the seventh and eighth centuries completed Europe's isolation. During these centuries the old stories inherited from the Greeks were retold and embellished with little effort to distinguish fact from myth. To these were added three legends of more recent origin: the stories celebrating the heroic exploits of the mythical Alexander; those rehearsing Saint Thomas the Apostle's missionary journey to India and his subsequent martyrdom; and those describing the rich, powerful, Christian kingdom of Prester John somewhere to the east of the Islamic world with which European rulers dreamed of allying against the Muslims. Even the trickle of precious Asian products brought to Europe by intermediaries seemed only to confirm the image of Asia as an exotic and mysterious world, exceedingly rich and exceedingly distant.

The rise of the Mongol empire in Asia during the thirteenth century resulted in direct overland travel between Europe or the West, and China. The Mongols' success also revived hopes among European rulers of finding a powerful ally to the east of the Muslims. Even the devastating Mongol incursions into Poland and Hungary in 1240 and 1241 scarcely dampened their enthusiasm. Already in 1245 the pope sent an embassy led by John of Plano Carpini to the Mongol headquarters near Karakorum. He was followed during the ensuing century by a fairly large number of envoys, missionaries, and merchants, several of who wrote reports of what they saw and did in Eastern Asia. Marco Polo's was the most comprehensive and reliable, and the most widely distributed of the medieval reports. By the time the Polos first arrived at Kublai Khan's court in 1264 it was newly established at Cambaluc (Beijing), from which the khan ruled the newly conquered Cathay (China). Like many other foreigners during the Mongol period (the Yuan Dynasty in China, 1260–1368) the Polos were taken into the khan's service. They were employed in the Mongol administration for seventeen years during which time Marco travelled extensively throughout China. On his return to Europe he produced the first detailed description of China in the West based primarily on first-hand observation and experience. No better account of China appeared in Europe before the middle of the sixteenth century. Marco Polo described China as the wealthiest, largest, and most populous land in the thirteenth-century world. While his understanding of Chinese culture was minimal he accurately and admiringly described cities, canals, ships, crafts, industries, and products. He noted the routes, topography, and people encountered in his travels, including his voyage home through South-east Asia to Sumatra, Ceylon and along the west coast of India.

The decline of the Mongol empire and the establishment of the Ming Dynasty in China in 1368 severed the direct connection between Europe and China. The fall of Constantinople in 1454 and establishment of the Turkish empire in the Near East disrupted the older connections between Europe and the near East and India. Europe's isolation from the outside world was complete, not to be restored until the opening of the sea route around the tip of Africa in the waning years of the fifteenth century. During this period no European appears to have travelled to China. Some few travel reports refer to India and South-east Asia. Of them only that written by the humanist Poggio Bracciolini in 1441 and based on Nicolò de' Conti's travels added to the West's store of knowledge about India and confirmed some of the more accurate of the ancient Greek reports. In fact amid the Renaissance humanists' enthusiasm for the rediscovery of ancient Greek literature, the Greek reports of India received new respect and attention.

During the long era of cultural balance before CE 1500 many important technological and scientific inventions and innovations appear to have migrated to the West from the other centres of civilisation, more often from China than from the others. The migration of technology was usually gradual, involving one or more intermediaries, the inventions usually being established in the West without any clear ideas about their origins. Much of the basic technology that enabled the Europeans to sail directly to Asia in 1500 and later to begin their march towards global domination was known earlier in the Asian centres and only later adopted or separately invented in Europe.

Among the more important technological borrowings were gunpowder, the magnetic compass, printing, and paper, all apparently originating in China. For none of them is the path of migration entirely clear, and thus for none of them can the possibility of independent invention be entirely ruled out. Gunpowder, for example, was known in China by 1040 and did not appear in Europe until the middle of the thirteenth century. The magnetic compass was fully described in an eleventh-century Chinese book, *Meng Qi Bi Tan* (Dream Pool Essays), written by Shen Gua in 1088. It began to be used in Europe during the late twelfth or early thirteenth century. Most likely Europeans learned about it from the Arabs. The case for moveable-type printing having been borrowed from the Chinese is more hotly debated than that for gunpowder or the compass. Woodblock printing was used in China by the seventh century, and paper was invented much earlier; the first printed books appeared there during the ninth century, six centuries before the invention of printing in the West by Johannes Gutenberg in 1445. Block printing probably became known in Europe through the introduction of printed playing cards and paper money during the Mongol period; medieval travellers frequently mentioned these. Because of the large number of Chinese characters the Chinese continued to prefer printing from page-sized blocks of wood carved as a single unit; European printing almost immediately employed moveable type, thus convincing some scholars that it was a separate invention. However, while they may have preferred block printing the Chinese also developed moveable type as early as the eleventh century. For none of these basic inventions taken separately – gunpowder, the compass, and printing – is the case for its diffusion from China to Europe indisputably demonstrated. Taken together, however, along with a rather large number of other technological and scientific innovations such as paper, the stern-post rudder, the segmented-arch bridge, canal lock-gates, and the wheelbarrow, which all appear to have migrated from China to the West, it becomes apparent that the general flow of technology and science in pre-modern times was from East to West. This would seem to increase the likelihood that these basic innovations also migrated to Europe from Asia. Those who used the technological innovations probably cared little about their ultimate origin and apparently did not seek it out. Nevertheless European mariners after 1500, confronted first hand with evidence that printing, gunpowder, the mariner's compass, and the like had been in use much longer in Asia than in Europe, frequently suggested that they had been borrowed from the Asians.

Even before gunpowder a group of military innovations found their way to Europe from China and India, again through intermediaries and apparently without Europeans being aware of their origins. The Chinese form of the Indic stirrup was the most important of these and may have been as important to military development in the eighth century as gunpowder was later. The Javan fiddle bow and the Indian Buddhist pointed arch and vault were acclimated in Europe before 1100. The traction trebuchet along with the compass and paper appeared in the twelfth century. Still more important for subsequent Western scientific achievements was the adoption of Hindu–Arabic mathematics in the twelfth and thirteenth centuries: the Indian system of arithmetical notation, trigonometry, and the system of calculating with nine Arabic numbers and a zero were all practiced in India as early as CE 270. Some components of Indian mathematics may have come from Babylonia or China. They came to Europe, however, through the translation of Arabic writings, and the European borrowers usually credited India rather than China, Babylonia, or the Arabic intermediaries as the source of the new mathematics. Along with Indian mathematics, Europeans learned some elements of Indian astronomy and also became fascinated with the Indian idea of perpetual motion.

Also before 1500, Europeans sometimes attempted to imitate desirable Asian products, not always successfully. Already in the sixth century the Byzantine emperor Justinian monopolised the silk trade in his realm and expressed his determination to learn the secret of its manufacture. In 553 a monk supposedly smuggled some silkworm eggs into Constantinople carrying them in a hollow stick, perhaps of bamboo. Nothing is said in the story about the importation of silk technology, but less than a century later sericulture had obviously taken root in Syria. From there it spread to Greece, Sicily, Spain, Italy, and France. In Italy during the fourteenth century, water power was used in silk spinning, as it had been used in China much earlier. Attempts to imitate Chinese porcelain, however, were unsuccessful. The best attempts to do so were made in northern Italian cities during the fifteenth century. None, however, approached Chinese porcelain in composition, colour, or texture. Nor did the Dutch Delftware of the seventeenth century, another attempt to imitate the Chinese product. Not until the eighteenth century were European craftsmen able to produce a hard-paste porcelain to rival that of China. Also appearing in Europe before 1500 were less important devices or techniques such as the Malay blowgun, playing cards, the Chinese helicopter top, the Chinese water-powered trip-hammer, the ball and chain governor, and maybe Chinese techniques of anatomical dissection.

While impressive, the West's importation of Asian science and technology before 1500 in no way deflected Western culture from its traditional paths. Seldom was the provenance of the new inventions or techniques known, and they could all rather easily be incorporated into the traditional Christian European world-view. They did not provoke any serious questions about the European way of life, its religious basis, its artistic and cultural traditions, or even its traditional scientific views. This is also true for the artistic and cultural borrowings from Asia prior to 1500. They too were often unconscious, and even when they were not, they were regarded as embellishments or decoration, rather than in any way a challenge to traditional themes. For example, the incorporation into the Christian calendar of Saints Baarlam and Josephat, derived as they were from stories of the life of Buddha, resulted not in a Buddhist challenge to the Christian faith but simply in the addition of two new saints to the growing Christian pantheon.

Nevertheless, the borrowing and adaptation of Asian science and technology by the West before 1500 was indispensable in making the long overseas voyages to Asia and the protection of the ships and shore installations possible. Without gunpowder, cannons, the compass, the stern-post rudder, etc. there would have been no European expansion. While they had lagged well behind the other centres of civilisation through most of civilised history, by 1500, European marine and military technology were beginning to equal that of the Near East, India, and China, and were obviously superior to that of the peripheral areas of Africa, South-east Asia, and the Americas. The Portuguese voyages down the coast of Africa and Columbus' voyage across the Atlantic attest to Europe's rapidly improving technology.

A small Portuguese fleet under Vasco da Gama reached Calicut on the Malabar Coast of India in 1498, thus establishing direct contact between Europe and Asia by sea and also inaugurating a new era in the relationships between the four major centres of civilisation on the globe. Soon after 1500 the Europeans began to dominate the seas of Asia, the Portuguese in the Indian Ocean being the first. The Portuguese, and later the Dutch and the English, moved along sea lanes, visited seaports, and fitted into a trading world which had been developed earlier by Muslim traders and which stretched from eastern Africa to the Philippines. As the Europeans at first tried to compete in that world and later tried to dominate and even to monopolise it, they found Muslim merchants and merchant-princes to be their most formidable opposition. That they were able to move so rapidly and effectively into that international trading system was due not so much to their superior technology and fire-power as to the fact that the great Muslim empires of the sixteenth century – the Ottoman Turkish, the Safavid Persian, and the Mughul in northern India – seem to have been too busy consolidating their newly won empires to contest the European intrusion. These Muslim empires, as well as the South-east and East Asian empires of Siam, Vietnam, China, and Japan had all as a matter of governmental policy turned away from the sea and looked inward to the control of their land empires and to land taxes as the source of their wealth and power. Had any or all of these major Asian and Near-Eastern states seriously resisted the western incursion the story might have had a different conclusion. Apart from Muslim traders in the Indian Ocean those who formidably opposed European power in the sixteenth and seventeenth centuries were small Muslim commercial port-city states like Makassar on Celebes and Aceh on Sumatra. The Mughul emperors in northern India usually allowed the Europeans to trade freely in their ports, often exempting them from customs duties. Apart from the illegal but locally tolerated Portuguese settlement in Macao the Europeans were not permitted to trade in China at all. China briefly contested Dutch maritime power only in 1624 after the Dutch had spent two years raiding the coast of Fukien Province, burning villages, seizing junks, enslaving their crews, and constructing a fort on one of the Pescadores Islands; all in an effort to force the Chinese to allow them to trade freely at some port along the Chinese coast. Confronted with a full scale Chinese war fleet the Dutch commander hastily sued for peace and gratefully accepted the

Chinese admiral's offer to let the Dutch trade on Formosa, which was not yet considered Chinese territory. From 1500 until about the middle of the eighteenth century the Europeans were able to carve out empires in the Americas, insular South-east Asia, and the Pacific Islands, but they did not threaten the major centres of civilisation in Asia.

During the first three centuries of direct maritime contact between Europe and Asia (1500–1800) the Westerners showed little sense of cultural superiority toward the high cultures of Asia. If anything they tended to exaggerate the wealth, power, and sophistication of the other centres of civilisation. Most Europeans were confident that Christianity was indeed the true religion, and they quickly began to send missionaries to convert Asian peoples. By 1600, if not earlier, they were also justifiably confident that European mathematics, science, and technology were superior to that of the Asians. Those areas aside, however, Europeans were endlessly fascinated with what they discovered beyond the line and realised that they still had much to learn from the high cultures of Asia. Between 1500 and 1800 the currents of cultural influence continued to flow mainly from east to west, but during this era the impact of Asia on the West was usually more conscious and deliberate than previously, and it was primarily in areas other than science and technology.

The new seaborne commerce with Asia almost immediately brought greatly increased quantities and varieties of Asian products into Europe. Pepper and fine spices were the first to appear on the docks in European ports, but they were soon followed by such goods as Chinese porcelain and lacquerware, tea, silks, Indian cotton cloth, and cinnamon. Following these staples of the trade came also more exotic products: Japanese swords, Sumatran or Javanese krisses, jewellery, camphor, rhubarb, and the like; the list is very long. Some of these products, such as tea, provoked striking social changes in Europe. Attempts to imitate others resulted in new industries and in new manufacturing techniques, Delft pottery in imitation of Ming porcelain, for example. Attempts to compete with cheap Indian cotton cloth seem to have touched off a technical revolution in the British textile industry which we customarily regard as the beginning of the industrial revolution.

Along with the Asian products came descriptions of the places and peoples who had produced the products. The earliest sixteenth-century descriptions usually seem designed to inform other Asia-bound fleets about the conditions of trade. After Christian missions were established they were often intended to elicit support for the missionaries. However, before long the travel tales and descriptions became popular in their own right and profitable to publish. During the seventeenth century what had been a sizeable stream of literature about Asia became a veritable torrent. Hundreds of books about the various parts of Asia, written by missionaries, merchants, mariners, physicians, soldiers, and independent travellers were published during the period. For example, during the seventeenth century alone there appeared at least twenty-five major descriptions of South Asia, another fifteen devoted to mainland South-east Asia, about twenty to the South-east Asian archipelagos, and sixty or more to East Asia. Alongside these major independent contributions stood scores of Jesuit letterbooks, derivative accounts, travel accounts with brief descriptions of many Asian places, pamphlets, news sheets, and the like. Many of the accounts were collected into the several large multivolume compilations of travel literature published during the period. In addition to the missionaries accounts, travel tales, and composite encyclopaedic descriptions, several important scholarly studies pertaining to Asia were published during the seventeenth century: studies of Asian medicine, botany, religion, and history; and translations of important Chinese and Sanskrit literature.

The published accounts range in size from small pamphlets to lavishly illustrated folio volumes. They were published in Latin and in almost all of the vernaculars, and what was published in one language was soon translated into several others, so that a determined enthusiast could probably have read most of them in his own language. They were frequently reprinted in press runs which ranged from 250 to 1000 copies. Five to ten editions were not at all uncommon, and some of the more popular accounts would rival modern "best sellers". In short the Early-Modern image of Asia was channelled to Europe in a huge corpus of publications which was widely distributed in all European lands and languages. Few literate Europeans could have been completely untouched by it, and it would be surprising if its effects could not have been seen in contemporary European literature, art learning, and culture.

From this literature European readers could have learned a great deal about Asia and its various parts. Perhaps most obviously their geographic horizons would have been continually expanded. Gradually Europeans gained accurate knowledge about the size and shape of India, China, and South-east Asia. During the seventeenth century, for example, several puzzles which had plagued earlier geographers were solved: for example, the identification of China with Marco Polo's Cathay, the discovery that Korea was a peninsula and Hokkaido an island. By the end of the century Europeans had charted most of the coasts of a real Australia to replace the imagined antipodes as well as those of New Guinea, the Papuas, numerous Pacific islands, and parts of New Zealand. Interior Ceylon and Java, as well as Tibet were visited and accurately described by Europeans before the end of the century. By 1700 only areas of continental Asia north of India and China, the interior of Australia, New Zealand, and New Guinea and parts of their coastlines remained unknown to the Europeans. Most of these lacunae were filled during the eighteenth century. Even more impressive than the greatly expanded geographic knowledge available to European readers in early modern times was the rapidly increasing and increasingly detailed information about the interiors, societies, cultures, and even histories of Asia's high cultures. Already during the seventeenth century European readers could have read detailed descriptions and even viewed printed cityscapes and street scenes of scores of Asian cities, interior provincial cities as well as capitals and seaports, and they could have learned countless details about Asia's various peoples, their occupations, appearance, social customs, class structures, education, ways of rearing children, religious beliefs, and the like. Details regarding Asia's abundant natural resources, crafts, and arts were described as well as its commercial practices and patterns of trade. Asian governments were described in exceedingly close detail, especially for major powerful states such as China, the Mughul Empire, Siam, and Japan. Jesuit missionaries in China, for example, described the awesome power of the emperor, his elaborate court, the complex imperial bureaucracy and its selection through competitive written examinations, and the Confucian moral philosophy on which it was all based. They also described the frequently less orderly and less savoury practice of Chinese government, complete with detailed examples of officials' abuse of power and competing factions within the administration. Similar details were reported for the governments of all the major states as well as for countless smaller states. By the end of the seventeenth century European observers had published many sophisticated accounts of Asian religions and philosophies; not only the frequently deplored Hindu "idolatry" and widow burning, but also the Hindu world view which lay beneath the panoply of deities and temples, the various schools and sects of Hinduism, and the ancient texts of Hindu religion. Similarly sophisticated and detailed accounts of Confucianism and Buddhism were available, as well as descriptions of the beliefs of peoples like the Formosan aborigines, the Ainu of Hokkaido, and the inner Asian and Manchurian tribes. Seventeenth and eighteenth-century readers could also have learned much about Asian history; especially, but not exclusively, that of Asia's high cultures. By the mid-seventeenth century, for example, a very detailed sketch of China's long dynastic history culled from official Confucian histories by Jesuit missionaries had been published. Martino Martini wrote the *Sinicae historiae decas prima*, which was published in Munich in 1658. During the eighteenth century an important Chinese history, the *Tongjian Gangmu* (Outline and Details of the Comprehensive Mirror [for aid in government]) by Zhu Xi, was translated into French in its entirety by Joseph-Anne-Marie de Mailla and published in Paris (*Histoire générale de la Chine*, 1777). Not only history, however, but also news was reported to Early-Modern readers. Their image of Asia was surely not that of a static world far away. Among the more important events reported in almost newspaper-like detail during the seventeenth century alone were the Mughul emperor's successful campaigns in the south of India, the Maratha challenge to Mughul supremacy, the fall of the Indian states of Golconda and Vijayanagar, the Manchu Conquest of China in 1644, the feudal wars and the establishment of the Tokugawa shogunate in Japan (1600), and the internal rivalries and wars in Siam and Vietnam. Natural disasters such as earthquakes, fires, volcanic eruptions, and the appearance of comets were also regularly reported. Readers of this richly detailed, voluminous, and widely distributed literature may well have known relatively more about Asia and its various parts than do most educated westerners today.

The post-1500 literature on Asia also contains a great many descriptions of Asian science, technology, and crafts: such things as weaving, printing, papermaking, binding, measuring devices, porcelain manufacture, pumps, water-mills, hammocks, palanquins, speaking tubes, sailing chariots, timekeepers, astronomical instruments, agriculture techniques and tools, bamboo and other reeds for carrying water, as well as products such as musical instruments, wax, resin, caulking, tung-oil varnish, elephant hooks and bells, folding screens, and parasols. Some, such as Chinese-style ship's caulking, leeboards, and strake layers on hulls, lug sails, mat and batten sails, chain pumps for emptying bilges, paddle-wheel boats, wheelchairs, and sulphur matches were quickly employed or imitated by Europeans. Some provoked documented experimentation and invention: Della Porta's kite in 1589, and Simon Stevin's sailing chariot in 1600. The effects of the new information on the sciences of cartography and geography are obvious and profound. Simon Stevin, whose sailing chariot was inspired by descriptions of similar Chinese devices, also introduced decimal fractions and a method of calculating an equally-tempered musical scale: both of which might also have been inspired by Chinese examples. The sixteenth-century mariners' cross staff may have been inspired by the Arab navigators' *kamals*. The Western science of botany was profoundly influenced by the descriptions of Asian flora and by the specimens taken back to Europe and successfully grown in European experimental gardens; rice, oranges, lemons, limes, ginger, pepper, and rhubarb were among the most useful. More important for botany, however, the Asian plants provoked comparisons with familiar plants, the development of comprehensive classification schemes, and thus the beginnings of modern plant taxonomy. Some Asian cures (herbs and drugs) were borrowed, especially for tropical medicine. Chinese acupuncture, moxibustion, and methods of diagnosis by taking the pulse were minutely described and much admired by European scholars, but it is not yet clear to what extent they were actually used in Europe.

The flow of cultural influence was not exclusively from East to West between 1500 and 1800. Many Asians became Christian. Able Jesuit missionaries translated scriptures and wrote theological works in Chinese and other Asian languages; they also translated European mathematical and scientific treatises into Chinese. The Kangxi emperor himself studied Euclidian geometry, Western astronomy, geography, the harpsichord, and painting under Jesuit guidance. Like many Chinese he was fascinated by European clocks. Many Asians, including the Chinese, learned how to use western firearms and cast western-style cannon. However, the cultural consequences of these efforts were disappointingly small and of short duration. Before the end of the eighteenth century they seem largely to have disappeared along with the Christian mission. The Japanese, however, even during the closed-country period after 1640, were far more curious about Western science. Samurai scholars, for example, studied Dutch medicine and science and in the eighteenth century repeated Benjamin Franklin's kite-flying experiment. Nevertheless through most of the early modern period the Europeans were far more curious about and more open to influence from Asia than were any of the high cultures of Asia to influence from the West.

During the seventeenth and eighteenth centuries the new information about Asia influenced Western culture primarily in areas other than science and technology. The extent of this impact remains to be comprehensively studied, and even of that which is known only a few examples can be mentioned here. Asian events and themes entered European literature in scores of instances from Lope de Vega, Ariosto, Rabelais, and More in the sixteenth century to the several Dutch, German, and English plays and novels depicting the fall of the Ming Dynasty and the triumph of the Manchus in the seventeenth century, to Voltaire's literary and philosophical works in the eighteenth. Even popular literature, seventeenth-century Dutch plays and pious tracts, for examples, show surprising familiarity with the new information about Asia. Asian influences in European art, architecture, garden architecture and the decorative arts also began in the sixteenth century and culminated in the *chinoiserie* of the eighteenth century. Confrontation with China's ancient history challenged the traditional European Four-Monarchies framework of universal history and touched off a controversy among European scholars which by the mid-eighteenth century resulted in an entirely new conception of ancient world history. Some scholars, beginning with Pierre Bayle in the late seventeenth century, have detected a Neo-Confucian influence in the thought of Spinoza and in some aspects of Leibniz's philosophy. Chinese government and especially the examination system were frequently held up for emulation by European states during the seventeenth and eighteenth centuries. It might well be that the institution of written civil-service exams in Western states, beginning with eighteenth-century Prussia, was inspired by the Chinese example. Of more general importance that any single instance of influence, however, was the challenge presented by long-enduring, sophisticated, and successful Asian cultures to traditional European assumptions about the universality of their own. Perhaps here can be found the beginnings of cultural relativism in the West.

By 1600 European science and technology generally and especially marine and military technology outstripped that of any Asian society, whatever had been borrowed earlier from them. Also, while the fascination with and appreciation of the high cultures of Asia continued through most of the eighteenth century, the West between 1600 and 1800 experienced the radical transmutation of its traditional culture which resulted in the development of a rational, scientific approach to the use of nature, and society – to agriculture, business, industry, politics, and above all warfare – which we have traditionally associated with the Scientific Revolution, Enlightenment, and Industrial Revolution. This transmutation has enabled Western nation-states during the past two centuries to establish the world-wide competitive empires which came to dominate all the other centres of civilisation and has resulted in the global dominance of Western culture. The triumph of this transmuted Western culture has reversed the centuries-long East-to-West flow of cultural influence and threatens all of the world's traditional cultures. It should be remembered however, that this rational, scientific, industrialised Western culture was not an obviously natural outgrowth of traditional Western culture, that in its early development it received important basic components from Asia, and that its triumph threatens traditional Christian Western culture almost as seriously as it threatens those of Asia.

Edwin J. Van Kley

References

Chaudhuri, K. N. *Trade and Civilization in the Indian Ocean: An Economic History from the Rise of Islam to 1750*. Cambridge: Cambridge University Press. 1985.

Cipolla, Carlo M. *Guns, Sails, and Empires: Technological Innovation and the Early Phases of European Expansion,* 1400–1700. New York: Pantheon. 1965.

Hodgson, Marshall. *The Venture of Islam: Conscience and History in a World Civilization*. Chicago: University of Chicago Press. 1974.

Lach, Donald F. *Asia in the Making of Europe*. Vol. I: *The Century of Discovery*. Bks. 1–2. Chicago: University of Chicago Press. 1965. Vol. II: *A Century of Wonder*. Bks. 1–3. Chicago: University of Chicago Press. 1970, 1977.

Lach, Donald F. and Van Kley, Edwin J. *Asia in the Making of Europe*. Vol. III: *A Century of Advance*. Bks. 1–4. Chicago: University of Chicago Press, 1993.

McNeill, William H. *The Rise of the West: A History of the Human Community*. Chicago: University of Chicago Press. 1963.

McNeill, William H. *The Pursuit of Power: Technology, Armed Force, and Society Since* CE 1000. Chicago: University of Chicago Press. 1982.

Needham, Joseph. *Science and Civilisation in China*. 9+ vols. Cambridge: Cambridge University Press. 1954.

Parker, Geoffrey. *The Military Revolution: Military Innovation and the Rise of the West, 1500–1800*. Cambridge: Cambridge University Press. 1988.

Wolf, Eric R. *Europe and the People without History*. Berkeley: University of California Press. 1982.

See also: Navigation – Military Technology – Mathematics – Technology – Zero – Colonialism and Science – Medicine

East and West : India in the Trasmission of Knowledge from East to West

The exhange of ideas was more balanced in the time before the European Scientific Revolution, after which the rapid growth of knowledge in the West dwarfed the interregional traffic that had taken place earlier. The Western tradition of the last few centuries has become the only system studied in universities and practiced in centres of science and technology worldwide. Quite often there is no interaction between this new tradition and the earlier knowledge from regions such as South Asia, even though there are many areas of learning that could enrich the Western tradition.

A study of the growth of the European and the South Asian scientific traditions shows considerable areas of overlap and mutual influence from very early times. When Europeans in the Renaissance looked back to Greek sources for new inspiration, they were in fact looking to Greek sources partly influenced by the South Asians.

Generally speaking, India was outside the world of shared ideas and values of pre-classical Greece. After the wars with the Persian empire the myth of a division into East and West was born, as was the concept of Europe. The conditions for a large-scale traffic of culture and ideas between Greece and Asia were created when the Persian Empire became a bridge from the Mediterranean to the Indus. One sees South Asian concepts that arose between 700 BCE and 500 BCE in the later Vedic hymns, the *Upaniṣads*, and among the Buddhists and the Jains, being echoed in Greek thought.

There are striking parallels between the two traditions. The *Upaniṣads* seek one reality; this has its echoes in Xenophanes, Parmenides, and Zeno. Pythagoras is thought to have been influenced by the Egyptians, Assyrians, and Indians. He believed in the possibility of recalling previous lives, which is also typical of South Asian philosophy. Pythagoreans abstained from destroying life and eating meat, as do Jains and Buddhists. They expounded many theories in the religious, philosophical, and mathematical sphere that were known in sixth century BCE India.

In Plato's philosophy, the 'cycle of necessity', a concept similar to Karma, was central. Humans were reborn as animals or other humans. The Indian elements *pṛthvī* (earth), *ap* (water), *tejas* (fire or heat), *vāyu* (air), and *ākāśa* (ether, or a non-material substance) have their counterpart in Empedocles, who belived that matter had four elements: earth, water, air, and fire.

After Alexander's encounter there was explicit dialogue with India. Several who travelled with Alexander are said to have met with Indian sages. In late antiquity India was seen in some debates as the origin of philosophy and religion. In the second century CE Lucianus stated that before philosophy came to the Greeks, Indians had developed it. It has also been suggested that Gnostic thought was influenced by Buddhist literature. Gnostic Carpocratians strongly supported the idea of transmigration. At least one Gnostic philosopher, Bardesanes of Edessa (ca. CE 200) had travelled extensively in India. Mani, the Persian Gnostic of the third century CE, incorporated several Buddhist ideas into Manichaeism. By the second century CE India had almost replaced Egypt as the presumed origin of Greek thought and learning.

At a later period, Plotinus, the father of the Neoplatonic school, took part in the military campaign against the King of Persia. Neoplatonism recommended abstention from sacrifice and meat eating. Neoplatonism, *vedānta*, yoga systems, and Buddhism all have strong similarities. In the second century CE, Clement of Alexandria spoke often about the existence in Alexandria of Buddhists, being the first Greek to refer to the Buddha by name. He was aware of the belief in transmigration and the worshipping of *stupas*.

During the Roman Empire, contacts between the two places continued. There was heavy trade in luxuries with South India and Sri Lanka by the Romans, and ambassadors were sent to Rome. An Indian delegation visited Europe in Emperor Antoninus Pius' reign. In the reverse direction, Apollonius of Tyana travelled to India. These repeated interactions between the two regions probably resulted in the exchange of ideas from South Asia to Greece. The Buddhists had sophisticated discussions prior to Heraclitus around the concept of being in a state of flux. Buddhists and Ājīvakas added joy and sorrow to the five elements, which precedes Empedocles's views that love and hate acted mechanically on the elements. The Buddhists and others taught a doctrine of the mean several centuries earlier than Aristotle (340 BCE). In medicine, the Hippocratic treatise *On Breath* deals in much the same way with the pneumatic system as we find in the Indian concept of *vāyu* or *prāṇa*. In his *Timaeus*, Plato discussed pathology in a similar way to the doctrine of *tridoṣa*.

The above examples should not suggest that there were no transmissions in the opposite direction. The ancient world had much cross flow of intellectual traffic. A well known example from Greece to South Asia concerns ideas on geometry and astronomy.

When the Classical age collapsed, European and South Asian contacts continued in the Middle Ages through Arab intermediaries. The Arabs performed the functions earlier performed by the Persian, Alexandrian, and Greek empires which brought together the ideas of East and West. It is useful to trace the transmission of Indian sciences to Europe as well as trace those that were not transmitted but remained in the region only to be rediscovered much later. This is done to some extent in the article on Indian mathematics in this Encyclopaedia.

The European Renaissance and Scientific Revolution brought about many changes that have been considered unique. However, the evidence indicates that many of the results were known, some albeit in an incipient form, in South Asia.

Alchemy was an important precursor to the development of chemistry. Greek alchemical texts do not show an interest in pharmaceutical chemistry, a marked contrast with China and India. In the *Atharva Veda* (eighth century BCE) there are references to the use of gold for preserving life. The transmutation to gold of base metals is discussed in the Buddhist texts of the second to fifth centuries CE by concoctions using vegetables and minerals.

In the West, iatrochemists, especially Paracelsus, were of the view that the human body consisted of a chemical system of mercury, sulphur, and salt. Sulphur and mercury were already known to the alchemists; salt was introduced by Paracelsus. This theory differed from the four humours theory of the Greeks advocated by Galen (CE 129–200). An Indian alchemist by the name of Ramadevar taught a salt-based alchemy in Saudi Arabia in the twelfth century.

In medicine, the work of Suśruta laid the foundations for the art of surgery. Suśruta emphasised observation and dissection, and described many instruments like those used in modern surgery, listed several kinds of sutures and needles, and classified operations into types. The operations described included those for hydrocele, dropsy, fistula, abscess, tooth extraction, and the removal of stones and foreign matter. The ancient Indian surgeons practiced laparotomy and lithotomy, plastic surgery, and perineal extraction of stones from the bladder. The region had considerable knowledge in dentistry including artificial teeth making. In CE 1194 the king Jai Chandra when beaten in battle was recognised by his false teeth. Suśruta describes details of operations for the conditions of obstructions in the rectum and for removal of a dead fetus without killing the mother, considered a very difficult procedure. He describes plastic surgery of the nose and cataract operations on the eye.

At the end of the eighteenth century the British studied Indian surgical procedures for skin grafting to correct for deformities of the face, which became the starting point for the modern specialty of plastic surgery. Dharmapal collected several illuminating accounts by Britons on Indian medical practices in the eighteenth century. This included one by

J.Z. Holwell, who gave a detailed report on the practice of inoculation against the smallpox. The smallpox epidemics in the nineteenth and early twentieth centuries have been attributed to the cessation of this practice before the vaccination system could become widespread.

In the West after Democritus, atomic theories were further expanded by Lucretius in the first century BCE but then virtually vanished from intellectual view for 1600 years. In the seventeenth century Gassendi, Boyle, Newton, and Huygens revived the atomic perspective. Atomic views of several schools such as the Buddhists, the Jains, and the Vaiśeṣika persisted. The Vaiśeṣika's theory of atomism considered atoms as eternal and spherical in form. The disintegration of a body results in its breaking down to constituent atoms. A solid block like ice or butter melts, and this is explained as a loosening of the atoms, giving rise to fluidity.

Evolution is one other element in the modern phalanx of scientific ideas. Evolutionary ideas had existed among pre-Socratic Greek and Indian thinkers. However, evolutionary thinking in Greek tradition was brought to a sudden end by the ideas of Plato and Aristotle. Plato viewed the real world as consisting of unchanging forms or archetypes; Aristotle viewed the physical world as a hierarchy consisting of kinds of things. For Aristotle the universe was unchanging and eternal. The idea of evolution is found in the *Upaniṣads,* the writings of the Buddhists, and others.

The *Encyclopaedia Britannica* lists three major innovative transformations in British agriculture in the Era of Improvement in the eighteenth century. They were the invention by Jethro Tull in 1731 of the drill plough, "whereby the turnips could be sown in rows and kept free from weeds by hoeing thus much increasing their yields"; the introduction of rotation of crops in 1730–38 by Lord Townshend, and the selective breeding of cattle introduced by Robert Bakewell (1725–95). There is evidence that all three were in existence in India, as reported by British scientists working there.

Roxburgh, generally recognised to be the "father of Indian botany" in the contemporary tradition, put this as follows: "the Western World is to be indebted to India for this system of sowing", meaning the implicit rotation of crops in the Vedic period where rice was sown in summer and pulses in winter in the same field. Other British works have attested to the use of "careful breeding of cattle", various kinds of drill ploughs, and rotation of crops and mixed cropping.

The Scientific Revolution had a deep impact on the philosophical underpinnings of Europe. There was for example the dethroning of human exclusivity as the special creation of God, as exemplified by the trials of Copernicus or the criticisms of Darwin's evolutionary theory. The discovery of the unconscious by Freud and others is also in this class. None of these events would have had the same impact on South Asia, whose cosmology allowed for a large number of worlds, for evolution and change, for humans as part of a larger living world, and for a subconscious.

Aside from these historical examples, are there innovations occurring even now which are drawing sustenance from the earlier South Asian tradition?

Helmut von Glasenapp observed that ancient South Asian ideas on fundamental issues had several parallels with those in modern science. Some of these concepts were: (1) an infinite number of worlds exists apart from our own; (2) worlds exist even in an atom; (3) the universe is enormously old; (4) there are infinitely small living beings parallel to bacteria; (5) the subconscious is important in psychology; (6) doctrines of matter in both Sāṃkhya and Buddhism are similar to modern systems; (7) the world that presents itself to the senses is not the most real; and (8) truth manifests itself differently in different minds giving the possibilty of a multiplicity of valid truths.

Following are examples of innovations based on the past taken from a few disciplines. The first is medicine. A recent study has documented the use of honey and sugar as treatment for wounds and ulcers in both Āyurvedic and contemporary biomedicine. The tranquiliser Reserpine was based on an ancient āyurvedic medicine. Hoechst, the West German pharmaceutical company, used Āyurvedic literature to help identify useful medicinal plants. By the early 1980s, over two hundred Indian medicinal plants were being tested every year in this program.

A recent study has evaluated the effect of *Rasāyana* therapy which aims at promotiong strength and vitality. The study covered six drugs from classical Āyurvedic literature. The clinical studies indicated that the drugs toned up the cardiovascular and respiratory systems and improved physical stamina. On the biochemical side, a significant drop in lipids was noticed.

References to curative plants in the Indian tradition go back to the Ṛgvedic period (3500–1800 BCE). The *Suśrutasaṃhitā* and *Carakasaṃhitā*, two compendia which are summaries of earlier works, dealt with about seven hundred drugs, some of them outside the subcontinental region. Clearly, a vast reservoir of explorable scientific knowledge exists.

One of the areas of study with a very long tradition in South Asia is psychology. There is also a very long tradition of sophisticated discussions on epistemology. There is potential for a fruitful interaction between these and the contemporary study of the mind, including the philosophy of language, methodology, ontology, and metaphysics.

Memory, motivation, and the unconscious are shown to have parallels in the theories of Freud and Jung, as well as in Patañjali. Similar parallels have been noted between the psychoanalytical theorists Heinz Hartmann and Erik Erikson, and the Hindu theory on the stages of life, as well as between Buddhism and early twentieth century analytical thought. Strong parallels between the concept of self-realisation used in subcontinental traditions such as Vedantic Hinduism, Theravāda, and Mahāyāna Buddhism and the concept of self-actualisation as developed in humanistic psychology by Arthur Maslow and Carl Rogers have been demonstrated.

Francisco Varela, a theoretical biologist and student of cognitive science and artificial intelligence, and co-workers have used Buddhist insights in extending the limitations of both the neo-Darwinian adaptation in biological evolution and of the current paradigm in cognitive sciences. Having noted that in Buddhist discourse, classical Western dichotomies like subject and object, mind and body, organism and environment vanish, Varela applies these discourses to several areas where these dichotomies had traditionally appeared. These include cognitive psychology, evolutionary theory, linguistics, neuroscience, artificial intelligence, and immunology.

Their position is that if cognitive science is to incorporate human experience, then it must have a means of exploring the dimension which is provided by Buddhist practice. Buddhist experiences of observing the mind are in the tradition of scientific observation. They can lead to discoveries about the behaviour and nature of the mind, a bridge between human experience and cognitive science.

Another area of interest is that of adaptation in evolutionary biology. In the conventional view it is assumed that the environment exists prior to the organism, into which the latter fits. This is not so. Living beings and the environment are linked together in a process of codetermination or mutual specification. In this light, environmental features are not simply external features that have to be internalised by the organism; they are themselves results of a long history of codetermination. The organism is both the subject and object of evolution. The processes of coevolution result in the environment's being brought to life through a process of coupling. Taking the world as pre-given and the organism as adapting can be categorised as dualism. Buddhism transcends this duality in its codeterminative perspective.

The standard arithmetic that we use today, based on the decimal place system and the use of zero, was transmitted through the Arabs from South Asia. It entails certain standard procedures, algorithms, to perform various operations. But are these the only such operations that exist and are these the ones that are computationally the most efficient? Could there be algorithms that did not get transmitted from India through the Arabs, or those that were developed after the transmission?

Indeed there are many such, as Ashok Jhunjhunwala, a professor of electrical engineering in the Institute of Technology, Madras, has discovered recently. He has examined everyday practices in arithmetic in areas not yet influenced by European techniques such as those used by artisans and businessmen in the non-Europeanised sector. He came across simple but fast methods of calculation. He described eight of these methods which are faster than conventional methods. These included means of finding area, multiplication, squaring, division, evaluation of powers, square roots of numbers, divisibility of numbers, and factorisation. They also included methods to catch errors. Jhunjhunwala has compared the speed of some of these old approaches with contemporary ones and found that some are faster. He is now applying these general methods to speed up calculations in computers.

Jhunjhunwala's collection of mathematics at the local level shows the proliferation of methods possible once the decimal system is understood. Local groups discovered new tricks, a process of grass roots creativity very much like the different responses to changing agroclimatic conditions across the world and the resultant variations in agricultural practices.

Time is yet another area to explore. There are many different philosophical discussions on *Saṁsāra* concerning what could be termed the nature of long duration processes. According to some Jain views, time was one of the causal factors in the evolution of nature; and Buddhism alone has a very large tapestry of conceptions of time.

One of these approaches developed an elaborate theory not only of atoms but also of moments, with some schools recognising four types of moments and others three. Other theories were also proposed by different schools to relate the theory of moments to the fact of continuity of temporal events.

Virtual Reality brings into question the constructor and the constructed. These types of questions are regularly dealt with in Buddhist and other South Asian philosophies. In the virtual realities that use visual representations, parallels also exist with visualisation techniques in certain branches of Buddhism. The author of a text on the topic, Howard Rheinhold, says that the Virtual Reality "experience is destined to transform us because it's an external mirror of something that Buddhists have always said, which is that the world we think we see 'out there' is an illusion."

The ethical and conceptual questions of the future brought about by modern science and technology could have many uses for South Asian perspectives.

There are many stores of valid information still to draw from. One authoritative estimate of manuscripts, roughly covering the areas of mathematics and astronomy, is about 100,000. Yet the recently published book *Source Book of Indian Astronomy* lists only 285. Of these only very few have been studied. They are mines of mathematical ideas and applications that have hardly been touched.

A passage from Suśruta stimulated the growth of modern plastic surgery in the nineteenth century in Europe. However, as Krishnamurty points out, that was only a stray reference in the many procedures described. It had the fortune of catching the imagination of a western expert. There could very well be many other descriptions that could be rediscovered for modern medicine. Under treatment for mental diseases Suśruta gives a very large list of plants. It is possible that screening of these plants could give rise to a much larger set of useful remedies.

Varela stated that the infusion of Eastern ideas into the sciences of the West would have as much an impact as did the Renaissance rediscovery of Greek thought. How far this may be true is for the future to decide. However, this would help make the present Western knowledge system more universal while still maintaining the rigour developed in the last few centuries. It would help both to enlarge the knowledge terrain covered by the present system as well as retrieve what is relevant from other traditions.

Susantha Goonatilake

Reference

Bose, D. M., S.N. Sen, and B.V. Subarayappa. *A Concise History of Science in India*. New Delhi: Indian National Science Academy. 1971.

Halbfass, Wilhelm. *India and Europe: An Essay in Understanding*. Albany: State University of New York Press. 1988.

Jain, S.K. *Medicinal Plants*. Delhi: National Book Trust of India. 1968.

Jhunjhunwala, Ashok. *Indian Mathematics: An Introduction*. New Delhi: Wiley Eastern Limited. 1993.

Krishnamurty, K.H. *A Source Book of Indian Medicine: An Anthology*. Delhi: B.R. Publishing Corporation. 1991.

Lach, Donald F. *Asia in the Making of Europe*, vol. 2. Chicago: University of Chicago Press. 1977.

Mohanty, J. N. "Consciousness and Knowledge in Indian Philosophy." *Philosophy East and West* 29(1): 3–11. 1979.

Rawlinson, H.G. "Early Contacts Between India and Europe." In *Cultural History of India*. Ed. A.L. Basham. Oxford: Clarendon Press. 1975.

Sangwan, Satpal "European Impressions of Science and Technology in India (1650–1850)." In *History of Science and Technology in India*, vol 5. Ed. G. Kuppuram and K. Kumudamani. Delhi: Sundeep Prakashan. 1991.

Varela, Francisco J., Evan Thompson and Eleanor Rosch. *The Embodied Mind: Cognitive Science and Human Experience*. Cambridge, Massachusetts: MIT Press, 1991.

von Glasenapp, Helmuth. "Indian and Western Metaphysics." *Philosophy East and West* 111(3): 1953.

See also: Mathematics – Knowledge Systems: Local Knowledge – Time

Eclipses

Eclipses of both the Moon and Sun are frequently recorded in the history of non-Western cultures throughout ancient and medieval times. These records originate almost exclusively from Babylonia and China in ancient times, and from East Asia (China, Korea, and Japan) and the Islamic world in the medieval period. As yet, virtually no eclipse observations have been uncovered from other major civilisations such as ancient Egypt, India, and Central America.

Eclipses were often noted on account of their spectacular nature, or because they were regarded as omens. However, early records provide many examples of a more scientific attitude, resulting in the careful measurement of times and other details. Ancient references to eclipses often prove of value in dating historical events, while the more careful observations have played a major role in present-day knowledge of variations in the rate of rotation of the Earth. Eclipses are thus among the most interesting of all celestial phenomena mentioned in history.

The oldest known allusions to eclipses in any civilisation are recorded on a series of astrological tablets (known as *Enuma Anu Enlil*) from the city of Ur, in southern Iraq. The earliest of these accounts may be translated as follows:

> "If in the month Simanu an eclipse occurs on day 14, the [Moon-] god in his eclipse is obscured on the east side above and clears on the west side below, the north wind blows, [the eclipse] commences in the first watch of the night and it touches the middle watch... The king of Ur will be wronged by his son, the Sun-god will catch him and he will die at the death of his father...".

Professor P.J. Huber of the University of Bayreuth, who provided the above translation, regards the historical details as relating to the murder of King Shulgi of the third dynasty of Ur by his son. Huber calculates the date of the eclipse as April 2 in 2094 BCE (Julian Calendar).

One of the earliest surviving reports of a solar eclipse is from Assyria. This is found in the *Eponym Canon*, which lists the names of the annual magistrates (*limmu*) who gave their names to individual years (similar to the Athenian archons or Roman consuls) under a year which is equivalent to 763–762 BCE.

> Revolt in the citadel. In (the month) Sivan the Sun was eclipsed.

Sivan corresponded to May–June. The only large eclipse visible in Assyria for many years occurred on June 13 in 763 BCE.

Between about 750 and 50 BCE (or possibly down to the first century CE), Babylonian astronomers systematically maintained a watch for eclipses (as well as other celestial phenomena) and carefully estimated the time of occurrence and magnitude (the maximum degree of obscuration of the Moon or Sun). Their aim was to use these observations to enable better predictions of future eclipses to be made, although the ultimate goal was astrological. Both genuine observations and failed predictions (the latter which were often described as eclipses which "passed by") are recorded on the late Babylonian astronomical texts which were recovered from the site of Babylon rather more than a century ago. Most of these clay tablets, many of which are badly damaged, are now in the British Museum. Dates are expressed in terms of a luni-solar calendar. Although earlier years were counted from the accession of each king, they began to be continuously numbered from the Seleucid Era (311 BCE) onwards.

Among Babylonian records of eclipses, the best known is the solar obscuration of April 15 in 136 BCE. This is reported on two separate British Museum tablets, which give overlapping details. A composite translation – based on work by Prof. H. Hunger of the University of Vienna – is as follows:

> "Year 175 (Seleucid), intercalary 12th month, day 29. At 24 degrees after Sunrise, solar eclipse. When it began on the south–west side, in 18 degrees of day in the morning it became entirely total. Venus, Mercury and the Normal Stars were visible. Jupiter and Mars, which were in their period of disappearance, became visible in its eclipse [...] It threw off [the shadow] from south-east to north-west. [Time-interval of] 35 degrees for onset, maximal phase and clearing".

Jupiter and Mars were too close to the Sun to be seen under normal circumstances. The unit of time here translated as "degree" (i.e. *us*) was equivalent to 4 minutes. Measurements were presumably made with a water clock, although direct evidence is lacking. The "Normal Stars" were certain reference stars in the zodiac belt. Many other detailed descriptions of both lunar and solar eclipses are preserved on the Late Babylonian astronomical texts.

Occasional allusions to eclipses are found in the Old Testament, notably in the Book of Joel (II, 31). However, the most direct reference to an eclipse in the history of ancient Palestine is recorded by Flavius Josephus, a Jewish historian of the first century CE. This event occurred only a few days before the death of Herod the Great and is thus of importance in dating the birth of Jesus Christ. Josephus gives the following account:

> "As for the other Matthias who had stirred up the sedition, he [Herod] had him burnt alive along with some of his companions. And on the same night there was an eclipse of the Moon. But Herod's illness became more and more severe..."

The events described occurred shortly before the Passover. Calculation shows that between 17 BCE and CE 3, the only springtime eclipses visible in Palestine occurred on March 25 in 5 BC (a total obscuration) and March 13 in the following year (a partial obscuration). The latter date is usually preferred by historians, although the two dates are conveniently close together.

For several centuries after this period, scarcely any eclipses are recorded outside East Asia. However, in the ninth century both Muslim astronomers and chroniclers began independently to note the occurrence of eclipses of both Moon and Sun. Between CE 830 and 1020, Muslim astronomers, largely based in either Baghdad or Cairo, regularly made careful observations of both local times and magnitudes. The intention was to test the reliability of contemporary eclipse tables and sometimes also to determine the difference in longitude between selected cities. Astronomers would first roughly predict when an eclipse was to occur and then assemble one or more parties of observers. Our principal source for such measurements is the handbook by the Cairo astronomer Ibn Yūnus, who died in CE 1009. This work, dedicated to Caliph al-Ḥakīm, is entitled *al-Zīj al-Kabīr al-Ḥakīmī*.

The following account of the lunar eclipse of CE 927 by Ibn Amājūr of Baghdad is fairly typical. As is customary, years are counted from the Hijra, which occurred in CE 622.

> (315 AH, a Friday – month and day of month not stated). "This eclipse was observed by my son Abū al-Ḥasan. Altitude of the star al-Shi'rā al Yamāniya [i.e. Sirius] at the start, 31 deg in the east; revolution of the sphere between sunset and the start of the eclipse, approximately 148 degrees or 9 hours and 52 minutes equal hours, which is 10 unequal hours. Estimated magnitude of the eclipse, more than one-quarter and less than a third, approximately $3\frac{1}{2}$ digits".

The only feasible date for the eclipse was CE 927 September 14, which was indeed a Friday in the year 315 AH. Medieval Muslim astronomers were in the habit of measuring eclipse times indirectly by determining the altitude of the Sun, Moon or a bright star (probably with a sextant or astrolabe) and then converting to local time with the aid of an astrolabe or tables. In estimating magnitude, they followed the practice of the ancient Babylonians and Greeks of using digits, each equal to one-twelfth of the diameter of the luminary.

After CE 1020, virtually no eclipse observations made by Muslim astronomers have survived. However, chroniclers continued to report eclipses – largely on account of their spectacular nature – for many centuries. An appealing eyewitness account of the total solar eclipse of CE 1176 Apr 11 was penned by Ibn al-Athīr, who was aged 16 when it occurred:

> (571 AH). "In this year the Sun was eclipsed totally and the Earth was in darkness so that it was like a dark night and the stars appeared. That was the forenoon of Friday, the 29th of the month Ramaḍān at Ibn 'Umar's island, when I was young and in the company of my Arithmetic teacher. When I saw it, I was much afraid; I held onto him and my heart was strengthened. My teacher was learned about the stars and told me, 'Now you will see all this go away', and it went quickly".

The island mentioned, now named Cizre, lies on the frontier between Turkey and Syria. This same eclipse also alarmed Saladin's army as the soldiers were crossing the Orontes River in Syria.

The very earliest observations of eclipses from China are recorded on inscribed bones dating from some time between 1350 and 1050 BCE. These are known as "oracle bones" since they are mainly concerned with divination. The following example provides a useful illustration:

> "The divination on day *guiwei* was performed by Zheng: 'Will there be no disaster in the next ten days?' On the third day *yiyu* an eclipse of the Moon was reported". (These events occurred) in the 8th month". *Guiwei* and *yiyu* are the 20th and 22nd days of a 60-day cycle.

Since the year is missing, the date of this event is uncertain; few bone inscriptions are intact, leading to considerable difficulties in dating.

Over the next 1500 years, little interest was shown in lunar eclipses in China, since they were regarded as of minor astrological consequence. By contrast, solar obscurations were considered to be unfavourable omens, especially for the ruler, and thus the records are much more complete. As many as thirty-six eclipses of the Sun are reported in a single ancient work, the *Chunqiu* – a chronicle covering the period from 722 to 481 BCE. This reports total eclipses in 709, 601 and 549 BCE, but without any descriptive details. Three further accounts (from 669, 664 and 612 BCE) which describe eclipse ceremonies in which drums were beaten and oxen were sacrificed underline the importance attached to solar obscurations at this early period.

Ancient records contain few mentions of eclipse times but from the fifth century CE, eclipses of both the Moon and Sun began to be carefully timed using water clocks. The main purpose was to verify the accuracy of astronomical tables. Solar eclipses were usually timed to the nearest *ke* (mark) – 1/100 of a day and night – roughly equal to 15 minutes. For lunar eclipses the time was often expressed to the nearest fifth of a night watch instead. The period from dusk to dawn was divided into five equal *geng* (night watches), each of which was in turn divided into five equal intervals, often termed "calls". Hence the night watches and their subdivisions varied in length with the seasons.

The earliest eclipse whose time is carefully recorded is noted in the *Songshu*, the official history of the Liu-Song Dynasty (CE 420–479); its date corresponds to CE 434 September 5:

> "Yuanjia reign period, 11th year, 7th month, 16th day, at full Moon... The Moon began to be eclipsed at the 2nd call of the 4th watch, in the initial half of the hour of *chou*. The eclipse was total at the 4th call".

The initial half of the *chou* hour was between 1 and 2 a.m. Many similar accounts are to be found elsewhere in the *Songshu* and in later official histories.

A few total solar eclipses are reported in vivid detail in Chinese history, and the onset of darkness and appearance of stars is described. Examples are found in CE 120, 454, 761 and 1275.

The very earliest Korean records of eclipses appear to be merely copied from Chinese history and probably do not become independent until around CE 700. Later accounts tend to be very brief, seldom giving more than the day of occurrence. These are mainly found in the *Koryo-sa*, the official history of Korea from CE 936 to 1392, and in later chronicles. The Chinese luni-solar calendar was closely followed in Korea, although years were numbered from the accession of Korean monarchs. Around CE 1200 there are occasional references to the king, dressed in white robes and accompanied by his closest ministers, attempting to rescue the Moon from its eclipse, but these may have been no more than ceremonies.

Japanese descriptions of eclipses tend to be more detailed than those of other East Asian countries and historical sources are much more diverse. The following account of the total solar eclipse of CE 975 August 9 is found in a privately compiled history known as the *Nihon Kiryaku*:

> "Ten-en reign period, 7th month, day *xinwei*, the first day of the month. The Sun was eclipsed. Some people say that it was entirely total. During the hours *mao* and *chen* (between 5 and 9 a.m.) it was all gone. It was the colour of ink and without light. All the birds flew about in confusion and the various stars were all visible. There was a general amnesty (on account of the eclipse)".

In this brief article it has only been possible to cite a minute proportion of the available records of eclipses from the non-Western world. However, it should be clear from the selection offered here that many detailed and important descriptions are available. Ancient and medieval observations often reveal a high level of sophistication, particularly in techniques of measurement. In particular, medieval Arab astronomers consistently determined the times of eclipses to within about five minutes while contemporary Chinese measurements were only a little inferior.

F. Richard Stephenson

REFERENCES

Beijing Observatory. *Zhongguo Gudai Tianxiang Jilu Zongji* (A Union Table of Ancient Chinese Records of Celestial Phenomena). Kiangxu: Kexue Jishi Chubanshe. 1988. sections 4 and 5.

Newton, Robert R. *Ancient Astronomical Observations and the Accelerations of the Earth and Moon*. Baltimore: Johns Hopkins University Press. 1970.

Said, Said S., F. Richard Stephenson, and Wafiq S. Rada. "Records of Solar Eclipses in Arabic Chronicles." *Bulletin of the School of Oriental and African Studies* 274(4): 170–183. 1989.

Stephenson, F. Richard. "Historical Eclipses." *Scientific American* 274(4): 170–183. 1982.

Stephenson, F. Richard. "Eclipses in History." In *Encyclopedia Britannica* vol. 16. 1993. pp. 872b–877b.

Stephenson, F. Richard. *Historical Eclipses and Earth's Rotation*. Cambridge: Cambridge University Press. 1996.

See also: Calendars

Environment and Nature

The philosophical and scientific ideas developed in India over the centuries are, on analysis, found to be deeply related to ecological issues, both generally and specifically. Like the ancient ideas of China and of the Hellenic world, the ancient Indian ideas of the comparable period are cosmological and comprehensive in character. The ideas presented in the *Vedas* and *Upaniṣads*, or in Laozi's *Dao* or in Parmenides' *Nature of Being* all were engaged in search of the first principle, One. They were all obliged to relate it to Many – many individual objects of knowledge. The One–Many relationship is pregnant with both cosmological and ecological implications.

Broadly speaking, the objects we know around us are biotic (living) or abiotic. However, to many thinkers, especially to the pluralists and evolutionists, this two-fold classification is simplistic and inadequate. They try to draw our attention to different grades of being or Reality – physical, chemical, paleontological, botanical, biological, psychological, and spiritual. The scientific philosophers of the Vedic insights, of the *Sāmkhya* persuasion, and also of latter times could discern different subgrades within each of these grades. These graded characteristics of different living and non-living Beings have to be recognised if we are to understand the complex and the interactive character of our environment. This insightful approach to ecology as an integral part of cosmology is evident in the tradition of Indian thought. It is very clearly available, for example in *Caraka-Saṁhitā*.

In ancient India the good of human life used to be thought of in the context of life's environment. The right relation between the individual and his environment received serious attention even at the levels of primary and secondary education. The aim was to impart basic knowledge of personal hygiene and medicine. This education was meant for all. In a way medical education was universal. It is interesting to note the five compulsory subjects for all high school students: Grammar *(Śabdavidyā)*, Art *(Śilpasthānavidyā)*, Medicine *(Cikitsāvidyā)d*, Logic *(Hetuvidyā)*, and Science of Spiritual philosophy *(Adhyātmavidyā)*.

Both abstract and concrete areas of knowledge, understood in their interconnection, are all relevant to the exercise of adjusting to our environment. Linguistic communication, artistic articulation, logical ratiocination, and medication are in different ways intended to awaken the best in us and to strike a balance with the large world around us. From the strength of body to the span of life, from good health to peace of mind, all are a unified function of food, personal hygiene, right actions, and character.

Life has a rhythm of its own, which is not necessarily manifest. The *Dharmaśāstra* are full of injunctions on how to attain that rhythm in terms of purity, diet, regulation, ablutions, behaviour, and physical and mental disciplines. These have to be followed as part of daily *(dinacaryā)* as well as seasonal routines *(ṛtucaryā)*. Food and drink habits and fulfilment of natural urges, avoidance and indulgence in sexual acts, and eating some ordinary things like curds, buttermilk, and honey are all necessary to define our correct relation with the environment. However, this normative rhythm can hardly be generalised, for it is integrally related to the individual constitution *(prakṛti and svāsthya-vṛtti)*.

The traditional Indian medical system *(Āyurveda)* takes a comprehensive view of a person. It neither encourages asceticism and mortification of flesh nor promotes unregulated sensualism. It highlights the importance of a "sound mind in a sound body". It points out that *svāsthya-vṛtta* or philosophy of hygiene is to be supplemented by *sadvṛtta* or the right life. Rightly understood, the philosophy of hygiene is a way of life containing in it the principles of eugenics, ethics, and healing.

In ancient Indian thought the basic principles of ecology are ontologically oriented and cosmological in implication. The three main concepts of the Vedas which are evident in *Āyurveda* are *satya* (truth, right, reality, or being), *svadhā* (self-position, self-power, or spontaneity) and *ṛta* (proper, suitable, and settled order). The basic point which is being emphasised here is that the true nature of reality cannot be changed by human will which is not informed of that reality. No technological skill, no arbitrary will of this or that person, can go indefinitely against the true nature of reality without causing harm to those who try to follow this wrong path. Even the most sophistical biotechnology cannot tamper with the essential nature of reality, the true Nature or *Prakṛti*. This role of the concept of truth has both ontological and axiological implications. Humans are required both to know reality in its true nature and to live and shape their lives accordingly.

Humans are also aided by the principle of becoming or the dynamic nature of reality in shaping their best possible life. This possibility is contained in the concept of *svadhā*. Reality has within itself the impetus or power for self-unfolding or gradual disclosure. It is not static, fixed, or self-enclosed. The world as reality and as a whole is perpetually expanding. This macrocosmic truth of *svadhā* is at work also in human nature. Consequently, human freedom knows no bounds. Nature and its laws are not antagonistic to the human will to be free. When we can discover the laws governing the true nature of the relation between individuals and the world, our knowledge of the world helps us to live in harmony with it. That harmony creates a favourable environment which not only nourishes our bodies but also enables us to be free from social conflict and tension. When we are free from ill health, we are better placed in our relation to nature. When we fail to strike the balance with nature, we are not only likely to be poorer in health but also less capable of getting the best out of our environment.

Finally, besides the true nature of reality (*satya*) and the power of self-disclosure (*svadhā*) what helps us to have a right environment is suggested by the concept of *ṛta*. It is orderliness or the law-governed character of reality or nature. From the change of seasons to the changing periods of life, *ṛta* is clearly perceptible. Even human cultures are found to exhibit certain rhythms. Cultures are characterised simultaneously by fragile and stable features. The rhythm of nature is not antagonistic to the spirit of human freedom. Our lives, both individual and collective, are marked by a sort of dynamic equilibrium.

The above three concepts embodying certain abstract principles may appear irrelevant to our actual lives. However, to the thinkers of *Āyurveda* the principles of philosophy and ethics, or those of good living, are inseparable from the basic characteristics of reality.

The same principle is illustrated in defining the ideal principles of town-planning and village-planning. Aśoka, the Buddhist emperor of ancient India, said that trees, plants, and shrubs of medicinal value had to be planted around every village and along the roadsides. The people were allowed to use the leaves, fruits, and bark of the trees. This practice survives today. Like geography, history, science, and arithmetic, the general principles of hygiene and physiology and simple methods of curing cuts, wounds, and everyday ailments were prescribed for everyone's general education. The underlying belief was that individuals had to be able to take care of their elementary medical needs.

The Indians believed that every village should be so constructed that its population must have the professional service of a medical practitioner (*vaidya*). People were advised to reside in a place with plenty of water, herbs, sacrificial sticks, flowers, grass, and firewood, and which yielded abundant food. They were also advised to live where there was safety of property and person, where the outskirts were beautiful and pleasing, and where there was a strong presence of learned people.

Ancient Indian thinkers paid attention to the development and preservation of the right type of environment for human life. The state was assigned a very important role for the purpose. It was expected to lay down rules and regulations for rubbish disposal and drainage systems. The state also determined where and how the quality of food and drink should be preserved. Those who violated these rules and thereby polluted the environment were punished. Equally conscious were the decision-makers in charge of public health. The state had a very important role to perform for preservation and promotion of a healthy environment. Various types of punishment were prescribed for cutting the tender sprouts of trees.

Kauṭilya mentioned elaborate laws for the promotion of agricultural activities. For example, he prohibited high agricultural taxes and bonded agricultural labour and attached much importance to animal husbandry and soil preservation.

D.P. Chattopadhyaya

References

Kauṭilya. *Kauṭilya's Arthaśāstra*. Trans. R. Shamasastry. Mysore: Mysore Print and Publishing House. 1961.

See also: Medicine in India: Āyurveda

Epilepsy

Āyurveda is the ancient Indian medical system (4500–1500 BCE). Its name is derived from the Sanskrit *Āyu*, meaning life-combined state of body, senses, mind and soul, and *veda*, meaning science. Āyurveda is therefore the science of life. It is considered by many to be the oldest system of medicine in the world. Epilepsy was referred to as *Apasmara* in Āyurveda. The prefix *Apa* means negation or loss of, and *smara* means consciousness or memory. Epilepsy was considered to be caused by both endogenous and exogenous factors. Among the exogenous factors were internal haemorrhage, high fever, excessive sexual intercourse, disturbances of the body due to fast running, swimming, jumping, or leaping, eating of foods that are contaminated, nonhygienic practices, and extreme mental agitation caused by anger, fear, lust, or anxiety. An endogenous disturbance refers to a metabolic derangement in the form of a disturbance of *doṣas* (humours) which are aggravated and lodged in the channels of *hṛt* (brain). It was also recognised that epilepsy could arise secondary to other diseases.

Aura was recognised and was called *Apasmara Pūrva Rūpa*. An actual attack of seizures was said to occur when the patient saw non-existent objects (visual hallucinations), fell down, and had twitching in the tongue, eyes, and eyebrows with jerky movements in the hands and feet. In addition, there was excessive salivation. After the paroxysm was over, the patient awakened as if from sleep.

Apasmara was classified into four types. The *Vatika* type was characterised by frequent fits with uncontrollable crying, unconsciousness, trembling, teeth gnashing, and rapid breathing. Upon regaining consciousness, the patient had a headache. In the *Pattika* type, the patient became agitated, had sensations of heat and extreme thirst and an aura of the environment being on fire followed by frequent fits accompanied by groaning and frothing at the mouth and finally falling on the ground. In the *Kaphaja* type the onset of convulsions was delayed and was preceded by an aura during which the patient felt cold and heavy and saw objects as white. The seizure was accompanied by falling and frothing at the mouth. The fourth type, *Sannipatika*, was due to a combination of all of the above. This type was considered incurable, occurred in older people, and resulted in emaciation.

The first step in the treatment was the "awakening of the heart" (meaning to wake the patient from unconsciousness) by using drastic measures to clear those doṣas that block the channels of the mind. After the patient was awake, drug formulations to alleviate epilepsy were administered. The ingredients of these preparations included sulphur, aged ghee (butter fat), and many herbs such as *Achyranthes aspena, Holanthena antidysenterica, Alstonia scholaris*, and *Ficus carica*. Blends of herbal formulations such as Pancamula and Triphala were also used. Bhagwan Dash describes pharmaceutical processes and preparations which involved fermenting, extracting, preparing inhalable substances, filtrating, heating in a closed cavity, purifying, and pill-making. General measures to correct exogenous factors, such as proper hygiene and a balanced diet were recommended. Epileptics like the insane were kept away from dangerous situations, namely, water, fire, treetops, and hills.

Bala V. Manyam

REFERENCES

Dash, Bhagwan. *Materia Medica of Ayurveda*. New Delhi: Concept Publishing. 1980. Kurup, P.N.V. *Birds-Eye-View on Indigenous Systems of Medicine in India*. Delhi: Depak. 1977.

Sharma, Priyavrat. *Caraka-saṃhitā* text with English translation, vols. I and II. Varanasi: Chaukhambha Orientalia. 1981.

Singhal, G.D., S.N. Tripathi, and K.R. Sharma. *Ayurvedic Clinical Diagnosis Based on Madhava- Nidana, Part 1*. Varanasi: Singhal Publications. 1985.

See also: Medicine in India: Āyurveda – Caraka

Ethnobotany

Ethnobotany deals with the relationship between humans and plants. Archaeological or paleobotanical evidence about the use and cultivation of plants for food, medicine, housebuilding, etc. suggests a long history, yet the word 'ethnobotany' was coined by J. W. Harshberger only in 1895. Human–plant relationships can be classified into two topics, material and abstract. Material relationships include use in food, medicine, building, painting and sculpture, and in acts of domestication, conservation, and improvement or destruction of plants. Abstract relationships include faith in the good or evil powers of plants, sacred plants, worship, taboos, and folklore. Ethnobotanical enquiry extends beyond botany, and has significant input of other branches of science such as archaeology or medicine.

The tribal people of India mostly live in forests, hills, and relatively isolated regions. They are variously termed *Ādivāsī* (original settlers), *Ādim Niwāsī* (oldest ethnological sector of the population), *Ādimjāti* (primitive caste), Aboriginal (indigenous), *Girijan* (hillsmen), *Vanyajāti* (forest caste), *Vanavasī* (forest inhabitants), *Janajāti* (folk communities), *Anusūchit Janajāti* (scheduled tribes), and other such names signifying their ecological, historical, or cultural characteristics. While the constitutional term is *Anusūchit Janajāti* (Scheduled Tribes), the most popular term is *Ādivāsī.*

Certain characteristics make India ethnobotanically rich. The region supports very varied and rich flora, from desert to tropical rain forests. Of deeper significance is the tradition of the Vedas and other ancient literature, an ethnobotanical continuum that enables contemporary investigators to delve into the distant past and often to link modern folklore with that of ancient cultures.

Ethnobotanical work in India can be broadly classified. First there is the ethnobotany of distinct ethnic groups such as the Mikir of Assam, Bhils of Rajasthan, and Thārūs of Uttar Pradesh. Next there is the ethnobotany of a specific region. Most ethnobotanical research is aimed at one utility group, e.g. the study of famine foods, special diets for festivals, beverages, spices, medicines in general, medicines for particular diseases, perfumes, oils, dyes, narcotics, and plants for tools, worship, ceremonies, or personal adornment.

People in remote areas cultivate numerous vegetables representing distinct genetic stocks adapted to local conditions. Some examples are *Piper peepuloides* (a condiment in the Khasi hills), *Digitaria cruciata* var *esculenta*, (a minor millet of Khasi Hills), and species of *Alocasia, Amorphophalus, Colocasia* and *Dioscorea* (cultivated by tribes in the north-east). In remote mountains or desert regions, wild plants (tubers, leaves, flowers, fruits, seeds, and grains) still provide considerable quantities of food. Examples of notable and less known plants in ethnomedicine are Yarrow (*Achillaea millefolium*), an anthelmintic; Cutch (*Acacia catechu*), and Indian Stinging Nettle (*Tragia involucrata*) for intestinal diseases; and Prickly Chaff Flower (*Achyranthes porphyristachya*) and Nutgrass (*Cyperus rotundus*) for liver complaints.

Many ethnobotanists have given attention to the diversity of traditional tools, gadgets, and articles of personal adornment. Good examples are the single-pan balance among the Mikirs of Assam, the variety of cattle-traps among the Goṅḍs and Bhils, and containers, utensils, bridegroom's hats, and other articles of personal adornment and agricultural or other tools. Researchers have found much ingenuity in the choice of raw materials and in the design or art of decorating these articles.

Vegetable dyes for colouring hair, teeth, palms, and other parts of the body have been used since ancient times. Hair, palms, and feet are usually dyed with Henna (*Lawsonia inermis*). Teeth are blackened by chewing walnut bark. The Mikirs of Assam burn green stems of *Murraya koenigii* and certain other plants and apply the gum exudates to their teeth. Tattooing on arms, legs, chest, cheeks, etc. is still popular in some societies; common designs include figures of gods and goddesses, flowers, or animals. Mikir women (Assam) make a dye from the juice of *Baphiacanthus cusia* with which they draw a perpendicular line from their foreheads, down over their noses to their chins. The aboriginals make indigenous musical instruments from naturally available materials. The Santals in Bihar and the Gonds in central India make drums from the wood of mango, Toddy Fish-tail Palm (*Caryota urens*), Bengal Padauk (*Pterocarpus marsupium*), Malay Bushbeech (*Gmelina arborea*), and Little Flower Crepe Myrtle (*Lagerstroemia parviflora*) trees. Stringed instruments are common. Flutes are made from bamboo, and are played singly or in accompaniment with drums.

The role of ethnobotany in conserving plant resources has been studied in sacred groves and traditional agriculture.

Many cultivators have not taken to using improved varieties of crops; they continue to use traditional land races or wild relatives of crops, thereby maintaining their genetic material. Some characteristics such as hardiness, disease resistance, and adaptability to waterlogging, drought, or cold in land races have been utilised by breeders; these have averted famines.

The study of mythological associations with plants like tree worship, plants in offerings, carvings in temples, and plants in Indian epics has been fascinating. Work has been done on plants associated with deities, sages, origins from the bodies of gods, and planets; with taboos on sowing, plucking flowers or fruits, and eating wild fruits in certain seasons; with woods suitable (or unsuitable) for making idols of gods, and with driving away the evil eye.

Detailed interdisciplinary studies on the ethnobotany of any particular plant include people's first association with that plant, the impact of selective use on the biology of the plant or on the ecosystem, and the implications of usage on the life and culture of people. Studies have been done on *Bauhinia, Coix, Coptis, Ficus, Saussurea*, and *Selaginella*.

Among the chief tools of research in ethnobotany in India, mention may be made of the following:

- *Literature*. Ancient or unnoticed, published or unpublished literature is an important source of information. Though the identity of a number of plants referred to in ancient works has sometimes been in doubt, Vedic literature has been a valuable resource.
- *Field work*. Information is collected from local informants on the local name, parts used, processing, and dosage of the drug. Voucher specimens are collected for recording in an herbarium.
- *Herbaria and Museums*. Notes on herbarium and museum specimens are a good source of data. They are attached to an actual specimen, and identification of the plant is precise.
- *Archaeological remains*. Sometimes, archaeological remains provide useful data. In India, an attempt was made to describe plants from reliefs on the Great Stupa of Sanchi, and the Stupa of the first and second centuries BCE respectively. Plant remains from the neolithic period apparently used as food, fodder, and shelter, were also studied.

Ethnobotany has been recognised as a priority area of research by the government.

S. K. Jain

REFERENCES

Harshberger, J.W. "Some New Ideas." *Philadelphia Evening News*. 1895.

Jain, S.K. "Ethnobotany." *Interdisciplinary Science Reviews* 11(3): 285–292. 1986.

Jain, S.K., ed. *A Manual of Ethnobotany*. Jodhpur: Scientific Publishers. 1987.

Jain, S.K. *Ethnobotany – Its Concepts and Relevance*. Presidential Address. 10th All India Botanical Conference. Patna. 1987.

Jain, S.K., ed. *Methods and Approaches in Ethnobotany*. Lucknow: Society of Ethnobotanists. 1989.

Jain, S.K., ed. *Contributions to Indian Ethnobotany*. Jodhpur: Scientific Publishers. 1990.

Jain, S.K., ed. *Dictionary of Indian Folkmedicine and Ethnobotany*. New Delhi: Deep Publications. 1991.

Jain, S.K. and Roma Mitra. "Ethnobotany in India". In *Contributions to Indian Ethnobotany*. Ed. S.K. Jain. Jodhpur: Scientific Publishers. 1990. pp. 1–17.

Jain, S.K., et al. *Bibliography of Ethnobotany*. Howrah: Botanical Survey of India. 1984.

Janaki-Ammal, E.K. "Introduction to Subsistence Economy of India." In *Man's Role in Changing the Face of the Earth*. Ed. L.T. William Jr. Chicago: University of Chicago Press. 1956. pp. 324–335.

F

Forestry

The history of forestry in India is related to the history of civilisation. If we look at the development of human kind we find that with the evolution of the human race various concepts evolved, such as those of family, tribe, etc. as well as developmental sciences like forestry.

One of the earliest civilisations of the Indian subcontinent was that of the Indus Valley (third or fourth millennium BCE). It was here that cedar (*Cedrus deodara*) and rosewood (*Dalbergia latifolia*) used for coffins was found. A wooden mortar of ber (*Zizyphus mauritiana*) for pounding grain and charred timber of *Acacia* spp., *Albizzia* spp., teak (*Tectona grandis*), haldu (*Adina cardifolia*), and *Soyamida febrifuge* were also found. This shows that neolithic people not only made extensive use of wood but also understood its particular characteristics for different purposes. This concept today is known as forest utilisation.

Evidence of tree worship during the Indus Valley civilisation is exhibited by various seals of the Harappan culture which depict the pipal (*Ficus*) and weeping willow (*Salix*) trees. Trees were an essential and integral part of the life support system and considered existing agencies of the creator or God.

Just as forests played an important part in Vedic India, tree worship was also practiced by the Aryans. Because of the human dependence on trees, they were venerated and protected by religious injunctions; their planting was encouraged by a promise of eternal bliss in future life. *Ṛgv̄eda* and *Aṭharvav̄eḍa*, basic texts of the Vedic period, contain several hymns praising and endowing trees, plants, and vegetation with various divine qualities, highlighting their medicinal significance, and enunciating the policy of conservation or sustainable management. Various *Upaniṣads*, a later group of philosophical treatises that explain the theology of Hinduism, like *Bṛhadāraṇyaka*, *Chāndogya*, *Chulikā*, and *Muṇḍakopaniṣad*, conclude that trees had life akin to human life. *Mattsyapurāṇa* and *Varāhapurāṇa* describe the benevolence of trees along with the rituals of tree planting. *Skanddapurāṇa* contains a long list of trees which should not be cut except for the purpose of *yajñas* (holy rituals). Whereas planting of trees led to heavenly comforts (*Agnipurāṇa*) indiscreet felling of trees meant torture in hell.

Vedas also contain valuable information about various species of birds. In *Ṛgveḍa* for instance there is a mention of *Garuḍa* (eagle), *Mayūra* (pea fowl) along with various Himalayan pheasants, partridges, and other species of birds. Surprisingly *Ṛgveḍa* also mentions not only anatomical details of some common birds but details about their staple food too. For example *Vartika* (partridges) are said to have well developed bills, legs, and rounded wings; their food consisted of grain, grass, weed, seed, tender shoots, insects, and even white ants. These kindled the sparks of wildlife management during the later period of civilisation.

Great saints and sages who understood these texts, passed on their knowledge and wisdom to their disciples. Such education was given in the *Gurukuls* (schools) which were located in the forests. The students lived in the forests and continued their studies. Thus a sense of tolerance and coexistence with various forms of plant and animal life was infused in early childhood. This *Araṇya* (forest) culture provided early exposure to nature study and ecology, as well as the policy of development without destruction in current parlance. Forest and environment consciousness was thus ingrained into the educational system from the very onset.

Even epics and religious texts written later on describe the protective role of trees and their unlimited usefulness. The *Rāmāyaṇa* and *Mahābhārata* describe the rich biodiversity and multiplicity of flora and fauna. In the *Bhagavadgīṭā* Lord Krishna compared himself to the *Aśvattha* (ficus) tree in order to emphasise the importance of trees.

Other religions of that time also mentioned the importance of forests and forestry. According to Lord Buddha it was the obligation of every good Buddhist to plant and nurture at least one tree every five years.

Moving along chronologically we find that forest management practices were well documented during the reign of Chandra Gupta Maurya (321–296 BCE). Reliable historical documents, such as the *Indika* of the Greek ambassador Magasthenese, the *Arthaśāstra* of Kauṭilya and the *Mudrārākṣasa* of Viśākhadatta vividly depict various aspects of forestry and wildlife. Kauṭilya's *Arthaśāstra*, recognised as the pioneer work in economics in India, indicates the existence of a regular Forest Department headed by *Kūpādhyakṣa* with definite duties and responsibilities for various officers of the department. Some of the important duties of *Kūpādhyakṣa* were to increase the productivity of forests, classification, price fixation, and disposal of various types of forest produce, raising block plantations of important species (e.g. sandalwood), pasture development in saline–alkaline waste lands, and joint management of forests with people dependent on forests.

Forests were legally classified into three main classes: reserved forests, forests donated to eminent Brahmans, and forests for public use. They were classified into six categories on the basis of crown density, luxuriance, growth, and origin. Planted forests were called *Upavana*.

Forest and game laws were stringent and draconian, containing corporal punishment. The death sentence was envisaged for poaching of elephants. Awards were provided to a person who collected elephant tusks from dead elephants and deposited them with government officials. Kautilya's treatise also dealt at great length not only with the use of various trees and shrubs but also with the specific use of various parts. For example it indicated that flowers of *Palāś, Kusuma,* and saffron are used as dyes; *Munja* and *Love* grass are used for making ropes, and fruits of *Aonla, Harara,* and *Bahera* are used as medicines.

Emperor Aśoka adopted and improved these practices. Plantation of fruit-bearing shade trees for the benefit of travellers and common people was started on an ambitious scale. Aśoka was the *Abhayāraṇya,* the *sanctum sanctorum* of wildlife, now known as National Parks and sanctuaries. Aśoka's edicts at Sarnath Varanasi bear ample testimony to the above.

Forestry flourished during the Gupta period (320–800). *Śukranīti,* a well known work of that time, throws light on the improvements in forest management practices. Seeds of "Social Forestry" were sown during this era. *Śukranīti* dwells on the concept of village forests, choice of species, afforestation and maintenance techniques, fertilisation procedures, irrigation schedules, and measures to increase flowering and fruiting in trees. Names of various multipurpose trees such as *Kadamba, Sīśama, Peepal, Mango, Nīm, Coconut,* and *Imlī* are listed in this policy document for planting near villages. Incidentally many of these are also listed in the latest ICRAF (International Council for Research in Agroforestry, in Kenya) booklets. This shows the worth of these ancient publications.

It is thus clear that in ancient India trees were the best friend of people in a hostile environment. They were held sacred, worshipped, and studied in great detail for service to humanity. Forestry practices evolved gradually, in a scientific and rational manner, and are important even today, although in the modern parlance new names have been coined for them. As a matter of fact some of the knowledge which originated in India on various aspects of forestry later spread to various parts of the world.

Deepa Pande

REFERENCES

Bhagwat Geeta. Gorakhpur: Geeta Press. 1992.

Chattopadhyay, Debiprasad. *Science and Society in Ancient India*. Calcutta: Research India Publications. 1979.

Ghildiyal Vineet and Sharma R.C. "Some Himalayan Birds and their Conservation in Rigvedic India." *Himalayan Research and Development* 4(2): 24–25. 1985.

Kautilya's Arthaśāstra. Mysore: Wesleyan Mission Press. 1923.

Pande, Deepa and I.D. Pande. "Forestry in India through the Ages." In *History of Forestry in India*. Ed. Ajay S. Rawat. New Delhi: Indus Publishing Company. 1992. pp. 151–162.

Sinha, B.C. *Tree Worship in Ancient India*. New Delhi: Books Today. 1979.

G

Geography

The study of geography (*Bhūgol*) as a systematic literature was not in use in ancient India. But various names, such as *Bhuvanakośa* (Terrestrial Treasure), *Trilokya Darpana* (World's Mirror) and *Kṣetrasamāsa* (Combination of Countries), were used to denote geographic phenomena. Beginning with the *Ṛgveda* (1500 BCE), tribes, rivers, and mountains of India were mentioned. As the Aryans, who brought the *Ṛgveda* with them, first settled in north-western India (Punjab and the Indus plain) only the related regional rivers like Sarasvatī (extinct), Sarayū (Sutlej) and Sindhu (Indus) were mentioned. Later, by the beginning of the Christian era when the three other *Vedas* were developed and the *Purāṇas* were written, a more extensive description of Indian geography evolved. The Aryan settlements advanced not only eastward to Bengal and Assam, but also to South India, which though Dravidian in terms of linguistic and physiognomic characteristics, became Hinduised. Kauṭilya's *Arthaśāstra*, written around 300 BCE, prescribed a state-planned colonisation policy for undeveloped (primarily natural vegetation covered) areas with people from foreign lands and kingdoms' surplus labour. Each village was to have five hundred families, mostly *śūdra* cultivators or labourers; villages were to be grouped in a hierarchy of settlements of eight hundred, four hundred, two hundred, and ten units with a large town *(sthānīya)*, small town *(droṇamukha)*, smaller town *(kārvaṭika)*, and large village *(saṃgrāhara)* to serve each group respectively.

The river Ganges (*Gaṅgā*) is mentioned in *Vāyupurāṇa* as a purifier of sinners, passing through thousands of mountains and irrigating hundreds of valleys. It was in *Kauṣītaki* that the southern mountain – Vindhyas – was first mentioned, while both *Vāsiṣṭha Dharmasūtra* and the *Code of Manu* refer to Vindhyas by name. The two great epics – *Rāmāyaṇa* and *Mahābhārata* – give extensive descriptions of geographic features of both North and South India. When the king Daśaratha, father of Rāma, performed *Aśvamedhayajña* (Horse Sacrifice), to establish his supremacy over the world, both North and South Indian kings were referred to. In the *Bhīṣmaparva* section of *Mahābhārata*, Sañjaya, the chariot driver of the blind king Dhṛtarāṣtra, identified the nations, mountains, and rivers of India.

Beginning with the *Ṛgveda* through the *Purāṇas,* cosmological and cosmogonic interpretations of causes of wind, precipitation, day and night, seasons, and planetary movement have been made. Explanation of artistic, mechanical, instrumental, and philosophical origins of the universe has also been given. The artistic origin refers to god as an artist who skilfully constructed the universe. The mechanical origin is conceived as a sacrifice of the primeval body (*Ādipuruṣa*) who not only had the soul and the nucleus of the universe, but also embodied the Supreme Spirit, resulting in the formation of the Earth, sky, wind, the Moon, the Sun, and other terrestrial elements. The philosophical concept considered the beginning as an empty space with no atmosphere or sky, and the universe was born out of its own nature, possibly by its own inherent heat. *Māhāpurāṇa,* composed by a Jain teacher in the ninth century CE, further crystallises this idea that the world endures under its own nature and is divided into hell, earth, and heaven. The instrumental origin idea is reflected in the union of heaven and earth caused by the action of different gods, such as *Agni* (fire), *Sūrya* (sun), and *Indra.* "The central idea of various cosmogonic theories of the Vedic and post-Vedic period appears to be (1) the existence of water in the beginning, and (2) the creation of a cosmic nucleus – *Prajāpati*" (Ali, 1966). *Prajāpati* is an embodiment of propagation and hence maker of the universe.

Ancient literature described a cyclical human development on the Earth. The Hindu view, expressed in the *Code* of *Manu,* speaks of four ages: Kṛta, Tretā, Dvāpara, and Kali, with a sharp break at the end of each age; physical and spiritual deterioration occurs at the completion of the four-age period with the universe coming to an end and then beginning a new cycle. The Jains also believed in the cosmic cycles, but unlike Hindus they did not foresee any sharp break at the ends of the periods (ages). The Jains divided a full cycle into six periods, each with ascending and descending halves. At the end of the sixth period, designated as "very wretched" (*Duḥṣama-Duḥṣama),* human deterioration reaches its peak with a fierce storm wiping out almost all inhabitants. After this a new six-period cycle starts again.

In a general view of explaining the distribution of continents and oceans the world was divided into seven regions or islands (*dvīpas*). S.M. Ali identifies those dvīpas as Śālmali (East Africa), Kuśa (Middle East), Polaksa (Mediterranean), Krauñcha (Europe), Puṣkara (East Asia), Śaka (South-east Asia), and Jaṃbū (Northern, Central, and Southern Asia). The southern part of the Jaṃbūdvīpa was inhabited by Hindus in the land called Bhārata. At the centre of the Jaṃbūdvīpa was the Meru identified with Pamir knots, and at the northern extremity adjoining the Arctic ocean was the Uttarakuru. Jaṃbūdvīpa was surrounded by an "ocean", meaning a physical barrier. Yet another Puranic conception compares the earth with a lotus. The pericap of the earth-lotus was the Meru; each petal represented a continent of the *Māhādvīpa* (great island), situated equidistant from each other and surrounded by oceans. Jaṃbūdvīpa was the largest of the continents.

Bhāratavarṣa (India) has been described in the *Purāṇas* as having various shapes: a half moon, a triangle, etc. It was given the shape of a rhomboid by Eratosthenes in BCE 320 and that of four equal triangles in the *Mahābhārata* composed around BCE 100. Ptolemy (CE 100–170) in mapping India ignored the peninsula and drew an almost straight line from the Gangetic delta to the Makran coast in Baluchistan. Varāhamihira (CE 550) identified seven divisions of India surrounding the central division of Pañchāla (Punjab/Haryana area); they were Magadha (east), Kaliṅga (south-east), Avanta (south), Sindhu-Sauvīra (west), Ānarta (south-west), Madra (north), and Kauṇḍa (north-east).

Though only the southern part of Jaṃbūdvīpa was considered Bhāratavarṣa, the Indians referred to the latter as the whole Earth because the landmass of South Asia surrounded by mountains in the north and seas in the south was not only a physical–cultural cul-de-sac, secluded from other peoples but was so productive and large that to the people of India it constituted the world. Puranic legends claim that the king Bharata, whose name was identified with the country, ruled the entire earth, meaning all of India. Such was the case with the emperor Aśoka (272–232 BCE) as inscribed in his Fifth Rock Edict.

Geopolitics took a special form in ancient India. The *Code of Manu* and the *Arthaśāstra* developed a set of stratagems and practices based on the position of states which were parts of a circle (*Maṇḍala*). A kingdom was considered to be surrounded by four circles of states. The first circle consisted of friends and friends' friends. The three outer circles represented the states that belonged to the enemy, the middle king, and the neutral king. A sixfold interstate policy – peace, war, marking time, attack, seeking refuge, and duplicity – was evolved to maintain a balance of power among the circle of states. Policies related to extension or defence of a state depended on one's power in relation to the nearness and distance from other states. As in the Puranic *Bhāratavarṣa*, *Arthaśāstra* defined the sphere of influence of an imperial ruler (*Cakravartin*) by the conquest of all land between the Himalayas and the southern sea, meaning Cape Comorin. Bāṇa's *Kādambarī* and *Harṣacharita* while conforming with *Arthaśāstra's* boundary specified the eastern limit as the mythical sunrise or Udaya mountains (possibly the Blue Mountains or Pawnpuri of Mizoram) and the western limit by another mythical mountain-sunset or Mandara (possibly the Western Ghats or Sahayadri).

Ashok K. Dutt

References

Ali, S. M. *The Geography of the Purāṇas*. New Delhi: People's Publishing House, 1966.

Cunningham, Alexander. *The Ancient Geography of India*. Varanasi: Indological Book House. 1979.

De Bary, W. M. Theodore, ed. *Introduction to Oriental Civilisations: Sources of Indian Tradition*. New York and London: Columbia University Press. 1967.

Goyal, S. R. *Kautilya and Megasthenes*. Meerut: Kusumanjali Prakashan. 1985.

Majumdar, R.C. and A.D. Pusalker, eds. *The History and Culture of the Indian People: The Vedic Age*. London: George Allen and Unwin Ltd. 1951.

Majumdar, R.C. and A.D. Pusalker, eds. *The History* and *Culture of the Indian People: The Age of Imperial Unity*. Bombay: Bharatiya Vidya Bhavan. 1953.

Majumdar R.C. and A.D. Pusalker, eds. *The History and Culture of the Indian People: The Classical Age*. Bombay: Bharatiya Vidya Bhavan. 1954.

Sircar, D. C. *Studies in the Geography of Ancient and Medieval India*. Delhi: Motilal Banarsidass. 1971.

Thapar, Romila. *A History of India I*. Middlesex, England: Penguin Books. 1966.

Tripathi, Maya Prasad. *Development of Geographic Knowledge in Ancient India*. Varanasi: Bharatiya Vidya Prakashan. 1969.

Geometry

An examination of the earliest known geometry in India, Vedic geometry, involves a study of the *Śulbasūtras*, conservatively dated as recorded between 800 and 500 BCE, though they contain knowledge from earlier times. Before what is conventionally known as the Vedic period (ca. 1500–500 BCE), there was the Harappan civilisation dating back to the beginning of the third millennium BCE. Even a superficial study of the Harappan cities show its builders as extremely capable town planners and engineers requiring a fairly sophisticated knowledge of practical geometry. An interesting conjecture has been suggested by a drawing on a seal found from Harappa (ca. 2500 BCE): was there an awareness then that the area of a polygon inscribed in a circle approaches the area of the circle as the number of sides of the polygon keeps increasing? This is the basic idea behind techniques that were developed for the mensuration of the circle in a number of mathematical traditions including Indian.

The *Śulbasūtras* are instructions for the construction of sacrificial altars (*vedī*) and the location of sacred fires (*agni*) which had to conform to clearly laid down instructions about their shapes and areas if they were to be effective instruments of sacrifice. There were two main types of ritual, one for worship at home and the other for communal worship. Square and circular altars were sufficient for household rituals, while more elaborate altars whose shapes were combinations of rectangles, triangles, and trapezia were required for public worship. One of the most elaborate of the public altars was shaped like a giant falcon just about to take (*Vakrapakṣa-śyena*). It was believed that offering a sacrifice on such an altar would enable the soul of the supplicant to be conveyed by a falcon straight to heaven.

It is clear that if in the construction of larger altars they had to conform to certain basic shapes and prescribed areas or perimeters, two geometrical problems would soon arise. One is the problem of finding a square equal in area to two or more given squares; the other is the problem of converting other shapes (for example, a circle or a trapezium or a rectangle) into a square of equal area or vice versa. The constructions were achieved through a judicious combination of concrete geometry (in particular what would be known today as the principle of dissection and re-assembly), ingenious algorithms, and the application of the so-called Pythagorean theorem. The essence of the dissection and re-assembly method involves two commonsense assumptions.The first is that both the area of a plane figure and the volume of a solid remain the same under rigid translation to another place. The second says that if a plane figure or solid is cut into several sections, the sum of the areas or volumes of the sections is equal to the area or volume of the original figure or solid. The reasoning behind this approach was very different from that behind Euclidean geometry, but the method was often just as effective, as shown in the Indian (and Chinese) "proofs" of the Pythagorean theorem.

In the *Kātyāyana Śulbasūtra* (named after one of the authors) the following proposition appears: "The rope (stretched along the length) of the diagonal of a rectangle makes an (area) which the vertical and horizontal sides make together." (2.11). Using this version of the Pythagorean theorem, the *Śulbasūtras* show how to construct both a square equal to the sum of two given squares and a square equal to the difference of two given squares. Further constructions include the transformation of a rectangle (square) to a square (rectangle) of equal area and of square (circle) to a circle (square) of approximately equal area. The constructions "doubling the square" and "squaring the circle" lead naturally to devising algorithms for the square root of 2 and other numbers, for implicit estimates of π, and for constructing similar figures in required proportions of a given figure.

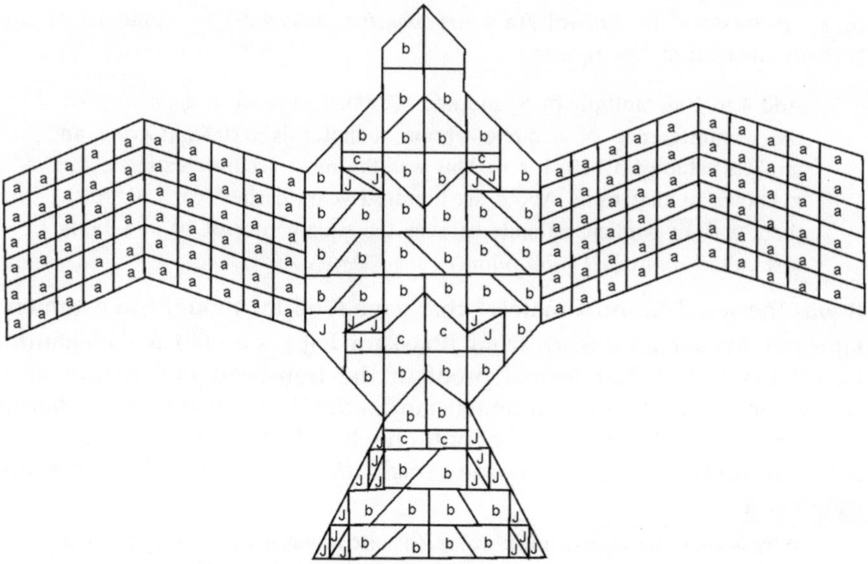

Figure 1 *The first layer of a Vakrapraksa-śyena alta. The wings are made from 60 bricks of type 'a', and the body, head and tail from 50 type 'b', 6 of type 'c' and 24 type 'd' bricks. Each subsequent layer was laid out using different patterns of bricks with the total number of bricks equalling 200.*

The composers of the *Śulbasūtras* made it clear that their work was not original but could be traced to earlier texts, notably the *Saṃhitās* and the *Brāhmaṇas* of which the most relevant text, *Śatapatha Brāhmaṇa,* is at least three thousand years old. In spite of its obscurities and archaic character the text is valuable for an early discussion of the technical aspects of altar construction. The instructions given in *Śatapatha Brāhmaṇa* (X.2.3.11–14) for constructing a falcon-shaped altar consisting of 95 layers of bricks are as follows (Figure 1):

Area of the body (Atman) $= 56 + \frac{12}{7}\sqrt{56}$;

Area of two wings $= 2\,(14) + \frac{3}{7}\sqrt{14} + \left(\frac{1}{5}\right)\left(\frac{1}{7}\right)(3)\left(\sqrt{14}\right)$;

Area of tail $= 14 + \frac{3}{7}\sqrt{14} + (\frac{1}{10})(\frac{1}{7})(3)(\sqrt{14})$.

The total area is about 116 square *purushas,* which is an over-estimate of the required 101.5 square *purushas,* arising in part from a rounding off error involved in taking 14 rather than 13 + 8/15.

A major strand running through the history of Indian geometry and also providing the main motivation for the development of the subject was a recognition of the impossibility of arriving at an exact value for the circumference of a circle given the diameter (i.e. the incommensurability of π). A passage in Āryabhaṭa's *Āryabhaṭīya* (CE 499) – Verse 10 of the section on *Gaṇita* – reads:

> Add 4 to 100, multiply by 8, and add 62,000. The result is *approximately* the circumference of a circle whose diameter is 20,000. (Giving an implicit value of 3.1416 for π. This was the most accurate estimate for π known at that time. About six hundred years earlier (ca. 150 BCE), there was an implicit estimate of π as the square root of 10 in a Jaina text called *Anuyoga Dwāra Sūtra*.)

It was the word "approximately" that gave food for thought to commentators of Āryabhaṭa's work from Bhāskara I (ca. CE 600) to Nīlakanṭha (b. CE 1445). The first formal proof of the transcendental nature of π was given by the Swiss mathematician Lambert in a paper to the Berlin Academy in 1671. However, about one hundred and fifty years earlier, Nīlakanṭha's commentary on *Āryabhaṭīya* contained the following statement:

> Why is only the approximate value (of circumference) given here? Let me explain. Because the real value cannot be obtained. If the diameter can be measured without a remainder, the circumference measured by the same unit (of measurement) will leave a remainder. Similarly the unit which measures the circumference without a remainder will leave a remainder when used for measuring the diameter. Hence the two measured by the same unit will never be without a remainder. Though we try very hard we can reduce the remainder to a small quantity but never achieve the state of 'remainderlessness'. This is the problem. (Adapted from Sarasvati Amma, 1979)

Once the incommensurability of π was accepted, the approach of the Indian mathematician was to obtain as accurate a value of this quantity as possible, and the strategy to be followed was expressed thus by Śaṅkara Variyār and Nārāyana Kriyākramakārī (ca. 1550)]:

> Thus even by computing the results progressively, it is impossible theoretically to come to a final value. So, one has to stop computation at that stage of accuracy that one wants and take the final result arrived at ignoring the previous results. (Adapted from Sarma, 1975)

The major breakthrough came from the revolutionary idea, most probably that of Mādhava (ca. 1340–1425), that it was possible to obtain an infinite series whose sum would be exactly equal to π and that an increasingly close rational approximation of the quantity could be obtained by taking partial sums successively of higher order. While the question of the slow convergence of this series was not explicitly discussed, the need for increasing rapidity of convergence was recognised and some remarkable corrections to be applied to truncated series were deduced. The work on infinite series for circular functions provided an impetus to derivation of other infinite series for trigonometric functions, namely arc tangent, sine, and cosine series. Often inductive reasoning (and intuition) built upon geometrical representation helped them to discover these results, but the proof of these results can withstand any rigorous criterion applied to it today. An implicit estimate for π based on infinite series expansion given by Mādhava around 1400 is correct to eleven decimal places.

Another area of geometry in which the Indian contribution was significant was in the study of the properties of a cyclic quadrilateral. In the *Brāhma Sphuṭa Siddhānta,* Brahmagupta (b. CE 598) gives these results:

(1) The area of a cyclic quadrilateral is given by the product of half the sums of the opposite sides, or by the square root of the product of four sets of half the sum of the sides (respectively), diminished by the sides.

(2) The sums of the products of the sides about the diagonal should be divided by each other and multiplied by the sum of the opposite sides. The square roots of the quotients give the diagonals of a cyclical quadrilateral.

The derivations of these results are first referred to in a tenth-century commentary on Brahmagupta's work, but find their full expression in the sixteenth century Kerala text *Yuktibhāṣā* by Jyeśṭhadeva. This makes use of Ptolemy's theorem that in a cyclic quadrilateral the product of the diagonals is equal to the sum of the products of two pairs of opposite sides.

Notable extensions in this area are contained in Nārāyaṇa Paṇḍita's *Gaṇita Kaumudī* in the fourteenth century and Parameśvara's *Līlāvatī Bhāṣya,* a detailed commentary on Bhāskaracharya's *Līlāvatī.* In the latter is found a new rule for obtaining the radius of the circle in which a cyclic quadrilateral is inscribed. The great interest in the cyclic quadrilateral in Indian mathematics arose from the fact that it was an important device for deriving a number of important trigonometric results which were in almost all cases used in astronomy.

In India geometry never became idealised in the way it did in Greece. Geometry was largely concrete and empirical in character. It did, however, have an algebraic character which is best seen in the genesis of trigonometry there. Because of their geometric emphasis, the Greeks used chords in their astronomical calculations, whereas the Indians developed the notion of sines and versines (i.e. 1 – cosine of an angle) as early as CE 500. Āryabhaṭa was perhaps the first Indian astronomer to give a special name to these functions and draw up a table of sines for each degree. Approximation formulae were developed for these functions, culminating in the construction of sine tables in Kerala during the fifteenth century where the values in almost all cases are correct to the eight or ninth decimal place, a remarkable degree of accuracy.

George Gheverghese Joseph

References

Colebrook, H.T. *Algebra with Arithmetic and Mensuration from the Sanscrit of Brahmegupta and Bhascara*. London: John Murray. 1817.

Datta, B. *The Science of the Śulbas: A Study in Early Hindu Geometry*. Calcutta: Calcutta University Press. 1932.

Gupta, R.C. "Paramesvara's Rule for the Circumradius of a Cyclic Quadrilateral." *Historia Mathematica* 4: 67–74. 1977.

Joseph, G.G. *The Crest of the Peacock: Non-European Roots of Mathematics*. London: Penguin. 1992.

Joseph, G.G. "What is a Square Root ? A Study of Geometrical Representation in Different Mathematical Traditions." In *Proceedings of 1993 Annual Meeting of the Canadian Mathematics Education Study Group*. Ed. M.Quigley. University of Calgary. 1994. pp. 1–14.

Kulkarni, R.P. *Geometry according to Śulba Sūtra*. Pune: Vadika Saṁśodhana Maṇḍala, 1983.

Varma, T. Rama and A. Aiyer, eds. *Yuktibhāṣā*. (in Malayalam). Trichur: Mangalodayam. 1948.

Sarasvati Amma, T.A. *Geometry in Ancient and Medieval India*. Delhi: Motilal Banarisidass. 1979.

Sarma, K.V., ed. *Lilavati of Bhaskaracarya with the Commentary Kriyākramakarī* . Hoshiarpur: Vishveshvaranand Institute. 1975.

Staal, F. *Agni: The Vedic Ritual of the Fire Altar*. Berkeley: Asian Humanities Press. 1983.

See also: *Śulbasūtras* – Nīlakantha – Āryabhaṭa – Nārāyana – Mādhava – Bhāskara – Jyeṣṭhadeva – Parameśvara

Gnomon

The gnomon is an instrument used widely in early astronomy. The shadow of a vertical rod on a horizontal plane determines the cardinal directions, the latitude of the place of observation, the celestial coordinates of the Sun, and the time of the observation.

A fairly complete account of its use in India was given by Varāhamihira in the *Pañcasiddhāntikā*. This was written in CE 505 and summarised the astronomical information current in India at that time. The *Āryabhaṭīya* of Āryabhaṭa also provides the main results of the theory of the gnomon, and these features appear again in the works of Bhāskara, Brahmagupta, and many later astronomers.

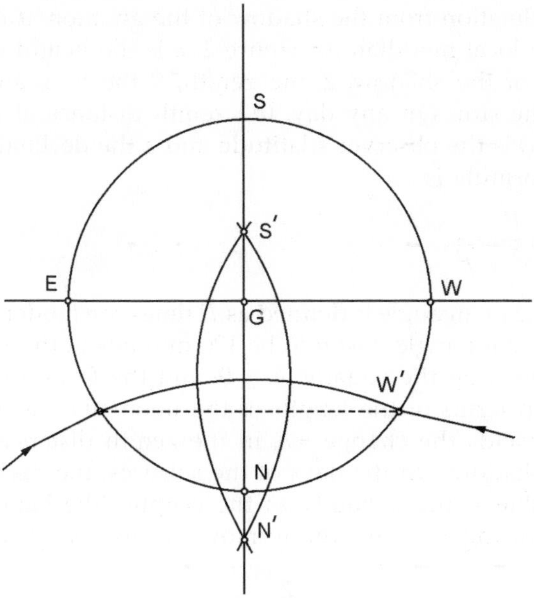

Figure 1 *Finding the cardinal direction (Neugebauer, 1971).*

The Cardinal Directions

The procedure is illustrated in Figure 1. G is the foot of the gnomon. The path of the end of the shadow enters and leaves a circle, centre G, at W′ and E′. Then the line E′W′ is in the east–west direction. With E′, W′ as centre, circular arcs are drawn intersecting at N′, S′. Then N′S′, the perpendicular bisector of E′W′, is in the north–south direction and intersects the circle at N and S, the north and south points. The east and west points, E and W, can be found by the same procedure since they are on the perpendicular bisector of NS.

This method depends on the symmetry of the shadow path about the north–south line. It does not take into account the small change in the declination of the Sun during the day. Brahmagupta prescribed a correction for this error in the *Mahābhāskarīya*. This method of finding the cardinal directions, described in the *Pañcasiddhāntikā*, is found in a much earlier treatise, the *Śulbasūtra*, which contains mathematical topics related to the construction of sacrificial altars.

THE NOON SHADOW

Trigonometric formulas enable us to find the latitude of the observer and the Sun's declination from the shadow of the gnomon at noon, when the Sun is on the local meridian. In Figure 2, g is the height of the gnomon, s the length of the shadow, Z the zenith, S the Sun, and z the zenith distance of the sun. On any day, the zenith distance at noon is $z = \phi \pm \delta$, where ϕ is the observer's latitude and δ the declination of the Sun. The Indian formula is

$$\text{Sin } z = \frac{Rs}{\sqrt{s^2 + g^2}},$$

where the Sine of an angle is defined as R times the modern sine function, R being a constant angle, taken to be 120 minutes in the *Pañcasiddhāntikā*. When the Sun is on the equator, $\delta = 0$, and the formula above gives us the latitude in terms of the length of the noon shadow. On other days, the formula yields the change $\pm\delta$, in the zenith distance, as a function of the noon shadow. On the days of the solstices, the declination has the maximum value ε, the obliquity of the ecliptic (the band of the zodiac through which the Sun apparently moves in its yearly course).

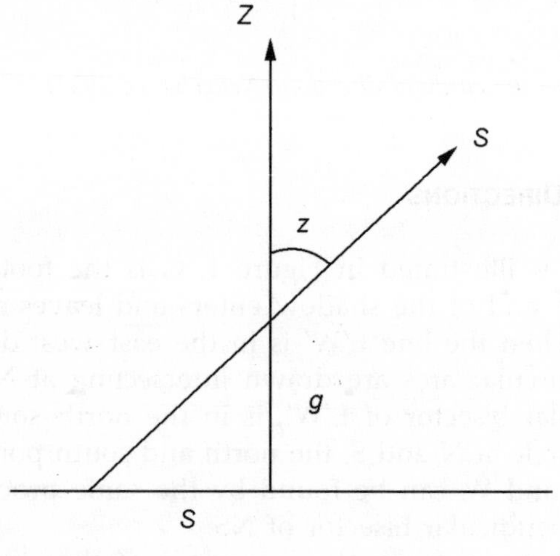

Figure 2 *The noon shadow.*

When the Sun is on the prime vertical (the great circle on the celestial sphere through the east and west points and the zenith), let z_1 be the zenith distance, and $a_1 = 90 - z_1$, the altitude of the Sun, and λ, the Sun's longitude. Then the *Pañcasiddhāntikā* formulae are

$$\text{Sin}a_1 = \frac{R\ \text{Sin}\delta}{\text{Sin}\phi} = \frac{\text{Sin}\lambda\text{Sin}\epsilon}{\text{Sin}\phi}.$$

With these two formulae, we can find the declination and longitude of the Sun from the shadow length, when the sun is on the prime vertical.

With the gnomon, it was also possible to find the time after sunrise, from the length of the shadow, using formulae which are equivalent to those used in modern spherical astronomy.

The second formula above gives the sun's longitude λ and declination δ, when it is on the prime vertical. λ and δ can be determined at any time, from the length of the gnomon's shadow and the distance of its endpoint from the east–west line. The *Pañcasiddhāntikā* also gives an approximate empirical algebraic formula which would have been useful in very early astronomy:

$$\frac{d}{2t} = \frac{s - s_0}{g} + 1.$$

This gives t, the time after sunrise in the morning or the time before sunset in the afternoon. s_0 is the noon shadow and d the length of daylight. This formula is derived from the following considerations:

(a) there is a linear relation between s and $1/t$,

(b) at noon $2t = d$,

(c) $s - s_0 = g$ at $4t = d$.

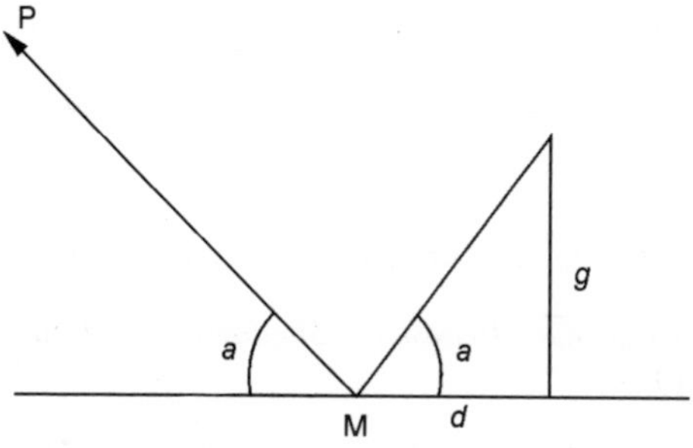

Figure 3 *Determining the altitude of the Moon or planets.*

The *Yavanajātaka* of Sphujidhvaja also has this formula. Chapter 20 of Kauṭila's *Arthaśāstra* gives the relation between the time after sunrise and the gnomon shadow. The *Arthaśāstra* also gives the rule for the uniform variation of the noon shadow from zero at the summer solstice to g, the gnomon height, at the winter solstice, a reasonable approximation for an observer on or near the Tropic of Cancer, for example at Ujjain. However, the rule for the uniform variation of the length of daylight from 12 to 18 *muhurtas*, also found in the two books above, implies a latitude of about 35 degrees, which suggests a Babylonian origin.

The theory of the gnomon presented above can be applied to the Moon and planets also. The altitude of the Moon or planet is determined in the following manner, illustrated in Figure 3.

The moon or planet (P) is seen reflected in a mirror M in the same horizontal plane as the foot G of the gnomon of height g at a distance d from the gnomon. Then the altitude, a, is given by the formula

$$\text{Sin}a = \frac{Rg}{\sqrt{g^2 + d^2}};$$

d is called the reversed shadow and takes the place of the shadow length s in the case of the Sun.

The eleventh-century Arabic scholar al-Bīrūnī wrote *Kitāb fī ifrād al-maqāl fī umr al-ẓilā* (The Exhaustive Treatise on Shadows), which contains a comprehensive account of the theory and applications of the gnomon shadow. Al-Bīrūnī refers to many Indian sources, for example:

(a) the method described above for finding the cardinal directions. He calls it the method of "the Indian circle";
(b) the algebraic formula; and
(c) the time from the shadow. Al-Bīrūnī follows the procedure of Brahmagupta.

George Abraham

References

Primary sources

Burgess, E. *Sūrya-Siddhānta*. Calcutta: University of Calcutta. 1935.

Kennedy, E.S. *The Exhaustive Treatise on Shadows by al-Biruni*, 2 vols. Syria: University of Aleppo. 1976.

Neugebauer, O., and D. Pingree. *The Pañcasiddhāntikā of Varāhamihira*. Copenhagen: Munksgaard. 1970–72.

Pingree, David. *The Yavanajātaka of Sphujidhvaja*. Cambridge, Massachusetts: Harvard University Press. 1978.

Sastri, T.S. Kuppanna. *Mahābhāskarīya*. Madras: Government Oriental Manuscripts Library. 1957.

Sastri, Ganapati. *The Arthaśāstra of Kautilya*. Madras: Kuppuswami Sastri Research Institute. 1958.

Sen, S.N., and A.K. Bag. *The Śulbasūtras*. New Delhi: Indian National Science Academy. 1983.

Sharma, R.S. *Brāhmasphuṭa Siddhānta*. New Delhi: Indian Institute of Astronomical and Sanskrit Research. 1966.

Shukla, K.S., and K.V. Sarma. *Āryabhaṭīya of Āryabhaṭa*. New Delhi: Indian National Science Academy. 1976.

Shukla, K.S. *Mahābhāskarīya*. Lucknow University. 1960.

Thibaut, G., and M.S. Dvivedi. *The Pañcasiddhāntikā*. Benares: Medical Hall Press. 1889.

Secondary sources

Abraham, George. "The Gnomon in Early Indian Astronomy." *Indian Journal of History of Science* 16(2): 215–218. 1981.

Pingree, David. "History of Mathematical Astronomy in India." In *Dictionary of Scientific Biography*, vol. 15. New York: Scribners. 1978. pp. 533–633.

Smart, W.M. *Text Book on Spherical Astronomy*. Cambridge: Cambridge University Press. 1977.

Somayaji, D.A. *Ancient Hindu Astronomy*. Dharwar: Karnataka University. 1971.

See also: Astronomy – Astronomical Instruments – Varāhamihira – Āryabhaṭa – Bhāskara – Brahmagupta – *Śulbasūtras* – Sphujidhvaja – al-Bīrūnī

H

Haridatta

Haridatta was the promulgator of the Parahita system of astronomical computation widely used in Kerala in South India, from where it spread to the neighbouring state of Tamilnadu. There are two basic texts of the system, the *Grahacāranibandhana* and the *Mahāmārganibandhana*, the latter of which is no longer extant. Haridatta inaugurated the system, as the legend goes, on the occasion of the 12-yearly religious festival held at the temple town of Tirunāvāy on the banks of the Bhāratappuzha River in CE 683. The system was called *Parahita* (suitable to the common man), because it simplified astronomical computation and made it accessible for practice even by ordinary people.

Haridatta based his system on the *Āryabhaṭīya* of Āryabhaṭa (b. 476), but made it simpler in several ways. First, he dispensed with the rather cumbersome and terse numerical symbolism used by Āryabhaṭa and substituted the facile, easily manipulated *ka-ṭa-pa-yā-di* system of notation. In this system, specific letters were used for representing digits, which could be arranged to form meaningful words and even sentences, which could be remembered with much less possibility of error.

Computations in Indian astronomy involved long numbers for planetary revolutions and other parameters for the aeon. To avoid these long numbers in multiplication and division, Haridatta ingeniously introduced a sub-aeon of 576 years or 210,389 days, and accurately determined the zero-correction for this sub-aeon for the mean motion of the several planets. In actual practice, Haridatta directed the aeonary days for any current day being divided by the sub-aeonary days. The quotient would then give the number of completed sub-aeons, and the remainder the days in the current sub-aeon. The above quotient multiplied by the sub-aeonary zero-corrections of the several planets would give the mean planets at the commencement of the current sub-aeon. In order to make calculation easier, when a large number of years passed, Haridatta gave these zero-corrections for chunks of six sub-aeons. To find the mean motion of the several planets for the completed days in the current sub-aeon, Haridatta gave certain sets of simple multipliers and divisors. The mean motion arrived at using these, when added to the mean position at the commencement of the current sub-aeon, would give the mean planet for the current date. In order to obviate the large numbers that would accumulate to mean planets when the Āryabhaṭan parameters were used, Haridatta prescribed a corrective called *Vāgbhāva* (Corrections to be Applied to the Different Planets from CE 523).

The Parahita system promulgated by Haridatta has been extremely popular in Kerala, and a very large number of texts and tracts based on the system have been produced in that part of India, both in Sanskrit and in the local language Malayalam. The system has also been regularly used for the computation of the daily almanac.

K. V. Sarma

References

Datta, B.B., and A.N. Singh. "Kaṭapayādi System." In *History of Hindu Mathematics*. Bombay: Asia Publishing House. 1962. pt. I, pp. 69–70.

Sarma, K.V. *Grahacāranibandhana: A Parahita Manual by Haridatta*. Madras: Kuppuswami Sastri Research Institute. 1954.

Sarma, K.V. *A Bibliography of Kerala and Kerala-based Astronomy and Astrology*. Hoshiarpur: Vishveshvaranand Institute. 1972.

Sarma K.V. "Parahita System of Astronomy." In *A History of the Kerala School of Hindu Astronomy*. Hoshiarpur: Vishveshvaranand Institute. 1972. pp. 7–9.

See also: Astronomy – Āryabhaṭa

I

Irrigation in India and Sri Lanka

Assessments by historians of Asia's irrigation systems and irrigation-related civil engineering techniques have been based on the scantiest of historical or empirical data. Naturally, they have ranged from one extreme to the other. Of these, the one most easily recognised and debated was provided by Karl Wittfogel whose theories led to the idea of "hydraulic civilisations".

A diametrically opposite assessment has been provided by some Indian historians who have concluded that there was no significant irrigation technology in use at all. Symbolic of this view is R. Majumdar and H.C. Raychaudhuri's *An Advanced History of India*, in which the authors make a categorical statement on the "comparative absence of artificial irrigation" in eighteenth century India.

However both views—fairly representative of the historiographical terrain—have had to be revised considerably because of the emergence of new historical materials and investigations during the last two decades. These are reflected in new literature specifically devoted to the subject. Illustrative of these materials is the report of Alexander Walker, an English specialist who toured India in the eighteenth century. Walker produced an elaborate treatise on Indian agriculture in which he drew the conclusion that "the practice of watering and irrigation is not peculiar to the husbandry of India, but it has probably been carried there to a greater extent, and more laborious ingenuity displayed in it than in any other country".

This display of ingenuity, however, is not restricted to eighteenth century India and has indeed found expression in a plethora of irrigation systems in Asia each designed to appropriate its own specific ecosystem potential. There is evidence of large-scale irrigation works in several Asian countries including China and Sri Lanka. The systems studied on the subcontinent include gigantic artificial lakes, large-scale and small-scale embankments, and diversion channels. They include schemes for taking water up a hill against gravity, elaborate canal distribution networks, innovations like the *khazans* on the west coast of India, where unmanned wooden sluice gates control the sway of salt and sweet water in low lying paddy fields adjoining the coastal or tidal rivers, and storage tanks with a bewildering variety of names.

"The irrigation history of Indian has been studied only in fragments" (Sengupta, 1993). But even this admittedly fragmentary picture that is emerging is far more fascinating than the simplistic or impressionistic scenarios of Wittfogel or later historians like Raychaudhuri and Majumdar. The most fundamental aspect of irrigation technology and civil engineering works to be noted is that almost all of them are related to monsoon precipitation in one way or another. Over ninety per cent of the annual run-off in the peninsular rivers and eighty per cent in the Himalayan rivers occurs during the four months of the monsoon. Thus, unlike the case of temperate ecosystems, irrigation becomes extremely crucial: in the wet season which stretches approximately four months, in several places less, there can be too much precipitation over intense bursts. This is followed by a dry season during which there is no precipitation at all. The result is predictable: periods of excess water followed by drought.

In this context, the basic design of irrigation technology is intimately related to precipitation: how to save it, store it, and divert it, so that spatially it reaches areas where there is no water (diversion techniques) or temporally makes it available during the dry months (storage techniques). Thus, rain-fed rivers are diverted into channels, or river basins are interconnected. Or the rainfall is directly collected in huge storage facilities on the land.

If this is the scenario, (and it is as valid today as it was in 3000 BCE), one would normally expect a much richer history of irrigation techniques in Asian conditions—where rice is a basic crop adapted to growing largely in water—than in any other part of the planet, particularly the temperate zones. It would also follow that the irrigation designs evolved for coping with such situations would not readily be available in other ecosystems. For this reason, it has taken some time for engineers and historians trained in other culture areas to appreciate their worth and function.

The irrigation experience of China is documented in Joseph Needham's *magnum opus, Science and Civilisation in China*, and will not concern us here. We shall restrict ourselves in this essay to a consideration of the irrigation and civil engineering techniques that arose in the Indian subcontinent including Sri Lanka and which were the result of a close interaction and adaptation between overall environmental situations and human ingenuity.

Irrigation Technologies

In the circumstances related above, it stands to reason that the primary design objective of irrigation engineers would be predominantly in the direction of a water storage system. The following listing is given by Shankari and Shah.

There were storage systems designed purely for drinking water: *nadi*, tanks, *bowari, jhalara*, and *pokhar*. Some were reserved only for human beings; others were for human beings and animals. These we shall ignore here.

The second category of storage technologies relates to irrigation, and there is considerable evidence of the spread of such technologies through the length and breadth of the subcontinent. Though the structures here were all designed for irrigation, they also provided other useful functions of soil conservation and ground water percolation and recharge.

Irrigation water stored in such storages was conveyed through two methods: first, under the force of gravity, or gravity irrigation, and second, through extractive or lift techniques or devices of some kind, including for instance the Persian wheel.

There were three main classes of such irrigation-related storage systems. The first comprised tank and pond irrigation systems. These in turn were of two types: *above surface* storage works, where a reservoir was created above the ground through a fairly long embankment. The corresponding structures were called *Keri, Eri, Cheruvu, Kalvai* and *Kunta* in South India and *Ahar* in Bihar (northern areas). The second category involved *below surface* storage works, which included dug ponds from which water was lifted by some means, manual or mechanical: *pokhar, talab, jhil, beel*, and *sagar*.

The second class of irrigation systems comprised land inundation systems: the land was flooded and saturated before cultivation and then drained off prior to planting (another term used is flood irrigation). These were primarily above surface types or referred to (in India) as submergence tanks. Important variations of these are the *khadin* and the *johad* in Rajasthan and the *bundhies* of Madhya Pradesh.

The *bundhies* were built generally in a series and therefore captured every possible drop of rainwater that fell. If there were a surplus, a waste weir was provided. There was generally a sluice at the deepest part of the storage reservoir. A *stambh* or pillar would indicate the location of the sluice. Sluices could be of different types: pipe sluices or sluices made of masonry for the larger tanks. (Some sluices open, as in the *khazans*, with the pressure of the incoming tides and discharge water automatically when the tide has fallen; these are made from wood.) The crop which was grown in the *bundhies* after the water was drained did not require any irrigation until harvest.

The third class comprised *in situ* techniques through which storage facilities were created to retain precipitation and ground water infiltration. The difference between class two and three was that in the former, cultivation followed drainage; in the latter it occurred simultaneously.

The sizes of these storage tanks varied and the tanks themselves were generally known from the command areas they irrigated: the smallest ones irrigated around fifty acres, the medium ones, about a hundred mined and shipped to dye factories and market towns. Minerals such as copper, tin, and iron also came to be used for mordanting, and thus added to the importance of mining for these minerals. Tree-borne dyes led to forest cultivation and above all, to extensive and destructive logging industries. The dye-yielding attributes of plants such as indigo stimulated the creation of plantation complexes.

The production of natural dyes and their use in dyeing textiles were often complicated skills requiring apprenticeships and years of training. Certain towns became known as dye centres or markets, such as ancient Tyre or medieval Venice in the Mediterranean, or Oaxaca in colonial Mexico. Guilds of dyers and castes devoted to weaving and dyeing were of considerable importance in European and Indian societies from medieval times.

Before European intrusion reached sub-Saharan Africa, Asia, Australia, and the Americas, there were several dyes which were produced and distributed over large areas. Perhaps the best known of the early root dyes was madder (from the Rubiaceae, of which there are over thirty species). *Rubia tinctorum* roots were used to obtain red dye in India, and were also known to the ancient Persians, Egyptians, Greeks, and Romans. For centuries Baghdad was the centre of the madder trade, and it was cultivated extensively – rather than gathered as elsewhere – in Mesopotamia. After its use spread to Western Europe, the Dutch became the most systematic and scientific madder growers, combining it with their leading role in cloth making (wool) and cloth importing (silks and linens). The French began to compete in the eighteenth century, but the French revolution damaged the industry and it never revived.

Before the European expansion, the most widely used dye made from plant leaves was woad (*Isatis tinctoria* of the Cruciferae family). It yielded various blues and greys. Although easy to produce in the temperate areas of Europe and Asia, it was not a very brilliant or fast dye. Woad in medieval and early modern Europe was manufactured and marketed by powerful guilds which, by monopolies, boycotts, and powerful legislation, managed to prevent largescale intrusions of other blues, especially indigos, from America and South-east Asia, until the seventeenth century.

The leading dye made from woad has always come from brazilwood (*Caesalpinia echinata*), although many other trees such as lima, sapan, and peachwood, all soluble redwoods, are often lumped together as brazilwoods. These medium-sized trees have been widely used for dyestuffs in many parts of the world. They are cut into small logs, ground to powder, then soaked and fermented, often with an aluminium or other metal ore mordant. They yield reds and browns, except on silks. Before the sixteenth century, India, Sumatra, and Ceylon (Sri Lanka) were the main producers. Other woods, such as camwood from India, which imparted a rough feeling to cloth because of its resins, were also traded. Cutch and the major ones five hundred and above. The large tanks were clearly impressive in scale. The Veeranam Tank in Tamilnadu has a bund (embankment) 16 km in length. The Gangaikonda Cholapuram Tank, also in Tamilnadu, was constructed by a Chola king from CE 1012 to 1044 and survives even today with a 25 km embankment.

The construction of tanks was a widely dispersed skill. To create a tank reservoir, an earthen embankment, usually curved, was erected in a concave form across the flow of water. The water was retained in the belly of the curve from where it was drawn and directed through channels to irrigate plots at lower levels through gravity irrigation. After the tank was emptied, the tank bed itself could be used temporarily for raising a crop utilising residual moisture. Many of the tanks in an area were interlinked and functioned as parts of an integrated system. The tank at the higher level released its surplus water as run-off to tanks at a lower level and the next in turn. These were called chain tanks. There were also tanks which were fed by canals from a river. Chain tanks were generally created in the upper reaches of the river basin and river-fed tanks in the lower reaches. Their construction would have required detailed cooperation among several communities in the region.

In Mysore in 1966 Major R.H. Sankey, Chief Engineer, wrote, "Of the 27,269 square miles covered by Mysore, nearly 60% has, by the patient industry of its inhabitants been brought under the tank system. Unless under exceptional circumstances, none of the drainage of these 16,287 square miles is allowed to escape. To such an extent the principle of storage has been followed, that it would now require some ingenuity to discover a site within this great area suitable for a new tank." The profusion of such tanks was not a feature of Karnataka alone. Experience was similar in Tamilnadu, Goa, or Bihar. The area north of the Vindhya mountain range in middle India for instance had more than eight thousand submergence tanks. In one district of Rajasthan alone, there were more than five hundred *khadins*.

In Sri Lanka, dry areas were populated with what are known as "tank villages". "The one-mile to an inch topographical maps of the island", writes D.L.O Mendis, "show nearly 15,000 of these, of which over 8000 are in working condition today. Tradition has it that some 30,000 of these small tanks had been constructed down the ages and there is a reference in the chronicles to 20,000 in the ancient province alone in the 12th century."

Water Conveyance

Apart from the storage works, the subcontinent witnessed the emergence of competent and impressive water conveyance systems designed to divert waters of rivers and flowing streams. Some diversions were accomplished without a check or embankment across the river; in such instances, the flood waters of the river were drained through a natural diversion. The *Kuhls* or *Guls* in the Himalayan areas, the *dongs* of the North-eastern states, and the *pynes* of Bihar all reflect this feature.

The second category involved check dams as a basic feature: the river bed was first raised and the resulting raised water diverted into a channel as was the case with the *bandharas*. Some of these schemes were fairly small, like those in the hilly areas. Others could be extremely large scale and it is reports of the latter that probably gave Wittfogel material for his speculations. According to Major T. Greenway, these were works of "truly gigantic magnitude, vast embankments and drainage channels equal to ordinary English rivers in capacity."

Historical Development of Irrigation Techniques

The interesting question that is now being asked is whether one can talk in terms of an evolution of irrigation technologies and civil engineering techniques from the earliest times to the present? The question is important in view of the fact that many of these storage systems, diversion channels, and embankments are largely intact and still in use. There are still parts of the country where Persian wheels are operated. Tanks and storage vessels are once again being made functional and weirs continue to be constructed.

The answer seems obvious: while more complex technological mechanisms did emerge as time passed, the earlier and the later techniques have continued to co-exist. The only major new innovation seems to be the idea of dams; these are new in terms of function, since the generation of hydropower was not intended in earlier times. The idea of large dams, once considered the temples of modern India, has taken a severe beating in recent years. They are now considered unsuited to tropical ecosystems, since the reservoirs invariably lead to displacement of large numbers of people, submergence of forests, and destruction of wildlife, and in places like Sri Lanka, submergence of smaller functioning reservoirs.

This being said, it is still possible to identify certain periods as distinct historical events in a possible history of irrigation and civil engineering on the subcontinent. There is archaeological evidence of artificial irrigation from pre-Harappan and Harappan times (ca. 5500–3500 BCE): this took the form of a large number of wells. One well found had a brick lining going down twelve metres. Post-Harappa, the major irrigation find is Inamgaon on the west coast of India where under the influence of the Jorwa culture (ca. 3400–3000 BCE), one finds evidence of a major diversion scheme reflected in a massive embankment 240 metres by 2.2 metres wide to divert the river into a channel. The channel itself was 200 metres long and 4 metres wide.

The first storage tanks in their rudimentary form appear around 2500 BCE—also in Sri Lanka—with the invention of iron tools. Hereafter, there are increasing references in literary works of both Sanskrit and Tamil up to the Gupta (CE 350) period. There are tank related inscriptions which give details of tank construction, maintenance, sources of funds for maintenance, and so on. The word *eri* also comes into circulation by the seventh century as a term for tanks.

The *anicut* (weir) technique is probably older than the *eri* or tank. The most famous of the *anicuts*, the Kaveri Anicut on the River Kaveri, is linked to a Chola king of the second century CE. It involved a dam on the River Kaveri 300 metres long and 12–18 metres wide and 5 metres deep. There is a dispute about the age of the *anicut*, since the *anicut* technique itself bears a strong resemblance to the Sri Lankan technique of massive stone dams and sluices, a technique which developed, according to Sri Lankan historian R.A.L.H. Gunawardene, only in the seventh century CE.

Sir Arthur Cotton paid eloquent testimony to the engineering talent involved in the large-scale irrigation works. He wrote: "There are a multitude of old native works in various parts of India... These are noble works, and show both boldness and engineering talent. They have stood for hundreds of years ... it was from them that we learnt how to secure a foundation in loose sand of unmeasured depth. In fact, what we learnt from them made the difference between financial success and failure, for the Madras river irrigations executed by our engineers have been from the first the greatest financial successes of any engineering works in the world, solely because we learnt from them ... With this lesson about foundations, we built bridges, weirs, aqueducts, and every kind of hydraulic work ... we are thus deeply indebted to the native engineers...".

SOCIAL ARRANGEMENTS/RELIGIOUS SANCTION

A significant feature of these irrigation works related to their construction and maintenance. Since water availability could be problematic with monsoon failure, those associated with the emergence of these works could gain religious merit for their deed. Though large systems were often sponsored by the state—to include kings, queens, local chieftains, *zamīndars* (landowners)—village communities, temples, and even individuals are associated with their construction. Thus a public park in Pondicherry bears an inscription recording a tank built by a *dāsī*—a temple dancer/courtesan, while another inscription in Karnataka (CE 1100) records a tank and shrine constructed by a village watchman.

All the major dynasties including the Mauryas (Sudarshan Lake near Kathiawar), the Cholas, the Hoysalas and the Vijayanagar kings and Muslim sultans were associated with irrigation works. Of these, the most impressive schemes are associated with the Cholas. However, these kings depended upon a cadre of skilled hydraulic engineers. Dikshit records the performance of one such engineer in the fourteenth century:

"The Kalludi (Gauribidanur taluk) inscription of 1388 CE is well known. According to it, when Vira Harihara Raya's son Sri Pratapa Bukkaraya was in Penugonda city in order that all the subjects might be in happiness—water being the life of the living beings—Bukkaraya in open court gave an order to the master of ten sciences, the hydraulic engineer (Jalasutra) Singayya Bhatta that he must bring the Henne (Pennar) river to Penugonda. Accordingly Singayya Bhatta conducted a channel to the Siruvara tank and gave the channel the name Pratapa Bukka Raya Mandalanda Kaluve".

The day-to-day operation and maintenance of both large and small works were mostly in the hands of local communities and of special professionals like the *nirkattis* of Tamilnadu. In many areas, produce from certain lands was set aside specifically to meet the maintenance costs of tanks. During the installation of the colonial regime, these revenues were appropriated by the colonial power and consequently the maintenance of such irrigation works fell into bad times leading to declines in efficiency.

Claude Alvares

References

Dharampal. *Indian Science and Technology in the 18th Century*. Delhi: Impex India. 1983.

Dikshit, G.S., G.R. Kuppuswamy, and S.K. Mohan. *Tank Irrigation in Karnataka: A Historical Survey*. Bangalore: Gandhi Sahitya Sangha. 1993.

Majumdar, R.C. and H.C. Raychaudhuri. *An Advanced History of India*. 3rd edition. London: Macmillan. 1967.

Mendis, D.L.O. "Lessons from Traditional Irrigation and Eco-systems." In *The Revenge of Athena: Science, Exploitation and The Third World*. Ed. Ziauddin Sardar. London and New York: Mansell. 1988. p. 317.

Needham, Joseph. *Science and Civilisation in China*. Cambridge: Cambridge University Press. 1954.

Sengupta, N. *User-Friendly Irrigation Designs*. New Delhi and Newbury Park: Sage Publications. 1993.

Shankari, U. and E. Shah. *Water Management Traditions in India*. Madras: PPST Foundation. 1993.

Somashekhar, R. *Forfeited Treasure: A Study on the Status of Irrigation Tanks in Karnataka*. Bangalore: Prarambha. 1991.

Wittfogel, Karl A. *Oriental Despotism: A Comparative Study of Total Power*. New Haven, Connecticut: Yale University Press. 1957.

J

Jagannātha Samrāṭ

Paṇḍita Jagannātha (1652–1744), who bore the title "Samrāṭ", writer on astronomy and mathematics, and designer of astronomical instruments, was the religious preceptor and collaborator in astronomical pursuits of Jai Singh Sawai (1688–1744), the astronomer-prince of Jaipur in Rajasthan. Born into a Vedic family, the son of Gaṇeśa, Jagannātha was attached to the court of Jai Singh from an early age and assisted his patron in all his social, religious, and scientific activities.

At Jai Singh's behest, Jagannātha mastered Arabic and Persian, the two foreign languages prevalent in the Mughal court, which he utilised in the study of Islamic astronomy and put to beneficial use translating into Sanskrit texts in those languages for his patron. In this way, Jagannātha produced his *Rekhāgaṇita* and *Siddhāntasārakaustubha*, which are translations of Euclid's *Elements of Geometry* and Ptolemy's *Almagest*, respectively, from their Arabic versions by Naṣīr al-Dīn al-Ṭūsī. It is interesting that in the case of *Rekhāgaṇita*, Jagannātha himself coined more than a hundred Sanskrit equivalents of technical terms.

In his original work *Siddhānta-samrāṭ*, composed at the behest of Jai Singh, Jagannātha described the construction and application of a number of astronomical instruments. He also mentioned the reasons that his patron, who was obsessed with metallic instruments like the astrolabe for making celestial observations and reading out the results, later opted for huge outdoor observatories with stone and mortar. The reason was that Jai Singh found that the metallic instruments did not give minute readings, and were also susceptible to wear and tear, and to climatic conditions, which brick observatories were not. His *Yantraprakāra* is a more elaborate work on the subject of astronomical instruments, which includes descriptions of some more instruments, computations, a number of tables, and allied data.

Jagannātha's *Siddhānta-samrāṭ* and *Yantraprakāra* carry a number of recordings of celestial observations of different types, for periods short or long, which demonstrate how the instruments designed, and observatories constructed, as above, had been put to use for correcting parameters, preparing almanacs and the like. The part played by Jagannātha in these endeavours was considerable and significant.

K. V. Sarma

References

Pingree, David. *Census of the Exact Sciences in Sanskrit*. Philadelphia: American Philosophical Society, 1981. Series A. vol. 3, pp.86–88; vol. 4, p. 95.

The *Rekhāgaṇita* or *Geometry in Sanskrit composed by Samrāṭ Jagannātha*. Ed. K.P. Trivedi. Bombay: Nirnaya Sagar Press. 1901.

Samrāṭ-siddhānta. Ed. Ram Swarup Sharma. New Delhi: Indian Institute of Astronomical and Sanskrit Research. 1967.

Sharma, M.L. "Jagannāth Samrāṭ's Outstanding Contribution to Indian Astronomy in Eighteenth Century AD." *Indian Journal of History of Science* 17(2): 244– 51. 1982.

Sharma, Virendra Nath. "Sawai Jai Singh's Hindu Astronomers." *Indian Journal of History of Science* 28(2): 131–55. 1993.

Siddhāntasamrāṭ of Jagannāthasamrāṭ. Ed. Muralidhar Chaturveda. Sagar: Samskrita Parishat, Saugar University. 1976.

Upadhyaya, B.L. *Prācīna Bhāratīya Gaṇita*. New Delhi: Vijnana Bharati. 1971.

Yantraprakāra of Sawai Jai Singh. Ed. and Trans. S.R. Sarma. New Delhi: Department of History of Medicine and Science, Jamia Hamdard. 1987.

See also: Jai Singh – Astronomical Instruments

Jai Singh

Jai Singh, or Jai Singh Sawai (Jaya Siṃha Savā'ī), the eighteenth century statesman-astronomer of India, was born on November 3, 1688 to the royal house of Amber, in the present state of Rajasthan, India. His ancestors were semi-autonomous rulers of their princely state under the Mughals and occupied important posts at the Mughal court. Jai Singh lived his life during one of the most troubled, uncertain, and critical periods of Indian history, and he was involved directly or indirectly in just about every political or military conflict of his time. With his diplomacy and political maneuverings, he acquired a great deal of authority and influence throughout the Mughal empire. Making full use of his prestige and power, Jai Singh embarked upon a program of reviving astronomy in his country.

Jai Singh displayed an early inclination towards astronomy and mathematics and soon acquired mastery over these two subjects. He realised that the astronomical predictions based on the Hindu, Islamic, or European books (which were available to him) did not agree with actual observations. He reasoned that the disagreements between predictions and observations were primarily due to the outdated parameters found in the astronomical books, and would not be alleviated until new parameters based on careful observations were made available. Consequently, he decided to obtain new parameters.

With the blessings of the reigning emperor, Muḥammad Shāh, Jai Singh initiated a multifaceted program in astronomy. He designed instruments, built observatories, compiled an excellent library, assembled competent astronomers of different scientific backgrounds, and sent a fact-finding scientific mission to Europe. His scientific career lasted for more than 20 years. He died in 1743, at the age of 54.

Jai Singh started out first with traditional instruments of brass built according to the designs given in the texts of the Islamic school of astronomy. However, the metal instruments did not measure up to his expectations; he discovered with disappointment that the instruments gave inaccurate results once their axes wore down, displacing their centres. The instruments were also unsteady during observing because of their portable nature. He discarded these instruments, therefore, in favour of the instruments of masonry and stone of his own design and tried to achieve the desired precision from their large sizes and steadiness from relatively inflexible structures. His instruments range anywhere from 1 m to 25 m in height.

Jai Singh built five observatories in cities of North India, at Delhi, Jaipur, Varanasi, Ujjain, and Mathura and equipped them with instruments of his own design. His observatory in Delhi was completed in 1724 and the others within a decade. His observatories, all except that of Mathura, are still extant in good to fair states of preservation. His observatory at Jaipur has the largest number of instruments and is in the best preserved state. The observatories of Delhi and Jaipur are big tourist attractions these days and visited by hundreds of thousands of people each year.

An inventory of Jai Singh's major instruments of stone and masonry is presented in Table 1.

Table 1 *Major instruments of masonry and stone of Jai Singh*

1	Jaya Prakāśa (hemispherical dial I)	2	Delhi, Jaipur
2	Samrāṭ yantra (equinoctial sundial)	6	Delhi, Jaipur, Ujjain, Varanasi
3	Rāma yantra (cylindrical dial)	2	Delhi, Jaipur
4	Rāśi valaya (ecliptic dial)	12	Jaipur
5	Ṣaṣṭhāṃśa yantra (sixty-degree instrument)	5	Delhi, Jaipur
6	Dakṣiṇottara Bhitti (meridian dial)	6	Delhi, Jaipur, Ujjain, Varanasi, Mathura
7	Digaṃśa yantra (azimuth circles)	3	Jaipur, Ujjain, Varanasi
8	Nāḍāvalaya (equinoctial dial)	5	Jaipur, Ujjain, Varanasi, Mathura
9	Kapāla A (hemispherical dial II)	1	Jaipur
10	Kapāla B (hemispherical dial III)	1	Jaipur

The Samrāṭ, Ṣaṣṭhāṃśa and Dakṣiṇottara Bhitti are Jai Singh's high precision instruments. With these instruments, he extended precision to the very limit of naked eye observing, i.e. 1′of arc.

Although the telescope had become common with European astronomers and had acquired refinements such as the micrometer and cross-hair, there is no evidence that Jai Singh benefited from it. His instruments do not use a telescopic sight, and with all their ingenuity of concept and design are no more than what may be called "naked eye tools" somewhat in the tradition of the medieval astronomers such as Ulugh Beg of Samarkand. It is reasonable to believe that the invention of the telescopic sight which had come into vogue with European astronomers only a few decades earlier did not come to his attention early enough. It should be pointed out, however, that Jai Singh was familiar with the telescope and had observed with it. His personal library inventory lists a telescope bought for him at a cost of 100 rupees.

Jai Singh's early training as an astronomer had been under Hindu *pundits*, and they remained the mainstay of his program until the very end. At one time there were at least twenty-two astronomers working at the observatory of Jaipur alone. Jagannātha Samrāṭ, Kevalarāma, and Nayanasukhopādhyāya were his principal astronomers. These astronomers constructed instruments, collected data, translated books and compiled original works in astronomy. The translated works included

Ptolemy's *Almagest*, Euclid's *Elements*, and De La Hire's *Tabulae Astronomicae*.

Figure 1 *Sawai Jai Singh (1688–1743). Courtesy of the Sawai Man Singh II Museum, Jaipur (used with permission of the author).*

Jai Singh was equally interested in the Islamic and the European traditions of astronomy. He collected astronomical works in Persian and Arabic and patronised Muslim astronomers of the Persian-Arabic school. The Muslim astronomers included Muḥammad Ābid, Sheikh Asad Ullah, Sheikh Muḥammad Shafī, and Dayānat Khān. These astronomers procured astronomical books for the royal library, constructed instruments, helped with the translations, collected data at the observatories of Delhi and Jaipur, and travelled to distant lands at the command of their patron.

By 1725, the involvement of the Muslim *nujūmūs* or astronomers in Jai Singh's astronomical program began to taper off and, in its place, the involvement of Europeans, primarily Jesuit priests, increased. The European astronomers included De Bois, Figuerado, Boudier, Gabelsberger, and Strobl. The Europeans played the role of conveyors of European knowledge to the Raja. Accordingly, they led a delegation to Europe, procured texts and instruments, translated De La Hire's tables, and carried out mathematical computations. However, the knowledge these Europeans brought to Jai Singh and his astronomers had already become outdated in Europe, for it did not include the theories of Galileo, Kepler, or Newton; nor did it include observational techniques such as those employed by Flamsteed in England.

In 1727, Jai Singh dispatched a scientific delegation to Europe after learning that "the business of the observatory was being carried out there." The delegation, first of its kind from the East, was led by Figuerado, and it reached Portugal in January 1729. The delegation stayed on in Portugal for over a year. It did not travel to Paris or London, however, where the most advanced work in astronomy was being done. In 1730, the delegation returned to Jaipur, the capital of Jai Singh'state at the time, with some instruments, books on mathematics, and the tables of De La Hire. The delegation did not bring any books elaborating the heliocentric world view, such as proposed by Newton, Kepler or Copernicus, since these publications were prohibited by the Catholic Church.

After collecting data for nearly a decade, Jai Singh succeeded in obtaining new astronomical parameters. With these parameters, he prepared a set of astronomical tables called a *Zīj*. The *Zīj*, completed sometime between 1731 and 1732, was dedicated to the reigning monarch, Muḥammad Shāh and is, therefore, called *Zīj-i Muḥammad Shāhī. Zīj-i Muḥammad Shāhī* is a 400-page long traditional work of astronomy similar to the *Zīj-i Sulṭānī* of Ulugh Beg. *Zīj-i Muḥammad Shāhī* may be considered Jai Singh's most important contribution to the astronomy of India. The *Zīj* remained a valuable resource for traditional astronomers of the country for nearly 150 years.

For the sake of rejuvenating astronomy in his country, Jai Singh expended a great deal of energy as well as his personal fortune, but he failed to initiate the new age of astronomy in India. He himself remained unaware of the Copernican revolution that had swept the intellectual circles of Europe. Lack of good communication systems, and a complex interaction of intellectual stagnation, religious taboos, theological beliefs, national rivalries, and the simple human failings of his associates share the blame for it. Jai Singh's scientific accomplishments were medieval in retrospect, but his scientific outlook was quite modern.

Virendranath Sharma

References

Bhatnagar. V.A. *Life and Times of Sawai Jai Singh* . Delhi: Impex India. 1974.

Garrett, A. ff. and Guleri, Chandradhar. *The Jaipur Observatory and its Builder*. Allahabad: Pioneer Press. 1902.

Kaye, G.R. *The Astronomical Observatories of Jai Singh*. New Delhi: Archaeological Survey of India, reprint. 1982.

Sarma, Sreeramula Rajeswara. "Yantraprakāra of Sawai Jai Singh." Supplement to *Studies in History of Medicine and Science*, Vols. X and XI, New Delhi, 1986 and 1987.

Sharma, Virendra Nath. *Sawai Jai Singh and His Astronomy*. New Delhi: Motilal Banarasidass, 1995.

See also: Jagannātha Samrāṭ – *Zīj* – Observatories – Astronomical Instruments in India

Jayadeva

Ācārya Jayadeva is an early Indian mathematician, known only through a long aphoristic (*Sūtra*) quotation in twenty verses from an unknown work of his. In this passage, Jayadeva sets out, step by step, an ingenious method for solving the indeterminate equation of the form $Nx^2 \pm C = y^2$. This quotation was extracted by Udayadivākara, an astronomer of Kerala, in his commentary called *Sundarī* on the *Laghu-bhāskarīya* of Bhāskara I (b. 629). Udayadivākara flourished about CE 1073. This means that Jayadeva lived before that date.

The above extract was made in the context of an astronomical problem involving two simultaneous equations: (1) $7y^2+1 = z^2$, and (2) $8x+1 = y^2$. Here, Udayadivākara states that the value of y in the first equation can be found by an ingenious method called *varga-prakṛti* (lit. "square-nature") enunciated by Ācārya Jayadeva, and the value of x in the second equation by the method of inversion.

The extract from Jayadeva forms an account of *varga-prakṛti* as he conceives it and solves problems through it. First he defines the term: "When (in an equation of the type $Ax^2 \pm C = y^2$) the square of an optional number is multiplied by a given number and then the product is increased or decreased by another number, and the result is in the nature of a square, such an equation is called *varga-prakṛti*." He then goes on to explain the technical terms which would be used in the course of his exposition, such as *Kaniṣṭha-mūlam* (lesser root), *jyeṣṭha-mūlam* (greater root), *kṣepa* (interpolator), *bhāvanā* (visualisation), and its two forms, *Samāsa-bhāvanā*, and *viśeṣa-bhāvanā* or *tulya-bhāvanā* and *atultya-bhāvanā*, all of which form the step by step processes for the solution of the indeterminate equation envisaged. Working through these processes is termed the *cakravāla* (cyclic method) through which any number of solutions can be found. The actual method of solving the equation is given in the last five verses, towards the close of which Jayadeva quips, "Thus have we identified a very ingenious method for solving the problem which is as difficult as it is for a flea to fly against the wind". Jayadeva is perhaps justified in making such a comparison, for his cyclic method was set out later by other authors like Bhāskara II (b. 1114) and Nārāyaṇa (1356). The historian of mathematics Hermann Hankel has remarked: "It is above all praise; it is certainly the finest thing which was achieved in the theory of numbers before Lagrange."

K. V. Sarma

References

Hankel, H. *Zur Geschichte der Mathematik in Alterum und Mittelalter*. Leipzig: Teubner. 1874, pp. 203–04.

Selenius, Clas-Olaf. "Rationale of the Chakravāla Process of Jayadeva and Bhāskara II." *Historia Mathematica* 2 : 167–84. 1975.

Shukla, Kripa Shankar. "Ācārya Jayadeva, The Mathematician." *Gaṇita* 5: 1–20 1954.

Udayadivākara's Commentary, Sundarī, on the Laghu-bhāskarīya of *Bhāskara I*. Mss. available in the Kerala University Oriental Research Institute and Manuscripts Library, Trivandrum, and a modern copy in the Tagore Library, Lucknow University.

K

Kamalākara

Kamalākara was one of the most erudite and forward-looking Indian astronomers who flourished in Varanasi during the seventeenth century. Belonging to Maharashtrian stock, and born in about 1610, Kamalākara came from a long unbroken line of astronomers, originally settled at the village of Godā on the northern banks of the river Godāvarī. Towards CE 1500, the family migrated to Varanasi and came to be regarded as reputed astronomers and astrologers. Kamalākara studied traditional Hindu astronomy under his elder brother Divākara, but extended the range of his studies to Islamic astronomy, particularly to the school of Ulugh Beg of Samarkand. He also studied Greek astronomy in Arabic and Persian translations, particularly with reference to the elements of physics from Aristotle, geometry from Euclid, and astronomy from Ptolemy. He wrote both original treatises and commentaries on his own works and those of others.

Kamalākara's most important work is the *Siddhānta-Tattvaviveka*, written in CE 1658. The work which is divided into fifteen chapters and contains over 3000 verses, faithfully follows the *Sūryasiddhānta* in the matter of parameters, general theories, and astronomical computation. However, in certain matters Kamalākara made original contributions and offered new ideas. Though he accepted the planetary parameters of *Sūryasiddhānta*, he agreed with Ptolemaic notions in the matter of the planetary system. He presented geometrical optics, and was perhaps the only traditional author to do so. He described the quadrant and its application. He proposed a new Prime Meridian, which is the longitude passing through an imaginary city called Khalādātta, and provided a table of latitudes and longitudes for twenty important cities, in and outside India, on this basis. Kamalākara was an ardent advocate of the precession of the equinoxes and argued that the pole star also does not remain fixed, on account of precession. Kamalākara wrote two other works related to the *Siddhānta-Tattvaviveka*, one a regular commentary on the work, called *Tattvavivekodāharaṇa*, and the other a supplement to that work, called *Śeṣāvasanā*, in which he supplied elucidations and new material for a proper understanding of his main work. He held the *Sūryasiddhānta* in great esteem and also wrote a commentary on that work.

Kamalākara was a critic of Bhāskara and his *Siddhāntaśiromaṇi*, and an arch-rival of Munīśvara, a close follower of Bhāskara. This rivalry erupted into bitter critiques on the astronomical front. Thus Ranganātha, younger brother of Kamalākara, wrote, at the insistence of the latter, a critique on Munīśvara's *Bhaṇgī* method (Winding method) of true planets, entitled *Bhaṅgī-vibhaṇgī* (Defacement of the *Bhaṇgi*), to which Munīśvara replied with a *Khaṇḍana* (Counter). Munīśvara attacked the theory of precession advocated by Kamalākara, and Ranganātha refuted the criticisms of his brother in his *Loha-gola-khaṇḍana* (Counter to the Iron Sphere). That in turn was refuted by Munīśvara's cousin Gadādhara in his *Loha-gola-samarthana* (Justification of the Iron Sphere). These kinds of astronomical and intellectual battles were typical of the philosophical and religious disputes which were common in ancient India.

K. V. Sarma

References

Dikshit, S.B. *Bhāratya Jyotish Śāstra (History of Indian Astronomy)*. Trans. by R. V. Vaidya. Pt. II. *History of Astronomy during the Scientific and Modern Periods*. Calcutta: Positional Astronomy Centre India. Meteorological Department. 1981.

Dvivedi, Sudhakara. *Gaṇaka Taraṅgiṇi: Lives of Hindu Astronomers.* Benares: Jyotish Prakash Press. 1933.

Loha-gola-khaṇḍana of Ranganātha and Loha-gola-samarthana by Gadādhara. Ed. Mithalala Himmatarama Ojha. Varanasi: Sañcālaka, Anusandhāna Saṃsthāna. 1963.

Pingree, David. *Jyotiḥśāstra—Astral and Mathematical Literature*. vol. VI, fasc. 4 of *A History of Indian Literature.* Ed. Jan Gonda. Wiesbaden: Otto Harrassowitz. 1981.

Siddhānta-Tattvaviveka. A Treatise on Astronomy by Bhaṭṭa Kamalākara, with *Śeṣavāsanā* by the same author. Ed. Sudhakara Dube. Benares: Benares Sanskrit Series, 5 vols., 1880–85 : Revised by Muralidhara Jha, Benares: Krishna Das Gupta for Braj Bhusan Das & Co, 1924–35.

See also: Astronomy – *Sūryasiddhāntha* – Precession of the Equinoxes – Bhāskara – Munīśvara

Knowledge System: Local Knowledge

The concept of local knowledge has recently come to the fore in the field of the sociology of scientific knowledge, where it is a common empirical finding that knowledge production is an essentially local process. Knowledge claims are not adjudicated by absolute standards; rather their authority is established through the workings of *local* negotiations and judgments in particular contexts. This focus on the localness of knowledge production provides the condition for the possibility for a fully-fledged comparison between the ways in which understandings of the natural world have been produced by different cultures and at different times. Such cross-cultural comparisons of knowledge production systems have hitherto been largely absent from the sociology of science. A necessary condition for fully equitable comparisons is that Western contemporary technosciences, rather than being taken as definitional of knowledge, rationality, or objectivity, should be treated as varieties of such knowledge systems. Though knowledge systems may differ in their epistemologies, methodologies, logics, cognitive structures, or in their socio-economic contexts, a characteristic that they all share is their localness. Hence, in so far as they are collective bodies of knowledge, many of their small but significant differences lie in the work involved in creating assemblages from the "motley" of differing practices, instrumentation, theories, and people (Hacking, 1992). Much of that work can be seen as strategies and techniques for creating the equivalences and connections whereby otherwise heterogeneous and isolated knowledges are enabled to move in space and time from the local site and moment of their production and application to other places and times.

In this view, all knowledge systems from whatever culture or time, including the contemporary technosciences, are based on local knowledge. However within the master narrative of modernism, local knowledge is an oxymoron. Exploring this contradiction and the manifold meanings of local requires a brief excursion into postmodernism as well as some of the arguments underpinning the sociology of scientific knowledge.

Though postmodernism eludes definition and is more likely a stage of modernism than a marked epistemological break, there has been a recent coalescence of strands of thought in a wide variety of areas that have questioned the assumptions underlying modernism. Postmodernism is most frequently equated with the collapse of the concepts of rationality and progress held to accompany the emergence of the post-industrial society and is consequently concerned with the rejection of universal explanations and totalising theories. But perhaps the strand that is most truly pervasive in the constellation of reformulated approaches to understanding the human condition is the emphasis on the local.

In physics, ecology, history, feminist theory, literary theory, anthropology, geography, economics, politics, and sociology of science, the focus of attention has become the specific, the contingent, the particular. This is the case whether it is a text, a reading, a culture, a population, a site, a region, an electron, or a laboratory. Within this diversity of uses of local there seem to be two broad and rather different senses being used. On the one hand there is the notion of a voice or a reading. The voice may be purely individual and subjective or may be a collection of voices belonging to a group, class, gender, or culture, but in all cases the notion captures one of the basic characteristic elements of postmodernism, courtesy of deconstruction, that all texts or cultures are multivocal and polysemous. That is they have a multiplicity of meanings, readings and voices and are hence subject to "interpretive flexibility" (Collins, 1985). On the other hand, local is used both in the more explicitly geopolitical sense of place and in the experiential sense of contextual, embodied, partial, or individual. A range of disciplines from meteorology to medicine now recognise the necessity of focusing on the particular conditions at specific sites and times rather than losing that specificity in unlocalised generalisations.

The sociology of scientific knowledge is one of the most classically post of all modernisms and is therefore an area in which the local is also a thematic presence which is only now coming into focus. Some philosophers of science have come to re-evaluate the role of theory and argue that scientists practicing in the real world do not deduce their explanations from universal laws but rather make do with rules of thumb derived from the way the phenomena present themselves in the operation of instruments and devices. Similarly philosophers and sociologists of science alike have recognised for some time the lack of absolute standards and the role of tacit knowledge in technoscientific practice, and have sought to display the context in which the practice of science is manifested as craft skills and collective work. However the recognition of the social and material embodiment of skills and work in the cultural practice of individuals and groups has only recently coalesced into the general claim that all knowledge is local. Knowledge, from this constructivist perspective can be local in a range of different senses. "It is knowledge produced and reproduced in *mutual interaction* that relies on the *presence* of other human beings on a direct, face-to-face basis. "(Thrift, 1985). It is knowledge that is produced in contingent, site, discipline or culture-specific circumstances (Rouse, 1987). It is the product of open systems with heterogeneous and asynchronous inputs "that stand in no necessary relationship to one another" (Pickering, 1992). In sum scientific knowledge is "situated knowledge" (Haraway, 1991).

Perhaps the most important consequence of the recognition of the localness of scientific knowledge is that it permits parity in the comparison of the production of contemporary technoscientific knowledge with knowledge production in other cultures. Previously the possibility of a truly equitable comparison was negated by the assumption that indigenous knowledge systems were merely local and were to be evaluated for the extent to which they had scientific characteristics. Localness essentially subsumes many of the supposed limitations of other knowledge systems compared with western science. So-called traditional knowledge systems have frequently been portrayed as closed, pragmatic, utilitarian, value laden, indexical, context dependent, and so on. All of which was held to imply that they cannot have the same authority and credibility as science because their localness restricts them to the social and cultural circumstances of their production. Science by contrast was held to be universal, non-indexical and value free, and as a consequence floating, in some mysterious way, above culture. Treating science as local simultaneously puts all knowledge systems on a par and renders vacuous discussion of their degree of fit with transcendental criteria of scientificity, rationality, and logicality. Now the multidisciplinary approaches to understanding the technosciences which together constitute the sociology of scientific knowledge can be made more fully anthropological by the addition of a new sub-discipline called comparative technoscientific traditions.

Emphasising the local in this way necessitates a re-evaluation of the role of theory which is typically held by philosophers and physicists to provide the main dynamic and rationale of science as well as being the source of its universality. Karl Popper claims that all science is cosmology and Gerald Holton sees physics as "a quest for the Holy Grail", which is no less than the "mastery of the whole world of experience, by subsuming it under one unified theoretical structure" (Allport, 1991). It is this claim to be able to produce universal theory that Western culture has used simultaneously to promote and reinforce its own stability and to justify the dispossession of other peoples. It constitutes part of the ideological justification of scientific objectivity, the "god-trick" as Donna Haraway calls it; the illusion that there can be a positionless vision of everything. The allegiance to mimesis has been severely undermined by analysts like Richard Rorty, but theory has also been found wanting at the level of practice, where analytical and empirical studies have shown it cannot provide the sole guide to experimental research and on occasion has little or no role at all. The conception of grand unified theories guiding research is also incompatible with a key finding in the sociology of science: "consensus is not necessary for cooperation or for the successful conduct of work". This sociological perspective is succinctly captured in Leigh Star's description:

> Scientific theory building is deeply heterogeneous: different viewpoints are constantly being adduced and reconciled... Each actor, site, or node of a scientific community has a viewpoint, a partial truth consisting of local beliefs, local practices, local constants, and resources, none of which are fully verifiable across all sites. The aggregation of all viewpoints is the source of the robustness of science (Star, 1989, 46).

Theories from this perspective have the characteristics of what Star calls "boundary objects"; that is they are "objects which are both plastic enough to adapt to local needs and constraints of the several parties employing them, yet robust enough to maintain a common identity across sites". Thus theorising is itself an assemblage of heterogeneous local practices.

If knowledge is local we are faced with a problem: how are the universality and connectedness that typify technoscientific knowledges achieved? Given all these discrete knowledge/practices, imbued with their concrete specificities, how can they be assembled into fields or knowledge systems; or in Star's terms "how is the robustness of findings and decision making achieved?" Ophir and Shapin ask, "How is it, if knowledge is indeed local, that certain forms of it appear global in domain of application?" The answers, considered here, lie in a variety of social strategies and technical devices that provide for treating instances of knowledge/practice as similar or equivalent and for making connections, that is in enabling local knowledge/practices to move and to be assembled.

Research fields or bodies of technoscientific knowledge/practice are assemblages whose otherwise disparate elements are rendered equivalent, general, and cohesive through processes that have been called "heterogeneous engineering" (Law, 1987). Among the many social strategies that enable the possibility of equivalence are processes of standardisation and collective work to produce agreements about what counts as an appropriate form of ordering, what counts as evidence, etc. Technical devices that provide for connections and mobility are also essential. Such devices may be material or conceptual and may include maps, calendars, theories, books, lists, and systems of recursion, but their common function is to enable otherwise incommensurable and isolated knowledges to move in space and time from the local site and moment of their production to other places and times.

Some of these devices have been revealed relatively un-problematically through direct observation. Others are less susceptible to investigation and analysis, being embodied in our forms of life. One way to catch a glimpse of these hidden presuppositions and taken for granted ways of thinking, seeing, and acting, is to misperceive, to be jolted out of our habitual modes of understanding through allowing a process of interrogation between our knowledge system and others. Such an interrogative process of mutual inter-translation can enable us to catch sight of the cultural glasses we wear instead of looking through them as if they were transparent.

This challenging of the totalising discourses of science by other knowledge systems is what Foucault had in mind when he claimed that we are "witnessing an insurrection of subjugated knowledges" and corresponds to an emphasis on the local that has emerged in anthropology at least since Clifford Geertz's *Interpretation of Cultures*. In his critique of global theories and in his emphasis on "thick description" Geertz pointed out that cultural meanings cannot be understood at the general level because they result from complex organisations of signs in a particular local context and that the way to reveal the structures of power attached to the global discourse is to set the local knowledge in contrast with it.

Equally there is the pervasive recognition characterised as postcolonialism that the West has structured the intellectual agenda and has hidden its own presuppositions from view through the construction of the other. Nowhere is this more acute than in the assumption of science as a foil against which all other knowledge should be contrasted. In the view of Marcus and Fischer we are at an experimental moment where totalising styles of knowledge have been suspended "in favour of a close consideration of such issues as contextuality, the meaning of social life to those who enact it and the explanation of exceptions and indeterminants". In this emphasis on the local we are postparadigm.

However we should not be too easily seduced by the apparently liberating effects of celebrating the local since it is all too easy to allow the local to become a "new kind of globalizing imperative" (Hayles, 1990). In order for all knowledge systems to have a voice and in order to allow for the possibility of inter-cultural comparison and critique, we have to be able to maintain the local and the global in dialectical opposition to one another. This dilemma is the most profound difficulty facing liberal democracies now that they have lost the convenient foil of communism and the world has Balkanised into special interest groups by genders, race, nationality, or whatever. By moving into a comparatist mode there is a grave danger of the subsumption of the other into the hegemony of western rationality, but conversely unbridled cultural relativism can only lead to the proliferation of ghettos and dogmatic nationalisms. We cannot abandon the strength of generalisations and theories, particularly their capacity for making connections and for providing the possibility of criticism. At the same time we need to recognise reflexively that theory and practice are not distinct. Theorising is also a local practice. If we do not recognise this joint dialectic of theory and practice, the local and the global, we will not be able to understand and establish the conditions for the possibility of directing the circulation and structure of power in knowledge systems. It is in the light of this recognition that I want to consider the ways in which the movement of local knowledge is accomplished in different knowledge systems and their consequent effects on the ways in which people and objects are constituted and linked together; that is their effects on power. The essential strength of the sociology of scientific knowledge is its claim to show that what we accept as science and technology could be other than it is. The great weakness of the sociology of scientific knowledge is the general failure to grasp the political nature of the enterprise and to work towards change. With some exceptions it has had a quietist tendency to adopt the neutral analyst's stance that it devotes so much time to criticising in scientists. One way of capitalising on the sociology of science's strength and avoiding the reflexive dilemma is to devise ways in which alternative knowledge systems can be made to interrogate each other.

Considerable advances in understanding the movement of local knowledge have been made possible through Bruno Latour's insightful analysis. For Latour the most successful devices in the agonistic struggle are those which are mobile and also "immutable, presentable, readable and combinable with one another". These immutable mobiles are the kinds of texts and images that the printing press and distant point perspective have made possible. These small and unexpected differences in the technology of representation are on his account the causes of the large and powerful effects of science. That which was previously completely indexical, having meaning only in the context of the site of production, and having no means of moving beyond that site, is standardised and made commensurable and re-presentable within the common framework provided by distant point perspective. Hence that which has been observed, created, or recorded at one site can be moved without distortion to another site. At centres of calculation such mobile representations can be accumulated, analysed, and iterated in a cascade of subsequent calculations and analyses.

Latour's account has been augmented by the work of Steven Shapin and Simon Schaffer in the *Leviathan and The Air Pump* (1985). They have shown that experimental practice in science is sustained by a range of social, literary, and technical devices and spaces that we take for granted but which had to be created deliberately to overcome the fundamentally local and hence defensible character of experimentally derived knowledge claims. In the seventeenth century, the problem for Robert Boyle, one of the earliest experimentalists, was to counter the arguments of his opponent Thomas Hobbes about the grounding of true and certain knowledge which they both agreed was essential in a country riven by dissent and conflicting opinion. Reliable knowledge of the world, for Hobbes, was to be derived from self-evident first principles, and anything that was produced experimentally was inevitably doomed to reflect its artifactual nature and the contingencies of its production; its localness would deny it the status of fact or law. Boyle recognised the cogency of these arguments and set out to create the forms of life within which the knowledge created at one site could be relayed to and replicated at other sites. In order for an empirical fact to be accepted as such it had to be witnessable by all, but the very nature of an experimental laboratory restricted the audience of witnesses to a very few. Boyle, therefore, had to create the technology of what Shapin calls virtual witnessing. For this to be possible three general sorts of devices or technologies had to be developed. Socially groups of reliable witnesses had to be formed. Naturally in the seventeenth century they were gentlemen. These gentlemen witnesses had to be able to communicate their observations to other groups of gentlemen so that they too might witness the phenomena. This required the establishment of journals using clear and unadorned prose that could carry the immutable mobiles, experimental accounts, and diagrams. The apparatus had to be made technically reliable and reproducible, but perhaps most importantly the physical space for such empirical knowledge had to be created.

Hobbes, of course, was right: experimental knowledge is artifactual. It is the product of human labour, of craft and skill, and necessarily reflects the contingencies of the circumstances. It is because craft or tacit knowledge is such a fundamental component of knowledge production that accounts of its generation, transmission, acceptance, and application cannot be given solely in terms of texts and inscriptions. A vital component of local knowledge is moved by people in their heads and hands. Harry Collins, a sociologist of science, has argued that this ineradicable craft component in science is ultimately what makes science a social practice. Because knowledge claims about the world are based on the skilled performance of experiments their acceptance is a judgment of competence not of truth. An example of the centrality of craft skill is the TEA laser, invented in Canada by Bob Harrison in the late sixties, which British scientists attempted to replicate in the early seventies. "No scientist succeeded in building a laser using only information found in published or other written sources" and furthermore the people who did succeed in building one were only able to do so after extended personal contact with somebody who had himself built one. Now TEA lasers are blackboxed and their production is routine and algorithmic. But in order to become routinised, Harrison's local knowledge had to be moved literally by hand.

Joseph Rouse (1987, 72) in considering the contemporary production process of scientific knowledge has summarised the implications of this understanding of science:

> Science is first and foremost knowing one's way about in the laboratory (or clinic, field site, etc.). Such knowledge is of course transferable outside the laboratory site into a variety of other situations. But the transfer is not to be understood in terms of the instantiation of universally applied knowledge claims in different particular settings by applying bridge principles and plugging in particular local values for theoretical variables. It must be understood in terms of the adaptation of one local knowledge to create another. We go from one local knowledge to another rather than from universal theories to their particular instantiations.

According to the historian of science Thomas Kuhn the way a scientist learns to solve problems is not by applying theory deductively but by learning to apply theory through recognising situations as similar. Hence theories are models or tools whose application results from situations being conceived as or actually being made equivalent. This point is implicit in the recognition that knowledge produced in a laboratory does not simply reflect nature because nature as such is seldom available in a form that can be considered directly in the lab. Specially simplified and purified artefacts are the typical subjects of instrumental analysis in scientific laboratories. For the results of such an artificial process to have any efficacy in the world beyond the lab, the world itself has to be modified to conform to the rigors of science. A wide variety of institutional structures have to be put in place to achieve the equivalences needed between the microworld created inside the lab and the macroworld outside in order for the knowledge to be transmittable. The largest and most expensive example of this is the Bureau of Standards, a massive bureaucracy costing six times the R&D budget. Without such social institutions the results of scientific research are mere artefacts. They gain their truth, efficacy, and accuracy not through a passive mirroring of reality but through an active social process that brings our understandings and reality into conformity with each other.

The result of the work of Latour, Collins, Shapin, Star, Hacking, Rouse, and others has been to show that the kind of knowledge system we call Western science depends on a variety of social, technical, and literary devices and strategies for moving and applying local knowledge. It is having the capacity for movement that enables local knowledge to constitute part of a knowledge system. This mobility requires devices and strategies that enable connectivity and equivalence, that is the linking of disparate or new knowledge and the rendering of knowledge and context sufficiently similar as to make the knowledge applicable. Connectivity and equivalence are prerequisites of a knowledge system but they are not characteristics of knowledge itself. They are produced by collective work and are facilitated by technical devices and social strategies. Differing devices and strategies produce differing assemblages and are the source of the differences in power between knowledge systems.

In conclusion, it has been argued that Western science, like all knowledge in all societies, is inherently local, and furthermore other non-Western societies have developed a variety of social and technical devices for coping with that localness and enabling it to move. Some of them are technical devices of representation like the mason's templates, and the Incan *ceques* and *quipus*. Some of them are abstract cognitive constructs, like the Anasazi and Incan calendars, and the Micronesian navigation system. All of them also require social organisation, rituals, and ceremonies. All of them have proved capable of producing complex bodies of knowledge and in many cases have been accompanied by substantial transformations of the environment. The major difference between Western science and other knowledge systems lies in the question of power. Western science has succeeded in transforming the world and our lives in ways that no other system has. The source of the power of science on this account lies not in the nature of scientific knowledge but in its greater ability to move and apply the knowledge it produces beyond the site of its production. However at the end of the twentieth century we can now perceive that there is a high cost to pay for science's hegemony. Much of that cost in terms of environmental degradation and ethnocide is due not so much to the totalising nature of scientific theories but to the social strategies and technical devices science has developed in eliminating the local.

The task of resisting and criticising science may now be addressed by reconsidering the causes of its dominating effects. Without the kinds of connections and patterns that theories make possible we will never be able to perceive the interconnectedness of all things. Without the awareness of local differences we will lose the diversity and particularity of the things themselves. Thus, rather than rejecting universalising explanations what is needed is a new understanding of the dialectical tension between the local and the global. We need to focus on the ways in which science creates and solves problems through its treatment of the local. Science gains its truthlike character through suppressing or denying the circumstances of its production and through the social mechanisms for the transmission and authorisation of the knowledge by the scientific community. Both of these devices have the effect of rendering scientific knowledge autonomous, above culture, and hence beyond criticism. Equally problematic is the establishment of the standardisation and equivalences required in order that the knowledge produced in the lab works in the world. The joint processes of making the world fit the knowledge instead of the other way round and immunising scientific knowledge from criticism are best resisted by developing forms of understanding in which the local, the particular, the specific, and the individual are not homogenised but are enabled to talk back.

David Turnbull

References

Allport, P. "Still Searching for the Holy Grail." *New Scientist* 132: 51–52. Oct 5, 1991.

Collins, Harry. *Changing Order: Replication and Induction in Scientific Practice*. London: Sage. 1985.

Foucault, Michel. *Power/Knowledge: Selected Interviews and Other Writings 1972–77*. New York: Pantheon Books. 1980.

Geertz, Clifford. *The Interpretation of Cultures: Selected Essays*. New York: Basic Books, 1973.

Hacking, lan. "The Self-Vindication of the Laboratory Sciences." In *Science as Practice and Culture*. Ed. Andrew Pickering. Chicago: University of Chicago Press. 1992. pp. 29–64.

Haraway, Donna. *Symians, Cyborgs and Women: The Reinvention of Nature*. London: Free Association Books. 1991.

Hayles, Katherine. *Chaos Bound: Orderly Disorder in Contemporary Literature and Science*. Ithaca: Cornell University Press. 1990.

Latour, Bruno. "Visualisation and Cognition: Thinking With Eyes and Hands." *Knowledge and Society* 6: 1–40. 1986.

Law, John. "Technology and Heterogeneous Engineering: The Case of Portuguese Expansion." In *The Social Construction of Technological Systems: New Directions in the Sociology and History of Technology*. Ed. W. Bijker, T. Hughes, and T. Pinch. Cambridge, Massachusetts: MIT Press. 1987. pp. 111–34.

Marcus, G.E. and M.M.J Fischer. *Anthropology as Cultural Critique: An Experimental Moment in the Human Sciences*. Chicago: University of Chicago Press. 1986.

Ophir, Adi and Steven Shapin. "The Place of Knowledge: A Methodological Survey." *Science In Context* 4: 3–21, 1991.

Pickering, Andrew, ed. *Science as Practice and Culture*. Chicago: University of Chicago Press. 1992.

Rouse, Joseph. *Knowledge and Power: Towards a Political Philosophy of Science*. Ithaca: Cornell University Press. 1987.

Shapin, Steven and Simon Schaffer. *Leviathan and the Air Pump: Hobbes, Boyle and the Experimental Life*. Princeton, New Jersey: Princeton University Press. 1985.

Star, Susan Leigh. "The Structure of Ill-Structured Solutions: Boundary Objects and Heterogeneous Distributed Problem Solving." In *Distributed Artificial Intelligence*. Ed. L. Gasser and N. Huhns. New York: Morgan Kauffman Publications. 1989. pp. 37–54.

Thrift, Nigel. "Flies and Germs: a Geography of Knowledge." In *Social Relations and Spatial Structures*. Ed. Derek Gregory and John Urry. London: Macmillan. 1985. pp. 366–403.

Knowledge Systems

Traditionally all knowledge in India has been traced to the Vedas. The Vedas are considered to be divine revelation. They were organised into four major branches: *Ṛgveda, Yajurveda, Sāmaveda*, and the *Atharvaveda*.

Various other branches of knowledge grew up as auxiliaries that were to be developed in order to interpret and put to practical use the material of the Vedas.

There were a total of fourteen *Śāstras* or branches of knowledge: the four Vedas, the four Upavedas (auxiliary Vedas), and the six Vedāṅgās (parts of the Vedas). The four Upavedas were (1) *Āyurveda*, literally "The Science of Life", which constituted the medical system; (2) *Arthaśāstra*, which constituted state craft and political theory; (3) *Dhanurveda*, literally, archery, but practically constituting the art of warfare in its varied aspects, and (4) *Gāndharvaveda* , constituting music, drama, and the fine arts.

Similarly, the knowledge systems required for understanding, interpreting, and applying the Vedas were organised into six branches called *Vedāṅgās*, literally " the limbs of the Vedas", with the Vedas personified in a human form. The six *Vedāṅgas* are *Vyākaraṇa* (Grammar), *Chandas* (Metrics), *Śikṣā* (Phonetics), *Nirukta* (Etymology), *Kalpa* (Ritual), and *Jyotiṣa* (Astronomy and Mathematics).

These *Vedāṅgās* were essential, since the Vedas had to be understood correctly (needing etymology and grammar), pronounced and chanted accurately (needing metrics and phonetics), and used properly in various contexts (needing ritual), and the times for these performances had to be computed correctly, requiring the knowledge of computation of the flow of time and of planetary movements (needing astronomy and mathematics). Even though the *Śāstras* originally evolved in the context of the Vedas, they also developed an independent identity and their own corpus of literature and applications that extended well outside the originally formulated requirements of the Vedic context.

In later periods the list of *Śāstras* became much larger and the area covered was much wider. For example, in his famous text *Kámasūtra*, the author Vātsyāyana provides a list of sixty-four arts with which any scholar should be familiar.

Since the various branches of Indian knowledge systems are extremely diverse, we will focus upon a few that can best illustrate some characteristic different features. These are: (1) the fact that linguistics occupied a seminal place even for exact sciences (unlike Western knowledge systems); (2) the nature of theorisation and theory building in Indian tradition; (3) the algorithmic nature of Indian computation, and (4) the sociology of organisation of knowledge—the "classical" and the "folk" streams and their interrelation.

In any scientific discourse it is essential to achieve precision and rigour. In the Western tradition, the geometry of Euclid is considered the paradigm of an ideal theory, and various other branches of knowledge tried to emulate Euclid by setting out their knowledge on the basis of a formal axiomatic system. In contrast, in Indian tradition, an attempt was made to use natural language and to refine and sharpen its potential by technical operations so that precise discourse was possible even in natural language. This is so, particularly in Sanskrit, where we find that even the most abstract and metaphysical discussions regarding grammar, mathematics, or logic are still written in natural language. In Indian knowledge systems, it is the science of linguistics that occupied the central place which, in the West, was occupied by mathematics.

LINGUISTICS

Linguistics is the earliest of Indian sciences to have been rigorously systematised. This set an example for all the other Indian sciences. Linguistics is systematised in *Aṣṭādhyāyī*—the text of Sanskrit grammar by Pāṇini. The date of this text is yet to be settled with any certainty. However, it is not later than 500 BCE. (The dates mentioned here are those based on Western scholarship. An indigenous Indian dating and chronology in this matter have yet to be established.) In the *Aṣṭādhyāyī*, Pāṇini achieves a complete characterisation of the Sanskrit language as spoken at his time, and also manages to specify the way it deviated from the Sanskrit of the Vedas. Given a list of the root words of the Sanskrit language (*dhātupaṭha*) and using the aphorisms of Pāṇini, it is possible to generate all the possible correct utterances in Sanskrit. This is the main thrust of the generative grammars of today that seek to achieve a purely grammatical description of language through a formalised set of derivational strings. It is understandable that until such attempts were made in the West in the recent past, to the Western scholars the Paninian aphorisms (*sūtras*) looked like nothing but some artificial and abstruse formulations with little content.

Science in India seems to start with the assumption that truth resides in the real world with all of its diversity and complexity. Thus for the linguist what is ultimately true is the language as spoken by the people. As Patañjali, a famous grammarian who wrote a commentary on Pāṇini's *Grammar* emphasises, valid utterances are not manufactured by the linguist, but are already established by practice in the world. Nobody goes to a linguist asking for valid utterances, the way one goes to a potter asking for pots. Linguists do make generalisations about language as spoken in the world, but these are not the truth behind or above the reality. They are not the idealisation according to which reality is tailored. On the other hand, what is ideal is real, and some part of the real always escapes our idealisation of it. It is the business of the scientist to formulate these generalisations, but also at the same time to be attuned to the reality, to be conscious of the exceptional nature of each specific instance. This attitude seems to permeate all Indian science and makes it an exercise quite different from the scientific enterprise of the West.

Astronomy and Mathematics

Indian mathematics finds its beginning in the *Śulbasūtras* of the Vedic times. Purportedly written to facilitate the accurate construction of various types of sacrificial altars of the Vedic ritual, these *sūtras* lay down the basic geometrical properties of plane figures like the triangle, the rectangle, the rhombus, and the circle. Basic categories of the Indian astronomical tradition were also established in the various *Vedāṅga Jyotiṣa* texts.

Rigorous systemisation of Indian astronomy begins with Āryabhaṭa (b. CE 476). His work *Āryabhaṭīya* is a concise text of 121 aphoristic verses containing separate sections on basic astronomical definitions and parameters; basic mathematical procedures in arithmetic, geometry, algebra, and trigonometry; methods of determining the mean and true positions of the planets at any given time; and descriptions of the motions of the Sun, Moon, and planets along with computation of the solar and lunar eclipses. After Āryabhaṭa there followed a long series of illustrious astronomers with their equally illustrious texts, many of which gave rise to a host of commentaries and refinements by later astronomers and became the cornerstones for flourishing schools of astronomy and mathematics. Some of the well known names belonging to the Indian tradition are: Varāhamihira (d. CE 578), Brahmagupta (b. CE 598), Bhāskara I (b. CE 629), Lalla (eighth century CE), Muñjāla (CE 932), Śrīpati (CE 1039), Bhāskara II (b. CE 1115), Mādhava (fourteenth century CE), Parameśvara (ca. CE 1380), Nīlakaṇṭa (ca. CE 1444), Jyeṣṭhadeva (sixteenth century CE), Gaṇeśa, and Daivajña (sixteenth century CE). The tradition continued right up to the late eighteenth century, and in regions like Kerala, original work continued to appear until much later.

The most striking feature of this tradition is the efficacy with which the Indians handled and solved rather complicated problems. Thus the *Śulbasūtras* contain all the basic theorems of plane geometry. Around this time Indians also developed a sophisticated theory of numbers including the concepts of zero and negative numbers. They also arrived at simple algorithms for basic arithmetical operations by using the place-value notation. The reason for the success of the Indian mathematician lies perhaps in the explicitly algorithmic and computational nature of Indian mathematics. The objective of the Indian mathematician was not to find "ultimate axiomatic truths" in mathematics, but to find methods of solving specific problems that might arise in astronomical or other contexts. The Indian mathematicians were prepared to set up algorithms that might give only approximate solutions to the problems at hand, and to evolve theories of error and recursive procedures so that the approximations might be kept in check. This algorithmic methodology persisted in the Indian mathematical consciousness until recently, so that Ramanujan in the 20th century might have made his impressive mathematical discoveries through its use.

Āyurveda: The Science of Life

The third major science of the classical tradition is Āyurveda, the science of life. Like linguistics and astronomy this finds its early expression in the Vedas, especially the *Atharvaveda*, in which a large amount of early medicinal lore is collected. Systemisation of *Āyurveda* takes place during the period from the fifth century BCE to the fifth century CE in the *Caraka Saṃhitā, Suśruta Saṃhitā* and the *Aṣṭāṅga Saṅgraha*, the so-called *Bṛhat-trayee* texts which are still popular today. This is followed by a long period of intense activity during which attempts are made to refine the theory and practice of medicine. This process of accretion of information and refinement of practice continued right up to the beginning of the nineteenth century.

Folk and Classical Traditions

There exists a vast amount of knowledge which represents the wisdom of thousands of years of observation and experience. While in any given area (such as medicine) there may be a body of experts or learned professionals, knowledge also prevails in more diffuse or scattered forms. In Indian tradition, it seems to be a general principle running through all types of learning that knowledge can and does prevail in various forms and also gets communicated in many ways.

The general picture that emerges seems to be that the "classical texts" in any area of learning may set out broad general principles as well as their application in a given context, say a particular region of the country. But in various different contexts or regions, knowledge gets expressed based on the given situation, and the generalities get adopted, modified, or even overridden sometimes based on the specificity. This can perhaps best be illustrated in the case of medicine, where classical medical texts themselves deal with this issue. A classical text of *Āyurveda* such as *Caraka Samhitā* expounds general principles of drug action on the six factors: *Dravya, Guṇa, Rasa, Vīrya, Vipāka*, and *Prabhava*. It also discusses remedies for several diseases and lists specific drugs. These may get modified to suit local conditions. In any recipe for a drug, one can substitute a non-principal component with an equivalent, which may be listed in the text or selected on the basis of the principle of *Rasa, Vīrya*, etc. From time to time traditional physicians produce texts and manuals which set out prescriptions for drugs in any given area based on what is available and suitable to the requirements of that area. For example, the text *Rājamṛgaṅka* lists 129 recipes, and in his foreword the editor states that it is a compilation that must have been made by a *vaidya* (physician) from Tamil Nadu, since it contains recipes based on herbs readily available in Tamil Nadu. Such recipes are not only easier to formulate, but they are also more suited to the area, in accordance with Caraka's dictum "For a person who belongs to a particular country or a region, herbs from the same region are most wholesome".

The fact that it is the particularity of the context that is the overriding consideration and that Sastric (i.e. scientific) principles are to be considered as precepts and guidelines and not applied in a mechanical or legalistic manner is clearly stated in many classical texts. "A Vaidya who comprehends the principles of *Rasa* , etc. would discard treatment if not wholesome to the patient in a given situation, even if it is prescribed in the texts; on the contrary he would adopt treatments that are helpful to the patient, even if they do not find a mention in the texts".

It is also interesting to note what the texts of *Āyurveda* say about folk knowledge. The *Caraka Saṃhitā* states that "the goatherds, shepherds, cowherds and other forest dwellers know the drugs by name and form." Similarly, the *Suśruta Saṃhitā* states "one can know about the drugs from the cowherds, tapasvis, hunters, those who live in the forest and those who live by eating roots and tubers".

This is an overview of Indian knowledge systems and does not go into the details of achievement in a variety of areas, particularly those pertaining to material sciences. Our attempt is to highlight basic characteristics of these knowledge systems, particularly in those respects where they differ from their modern counterparts.

A. V. Balasubramanian

References

Balasubramanian, A.V. and M. Radhika. *Local Health Traditions: An Introduction.* Madras: Lok Swasthya Parampara Samvardhan Samithi. 1989.

Cardona, *G. Pāṇini: A Survey of Research.* The Hague: Mouton. 1976.

Srinivas, M.D. "The Methodology of Indian Mathematics and its Contemporary Relevance." *PPST Bulletin-S* 12: 1–35. 1987.

See also: Medicine: Āyurveda – Mathematics – Astronomy

L

Lalla

Lalla, an eighth century Indian astronomer, was an exponent of the school of astronomy founded by Āryabhaṭa (b. CE 476). He was the son of Tāladhvaja and grandson of Sāmba alias Trivikrama, and hailed from Daśapura in Mālava in Western India.

Lalla was a popular astronomer who wrote both on astronomy and astrology. His most important work is the *Śisyadhīvṛddhida* (Treatise Which Expands the Intellect of Students), which, as he says, was composed to expatiate astronomy as set out by Āryabhaṭa. He uses the parameters enunciated in the *Āryabhaṭīya,* but propounds corrections to them every 250 years commencing from CE 498, the time of Āryabhaṭa. The first such correction falls in CE 748, which gives an indication of Lalla's date. The *Śiṣyadhīvṛddhida* is in two sections, entitled *Grahādhyāya,* dealing with planetary computations, and *Golādhyāya,* dealing with spherics, and theoretical and cosmological material. The first section, which is comprised of chapters I–XIII, treats of the mean and true planets, the three problems relating to diurnal motion, eclipses, rising and setting of the planets, the moon's cusps, planetary and astral conjunctions, and complementary situations of the Sun and the Moon. The second section (chapters XIV–XXII), deals with the graphical representation of the motion of the planets, the rationale of the rules enunciated earlier, rejection of popular false notions on astronomy, and astronomical instruments. Another work of Lalla, known from quotations by later authors on astronomy, is *Siddhāntatilaka.*

Lalla wrote a work on natural astrology, entitled *Ratnakośa* which is still in manuscript form. Lalla's verses on mathematical topics are frequently quoted by later writers, but the complete text from which these verses are taken has yet to be found. This provides the justification for Lalla's being referred to in later works as *Tri-skandhavidyākuśalaikamalla,* "the one stalwart versed in all three branches", that is, mathematics, astronomy, and astrology.

Though Lalla follows Āryabhaṭa in certain aspects, he follows Brahmagupta (b. CE 598), and Bhāskara I (fl. CE 629), in certain others. It is also interesting that some of his innovations are followed by later astronomers like Śrīpati (tenth century), Vaṭeśvara (ca. CE 900), and Bhāskara II (b. CE 1114). This makes Lalla an important link in Indian astronomical tradition.

K. V. Sarma

REFERENCES

Chatterji, Bina. *Śisyadhīvṛddhida Tantra of Lalla with the commentary of Mallikārjuna*. 2 vols. New Delhi: Indian National Science Academy. 1981.

See also: Astronomy – Āryabhaṭa – Brahmagupta – Bhāskara – Astronomical Instruments – Śrīpati

Lunar Mansions in Indian Astronomy

In Indian astronomy the 27 or 28 *nakṣatras* (constellations) with their *yogatārās* (junction stars), all situated in the zodiac, correspond to the lunar mansions called *xius* in the East Asian, and *manāzil* in the Islamic tradition. *Nakṣatra* (lit. *na-kṣatra*, non-moving, fixed; *nakta-tra*, 'guardian of the night'), meaning 'star', refers to the 27 asterisms or star groups that occur on or on the sides of the zodiac. It refers also to the 27 equal spaces into which the zodiac can be divided, each space being equal to: 360°/27 = 13°20′ or 800′, commencing from the First point of Meṣa, which is the starting point of the zodiac in Indian astronomy. Now, the zodiac is divided into 12 equal parts, each being 30°, called 'sign' or *rāśi*, to accommodate the 12 solar months. Hence, each *rāśi* holds, inside it, two and a quarter *nakṣatra*-spaces, each distinguished by a prominent star which is called *yoga-tārā*.

Since the performance of rituals and sacrifices at specified times, days, seasons, and years was obligatory for the Vedic Indians, they were interested in the preparation of a workable calendar to be able to ascertain the specified times. This required the study of the motions of the Sun and the Moon, which moved along or near the zodiac. The fixed stars and constellations provided the astronomers with a stellar frame of reference against which they could follow and measure the movements of the Sun, the Moon, and the planets. Hence Indian astronomy identified and concentrated on the study, from very early times, of these stars only, to the exclusion of the general array of stars that stud the heavens. Thus the *nakṣatra* system of the Hindus came into being.

The *nakṣatras* had been identified even during the time of the *Ṛgveda*, though only those which were relevant to specific Vedic prayers were mentioned therein. In the *Yajurveda* and the *Atharvaveda*, however, all the *nakṣatra* were mentioned in the order in which they appear on the zodiac, since in those texts contexts required their mention in a row. In the *Yajurveda* literature, the several *nakṣatras* were assigned presiding deities, pictured as male or female or neuter, and the plurality specified. Legends have also been narrated to explain some of the characteristics of the *nakṣatras*, besides specifying them as benefic or malefic, which aspect was elaborated in later astrological literature.

Yogatārā (Junction star) is the cardinal star in a *nakṣatra* which is made up of several stars. Normally, the *Yogatārā* would be the brightest star in the group, and the zodiacal signs would mostly be named after that star. The several constellations of Hindu astronomy, the details regarding them, and their *yogatārās* are listed in Table 1.

Table 1 *Constellations of Hindu astronomy*

No.	*Nakṣatra*	Presiding deity	Gender	Plurality	*Yogatārā*
1.	Kṛttikā	Agni	Feminine	Plural	η Tauri
2.	Rohiṇī	Prajāpati	Feminine	Singular	α Tauri
3.	Mṛgaśiras	Soma	Neuter	Singular	λ Orionis
4.	Ārdrā	Rudra	Feminine	Singular	α Orionis
5.	Punarvasu	Aditi	Masculine	Dual	β Geminorum
6.	Puṣya	Bṛhaspati	Masculine	Singular	δ Caneri
7.	Āśleṣā	Sarpa	Feminine	Plural	α Caneri
8.	Maghā	Pitṛ	Feminine	Plural	α Leonis
9.	Pūrvaphalgunī	Aryamā	Feminine	Dual	δ Leonis
10.	Uttaraphalgunī	Bhaga	Feminine	Dual	β Leonis
11.	Hasta	Savitā	Masculine	Singular	δ Corvi
12.	Citrā	Indra	Feminine	Singular	α Virginis
13.	Svātī	Vāyu	Feminine	Singular	α Bootis
14.	Viśākhā	Indrāgni	Feminine	Dual	α Librae
15.	Anurādhā	Mitra	Feminine	Plural	δ Scorpii
16.	Jyeṣṭhā	Indra	Feminine	Singular	α Scorpii
17.	Mūlā	Nirṛti	Feminine	Singular	λ Scorpii
18.	Pūrvāṣāḍhā	Āpaḥ	Feminine	Plural	δ Sagittarii
19.	Uttarāṣāḍhā	Viṣvedevāḥ	Feminine	Plural	α Sagittarii
20.	Śroṇā	Viṣṇu	Feminine	Singular	α Aquilae
21.	Śraviṣṭhā	Vasu	Feminine	Plural	β Delphini
22.	Śatabhiṣaj	Indra	Masculine	Singular	λ Aquarii
23.	PūrvaProṣṭhapada	Ajekapād	Masculine	Plural	α Pegasi
24.	UttaraProṣṭhapada	Ahirbudhnya	Masculine	Plural	γ Pegasi
25.	Revatī	Pūṣā	Feminine	Singular	ζ Piscium
26.	Aśvinī	Aśvin	Feminine	Dual	β Arietis
27.	Bharaṇī	Yama	Feminine	Plural	41 Arietis

It is interesting that besides the above details, the work *Nakṣatra-kalpa*, an ancillary text of the *Atharvaveda*, also provides, among other things, the number of stars making up each constellation.

K. V. Sarma

REFERENCES

Bose, D.M., S.N. Sen, and B. V. Subbarayappa. *A Concise History of Science in India*. New Delhi: Indian National Science Academy. 1971.

Chakravarty, S.K. "The Asterisms." In *History of Oriental Astronomy, IAU Colloquium, 91*. Ed. G. Swarup et al. Cambridge: Cambridge University Press. 1987. pp. 23–28.

Dikshit, S.B. *Bhāratīya Jyotish Śāstra (History of Indian Astronomy)* Trans. R.V. Vaidya. pt. I. *History of Astronomy During the Vedic and Vedāṇga Periods*. Delhi: Manager of Publications. Civil Lines. 1969.

Modak, B.R. "Nakṣatra-Kalpa." In *Proceedings of the Twenty-sixth International Congress of Orientalists*. New Delhi. 1964. vol. III, pt. I. Poona: Bhandarkar Oriental Research Institute. 1969. pp. 119–122.

Report of the Calendar Reform Committee, Government of India. New Delhi: Council of Scientific and Industrial Research. 1955.

Yano, M. "The Hsiu-yao Ching and its Sanskrit Sources." In *History of Oriental Astronomy, IAU Colloquium 91*. Ed. G. Swarup et al. Cambridge: Cambridge University Press. 1987. pp. 125–134.

See also: Astrology – Astronomy

M

Mādhava of Saṅgamagrāma

During the Muslim rule in north India, there was a decline in Hindu culture. This adversely affected the creative spirit in indigenous art, literature, and science. Southern India was comparatively less affected, and traditional culture and the sciences flourished there. The followers of Āryabhaṭa I made enormous contributions to the development of mathematical sciences. It was a golden age of Indian mathematics.

Mādhava of Saṅgamagrāma, who flourished about CE 1400, was the first great astronomer and mathematician of the Late Āryabhaṭa school, which he in fact founded. Saṅgamagrāma has been identified as the modern Irinjalakkuda, a town near Cochin in Kerala State. Mādhava belonged to the Emprantiri subcaste group of Kerala Brahmins. We have no knowledge about his parents and teachers, or of the exact dates of his birth and death. Various dates ranging from CE 1336 to 1418 are used in his works. Hence the period of activity of his life has been roughly fixed from CE 1340 to 1425.

There is no doubt that Mādhava was an extraordinarily brilliant man. He used his talent and sharp intelligence to acquire knowledge by private study, and could thus overcome the difficulty of finding a good Guru because of his inferior status in the dominant Nampūtiri Brahminic community. He was a self-taught genius and not a gifted pupil. He was generally referred to as *Golavid* (Master of Spherics) by subsequent scholars and followers of his School, such as Nīlakaṅṭa Somayājī (CE 1444–1545) and Acyuta Piṣāraṭi (CE 1550–1621).

Mādhava wrote all his works in Sanskrit, the classical language of India. One of his earliest works is the *Candra-Vākyāni* (Moon Sentences). This was composed as a revision of Kerala's ancient traditional astronomical work attributed to Vararuci, who lived a thousand years earlier. The *Candra-Vākyāni* gives 248 mnemonic phrases regarding the longitudinal position of the moon for each of the 248 days which comprise a period of nine anomalistic months.

The *Sphuṭacandrāpti* (Computation of True Moon) contains 51 verses and is a work of the *Karaṇa* category. A *Karaṇa* is a handbook on practical astronomy. In this one, he provides an ingenious method for finding the true position of the moon.

Mādhava's *Veṇvāroha* (Bamboo Climbing) is an elaboration of his *Sphuṭacandrāpti* and consists of 74 verses. In this work the author created a facile procedure to find the true lunar positions at intervals of about half an hour. It is dated as CE 1403, and is the most popular astronomical work of Mādhava. Acyuta Piṣāraṭi wrote a Malayalam commentary on it.

A recently identified astronomical work of Mādhava is *Agaṇita-grahacāra.* It is an extensive work on planetary computations using somewhat novel methodologies. It is a treatise of the *Karaṇa* category and must have been composed just after CE 1418 which is the latest date mentioned.

Among other unpublished works of Mādhava there are two short astronomical tracts. One is the *Madhyamānayanaprakāra* (Method for Computing Mean Positions) which is extant in a unique manuscript at the India Office Library. The other is *Lagnaprakaraṇa*, of which at least three manuscripts exist in South India. This work deals with computations of ascendants.

It is possible that Mādhava composed a work on *Golavāda* (Spherics) which earned him the appellation *Golavid.* But the reported manuscript from a private collection has been eaten by white ants. Mādhava was also the author of a number of stray or free verses which have been cited by later authors and commentators.

Scientific Contributions

The traditional "moon sentences" of Vararuci (fourth century), used in Kerala, gave daily longitudes of the moon only up to minutes of the arc or angle. Mādhava computed more sophisticated moon sentences which expressed the longitudes correctly up to seconds. By making use of the popular system of alphabetic numerals, called the *Kaṭapayādi Nyāya,* these mnemonic phrases were made short and aphoristic. Mādhava also provided a value of π using a system of word numerals *(bhūta-saṁkhyās).*

The knowledge of an accurate value of π enabled Mādhava to obtain a better value of the traditional *Sinus Totus* (Total Sine, or radius). Mādhava's sine table is quite precise and accurate. He may have used traditional methods for getting the table or the newly discovered power-series expansion of sine (see below). For computing sine for the argument intermediary between any two tabulated angles, he knew a formula which is equivalent to the modern Taylor series approximation up to the second order. Higher interpolation based on second order finite differences had been known in India since the time of Brahmagupta (seventh century CE).

For computing π to any desired degree of accuracy, Mādhava discovered a number of series including the one $\frac{\pi}{4} = 1 - \frac{1}{3} + \frac{1}{5} - \frac{1}{7} \cdots$, often called the Leibniz series after the German mathematician G.W. Leibniz, who rediscovered it in 1673 or so.

Another formula perhaps known to Mādhava is now called the Gregory Series, after the Scottish mathematician James Gregory (1638–1675). The Indian proof is found in the *Yuktibhāṣā* (CE sixteenth century) and other works.

One of Mādhava's major achievements was the discovery of the power-series expansions of sine and cosine, which were rediscovered in Europe at the time of Newton, and which are equivalent to

$$\sin x = x - \frac{x^3}{3!} + \frac{x^5}{5!} - \frac{x^7}{7!} + \cdots,$$

$$\cos x = 1 - \frac{x^2}{2!} + \frac{x^4}{4!} - \frac{x^6}{6!} + \ldots,$$

The two Sanskrit verses which embody the method of computing sine based on power series up to x^{11} are quoted by Nīlakaṇṭha in his commentary on the *Āryabhaṭīya* (II, 17b).

In this connection it is relevant to discuss a small tract called *Mahājyānayana-prakāra* (Method for the Computation of the Great Sines). It gives the power-series methods for computing *Mahājyās* (Great Sines). Unfortunately there is no mention of an author's name in it, although Mādhava's rule for computing sines up to x^{11} is mentioned. K.V. Sarma attributed the tract to Mādhava, but Gold and Pingree consider it to be the work of his follower(s). Perhaps Mādhava explained his theory during lectures to his pupils, whose lecture notes may be the basis of the above tract.

R.C. Gupta

REFERENCES

Primary sources

Candra-Vākyāni. Ed. K.V. Sarma as the appendix to his edition of *Sphuṭacandrāpti* as well as of *Veṇvāroha* (see below).

Mahājyānayana-prakāra. Ed. and trans. D. Gold and D. Pingree. *Historia Scientiarum* 42: 49–65. 1991.

Sphuṭacandrāpti. Ed. K.V. Sarma. Hoshiarpur: Vishveshvaranand Institute. 1973.

Veṇvāroha. Ed. K.V. Sarma, with a Malayalam commentary of Acyuta Piṣāraṭi. Tripunithura: Sanskirt College, 1956.

Secondary sources

Gupta, R.C. "Mādhava's Power Series Computation of the Sine." *Gaṇita* 27: 19–24. 1976.

Gupta, R.C. "South Indian Achievements in Medieval Mathematics," *Gaṇita Bhāratī* 9: 15–40. 1987.

Gupta, R.C. "The Mādhava-Gregory Series for $\tan^{-1} x$". *Indian Journal of Mathematics Education* 11(3): 107–110. 1991.

Gupta, R.C. "On the Remainder Term in the Mādhava–Leibniz's series." *Gaṇita Bhāratī* 14: 68–71. 1992.

Hayashi, T. et al. "The Correction of the Mādhava Series for the Circumference of a Circle." *Centaurus* 33: 149–174. 1990.

Pingree, D. *Census of the Exact Sciences in Sanskrit.* Series A. vol. 4. Philadelphia: American Philosophical Society. 1981.

Sarma, K.V. "Date of Mādhava, a Little-known Indian Astronomer." *Quarterly Journal of the Mythic Society* 49(3): 183–186. 1958.

Sarma, K.V. *A History of the Kerala School of Hindu Astronomy.* Hoshiarpur: Vishveshvaranand Institute. 1972.

Magic and Science

The concepts "magic" and "science" are products of Euro-American history; thus their use in other regions of the globe, from the colonial era to modern anthropological studies, is intertwined with Western intellectual history. Consequently, understanding the meaning of, and relationship between, magic and science in the context of western notions of rationality is essential when examining phenomena in non-Western societies placed in these categories. Magic and science are, in essence, labels used to exclude and include according to an intellectual value system rooted in European history. They are part of the cultural baggage taken abroad by Euro-American travellers, and used to identify "otherness" in foreign cultures. In non-Western cultures, similar practices of "magic" were not necessarily excluded or marginalised by the growth of science as they were in the West.

The European evolution of these two words from the classical (Greco-Roman) era to the twentieth century shows a growing gap between magic as occult or hidden knowledge on the one hand, and science and religion as public knowledge on the other. Increasingly in the European intellectual tradition, science was defined in narrower ways while pushing magic out of the realm of knowledge. This contributed to an evolutionary paradigm applied by Westerners to non-Western societies, of progress from magic to religion to science, a model now called into question by modern anthropologists. Because of this conceptual evolution in the intellectual history of Europe, "magic" and "science" can be used in a number of senses in modern English usage (see the *Oxford English Dictionary*). The gradual transformation of the word science as a distinctive rationality valued above and against magic is part of a uniquely European duality not generally found in non-Western societies, where magic can exist side-by-side with science and religion.

HISTORY OF THE TERMS IN THE WESTERN INTELLECTUAL TRADITION

European intellectual history is full of self-imposed oppositions: temporal–spiritual, natural–supernatural, pagan–Christian, devil–God, magic–religion, magic–science. All of these are subject to a moral scale of Good versus Evil, a distinction in Western thought that has its roots in the Judeo–Christian monotheistic system positing a single, omnipotent, all-good Deity. This way of thinking is very different from, for example, the Chinese world view embodied in *Yin* and *Yang*, opposites that create balance (positive–negative, active–passive forces) without the identification of Evil versus Good. Thus, the Western dualities are not, as some westerners visiting other cultures have assumed, universals found in all cultures; rather they are a particular product of the belief systems and intellectual history of Europe.

None the less, despite these polar oppositions in European thought, changes in definitions over time caused overlaps and grey areas. Thus magic, as the opposite of religion or science, has a history that complicates the way the word is used at different times and places by different classes of people. The self-defined shape of European history is one of progress from magic to religion to science: from root definitions in classical culture (Greco-Roman), through the medieval magical and religious mentality, through major religious and intellectual changes (renaissances) in the medieval and early modern world, to the development of science as a separate discipline in the modern world. At the same time, this chronological picture is muddied by the slippery definition of magic as it changes in relation to the growth of religion and science. Throughout the development of these distinctions, from the fourteenth through the nineteenth and twentieth centuries, Europeans and Americans went abroad and applied these differing notions to the peoples they met.

The root meaning of magic contains a sense of exclusion found throughout the history of the term: in the Roman world, the Latin *magia*, derived from *magi* (Persian astrologers like the Magi of the Christmas story), implied a foreigner, even when the practitioner was a Roman. This was someone who possessed secret and powerful knowledge both feared and respected, displayed in the ability to manipulate unseen or spiritual agencies, in such arts as divination, astrology, curses, oracles, and amulets to ward off evil (Luck, 1985). Thus the word magic has at its root a sense of marginality in its otherness and its paranormal, unknown, and supernatural associations, but also a strong sense of power held exclusively and secretly by the *magus.* The root word for science, on the other hand, is more normative: *scientia* includes knowledge, art, or skill. It derives from the Greek heritage a strong sense of human rationality, but comes to include divinely-revealed knowledge as well in the Christian era (from ca. 200).

In the progress model, the magic of the European past is associated with the medieval period (ca. 500–1350), in contrast to the rationality of the Renaissance (fourteenth and fifteenth centuries) and Enlightenment (eighteenth century) intellectual revolutions. "Medieval" thinking has earned the label "magic" from later generations on several grounds. The medieval otherworldly emphasis and reliance on divine revelation led to a lack of distinction between natural and supernatural and contributed to an allegorical way of thinking about nature, so that objects such as a flower and events such as storms or illnesses were read as divine messages. Medieval thinking rested on a belief in a wonder-working, ever-present God and also a magic-working evil presence in the devil; sometimes the miracles of God's saints and the tricks of the devil appear similar in method in the eyes of later thinkers (Flint, 1991; Thomas, 1971). Specifically, the belief in supernatural powers in words is found in both Germanic animism (worship of nature spirits) and the Christian tradition: in Germanic animism charms (ritual words and actions) invoke the inherent virtues of a plant. In the Christian eucharist, bread and wine are changed into flesh and blood through prayer. The two were joined in Christianised folk medicine in the production of charms using Christian prayers as the powerful words (Jolly in Neusner et al., 1989). To Protestants (after the sixteenth century) and anthropologists (late nineteenth century), these practices are all magical in their manipulation of nature through word-magic (as opposed to true religious prayer as supplicatory). Yet to the medieval mind, magic was defined by *who* – God or the Devil – not by *how* – supplication versus manipulation. Thus medieval thinking was rejected as backward by later rationalists and was used to describe cultures that had not advanced out of magical or superstitious thinking.

Religion, in the history of European "progress", moves away from magical thinking and opens the door for rationalism. The tradition of logic and deductive reasoning dates back to the Greeks and partially survived into the Middle Ages through Roman-Christian church leaders and thinkers; the recovery of Aristotle through Arab sources in the twelfth century helped spur a renaissance in learning among medieval scholars so that human reason was placed alongside divine revelation as a way of knowing truth. This interest in the potential of human reason to understand things in conjunction with divinely-revealed knowledge cleared a space for reason to function independently over the succeeding centuries. The separation of natural from supernatural, and reason from revelation, allowed thinkers to focus on the human study of natural phenomena. Magic in this context became things not in the category of the divine (miracles) and not subject to human reason either: black magic associated with the devil and witches, such as curses and evil spells, or the low magic of ignorant persons based on false reasoning, such as herbal charms and love potions. High, white magic was associated with the intelligentsia of the high Middle Ages and Renaissance (twelfth through fifteenth centuries). These early scientists dabbled in the occult, a grey area in between divine knowledge and human reason: occult phenomena were insensible (not subject to human sense perception), such as magnetism, gravity, or the pull of the stars, but might be intelligible (something to be reasoned about); these occult phenomena, classified as "natural magic", became the sciences of astrology, astral medicine, and alchemy, for example.

In the Scientific Revolution, science became a separate, and increasingly higher, discipline from religion; it came to mean exclusively the human (versus divine) study of natural (versus supernatural) phenomena. This secularisation of knowledge was the product of the Italian Renaissance of the fourteenth and fifteenth centuries and the Enlightenment of the seventeenth and eighteenth centuries. Simultaneously, Protestant ethical values contributed to this process a utilitarian view of the created order that effectively circumscribed religion into a rational system: God made the world to work by certain laws that humans could understand and systematise (Tambiah, 1990). Human study could produce a true understanding of reality independent of divine insight. Magic was now clearly marked as something not rational: it could be proven to be a hoax (prestidigitation or sleight-of-hand), and was relegated to the entertainment industry where it could be enjoyed as an illusion, a deception that could be scientifically explained. As science became the religion of the modern west, magic was being exorcised from modern consciousness not as demonic but as irrational and backwards.

This simple pattern of progress from magic to religion to science is misleading in two ways: it is anachronistic in applying later definitions of magic back on to earlier periods where the word had different meanings, and it does not take into account the overlaps and continuities whereby magic survives alongside religion and well into science. For example, the Scientific Revolution is compromised by intellectual dabbling in the occult: the great shift from a geocentric worldview (earth at the centre) to heliocentric (Sun at the centre) was founded not just on forward-looking developments in mathematics and astronomy but was motivated by a backward-looking interest in a supposedly Egyptian magical tradition, Hermeticism (Tambiah, 1990). Differences of class in relation to conceptions of magic further complicate this picture of Euro-American intellectual history: the older ways of folk belief, in medicine for example, as a viable, not magical, method of manipulating nature or spiritual agencies was retained among many classes long after the religious authorities or the intellectual elite had dispensed with it, and in some cases had begun investigating it as witchcraft (anti-religion) or fraud (unscientific). All of these divergent attitudes toward magic were carried abroad by Europeans and Americans: magic as demonic, evil, and fearful; magic as medieval or backward; magic as unscientific, irrational or uncivilised; but always as something "other."

Modern ethnography, the study of cultures, is a product of this western history and its intellectual legacy of magic versus religion or science. The earliest ethnographers were explorers and missionaries, some of whom made an effort to observe and document these "new" peoples. The paradigm of progress some missionaries used in meeting non-urban, preliterate peoples was to categorise them as children needing to be fostered into adulthood; other colonisers used a model to exploit the "Indians" as sub-human slaves. For the missionaries, conversion was one step in the maturing process necessary for the native peoples to reach the "level" of civilisation mastered by the Europeans.

Modern anthropology attempted to break with the religiously-biased view of these missionaries and take an objective observer stance which was, none the less, still coloured by an evolutionary model of progress from magic to religion to science (Herbert, 1991). This model is clearly evident in the works of early nineteenth-century founders Edward Tylor and Sir James Frazer and into the twentieth century in Jacob Bronowski and Bronislaw Malinowski (Tambiah, 1990). The anthropological definition focused on magic as unscientific manipulations of nature or supernatural forces and classified it according to its false premises (imitative magic, contagious magic, sympathetic magic). These notions of magic were assumed as a universally valid construct applicable cross-culturally. Consequently, observed peoples were placed into the spectrum of development from magic to science. This model is the subject of debate in late twentieth-century anthropological scholarship, by such authors as Francis Hsu and Stanley Tambiah, who question whether the European concept of magic can be used accurately to classify a set of phenomena in a non-European culture.

Indigenous Views in Non-Western Societies

In Western thought, then, magic has become something marginal, separate from or opposite to a mainstream tradition of religion or science. Non-Western practices of magic seen in their own cultural context are not the opposite of religion or science, but are complementary to their political, social, religious orders; magic is not the "other" in their worldview, but is part of the norm. Magical practices in non-Western societies can function as part of their cultural identity, alongside scientific development or as a sub-group of religion. In many parts of the world, syncretism is more prominent as a response to alternate worldviews, resulting in co-existing modes of rationality rather than competing ones.

The ancient civilisations of Asia offer examples of traditions developing their own modes of rationality with different dynamics than the European model, between the spheres of religion, science, and magic. In China, the traditions are as complex and overlapping as they are in Europe: ancestor-worship, Confucianism, Daoism, and Buddhism as belief-systems evolved and interacted amid the simultaneous development of science and technology. These belief-systems cannot be easily categorised as religion, philosophy, or magic along Western lines. The Buddhist emphasis on the world as illusory and the Daoist focus on metaphysics lent themselves more readily to practices resembling magic in a Western sense (appeals to supernatural aid, fortune-telling), as did ancestor worship. Confucianism, on the other hand, has both religious and philosophical elements; its concern with the social and political world resembles more the secular humanism of the western tradition. All of these co-exist as alternate, and sometimes complementary, modes of rationality in China.

Similarly in the Chinese world, science and technology do not necessarily replace magico-religious belief: villagers' responses to crisis (for example, a plague) incorporate both medicinal remedies such as serums proven through experimentation in the Western scientific tradition, and rituals seeking to appease the gods. While Westerners would be under some pressure to justify such magical practices with some rational or pseudo-scientific explanation, the Chinese do not feel compelled to argue about where or how the practice fits into some duality of true or false, natural or supernatural, orthodox or heretical, scientific or magical. Ancestor worship and recourse to geomancers (practitioners of *feng shui,* the art of finding spiritually-correct locations for buildings) and fortune tellers (Daoist priests or other spiritualists) continue in twentieth-century China and in Chinese communities in the West without shame or apology (Hsu, 1983).

Elsewhere in Asia, "magic" continues as part of mainstream culture because it is closely linked with cultural identity. In Korea, female shamans perform ritual cures and exorcisms *(kut)* as part of the uniquely Korean fabric of life, alongside imported traditions from China and the West; in this way shaman practice is part of a unique Korean identity (Kendall, 1985). Shintoism in Japan, like ancestor-worship in China, is part of the Japanese cultural heritage embedded in everyday life, so much so that it easily accommodated an incoming religious system such as Buddhism or the development of science and technology, with which it has no reason to quarrel. Thus, syncretism, rather than opposing dualities, is a common pattern in the dynamics of magic, science, and religion in Asian cultures.

In many of the ancient near eastern and south Asian cultures, practices resembling magic (fortune-telling, amulets, sorcery) form a sub-group of either religion or science/medicine. Magicians in India, for example, are closely linked with the religious traditions of Hinduism, Buddhism, and Islam. Indian thought emphasises the illusory nature of the physical world; religious masters (gurus, yogi, and other mendicants) are able to manipulate the natural world precisely because they have reached a state of enlightenment where the world is truly an illusion to them. Likewise street magicians and stage entertainers practice "deceptions" (sawing people, producing trees from a basket, pulling an egg or a bird out of a bag, making ropes rise) that echo Hindu or Islamic stories and thus embody certain truths about the world as a wondrous, deceptive, and illusory place that one looks *through* to find meaning. Although such street magicians are a separate, low (Muslim) caste in India, they are a prominent part of India's cultural landscape, popular as reflections of Indian values (Seigel, 1991).

The interconnection of belief, science, and magic is also seen in Islamic culture. Medieval Islam fostered scientific, technological, and medical research because of their belief in a monotheistic Deity who made all things rational for humans to study (a view that eventually sparked medieval European science). Yet Islam also has magical–mystical sub-groups, some of whom were condemned by Islamic law: their explorations of astronomy have astrological connections, as in Europe; their mathematical concepts have occult meaning to some; Sufi mystics distanced themselves from the Islamic intellectual heritage and sought knowledge through other, spiritual or interior, means. While these branches of Hinduism, Buddhism, or Islam may be minority groups, they are not marginal to their own cultures. Rather, they form an important extension of mainstream ideas that influences the whole culture, although not without conflict.

Societies in Africa, the Pacific, and South-east Asia that did not have the urban, literate characteristics of these older civilisations provide clear examples of how poorly the conceptual dichotomy of magic versus science works as a model for understanding cultures where so-called magical practices are part of the norm. In Mali (Africa), the *Sundiata,* a twentieth-century version of an oral tradition dating back to the thirteenth century, speaks of the war magic used by Malian sorcerers to make a tribe successful so that others feared them; their magic oracles dispensed wisdom for successful living, a combination of prophecy and character insight. Likewise, many Pacific island cultures practiced a kind of potent magic in love, war, and healing that relied on a sorcerer's ability to conjure nature and spirits.

These practices encompass both white and black magic in the European paradigm, to both heal and curse; such mastery is always dangerous to the practitioner because of the power of the sources he or she is using. In New Caledonia (Melanesia), sorcerers concoct love potions; in Samoa (Polynesia), chants ward off a headache caused by a god. In Pohnpei (Micronesia), tribal groups employ sorcery to win a war, calling on ghosts or natural forces such as tides; during the Spanish occupation, they used both the borrowed technology of guns and their own tradition of sorcery to hold off attacks (Hanlon, 1992). Marquesans (in French Polynesia) also practiced a kind of war "magic" closely linked to their tribal identity: chants of power specific to their people and location used words ritually to invoke natural/supernatural forces to aid them in battle. The concept of *mana* in Polynesian cultures embodies this sense of powerful forces that can be channelled through words and actions, and is a more authentic construct for understanding their practices than is "magic".

Such beliefs and practices, prior to European contact, were part of political and social value systems; the practitioners were feared for their power, but were not marginal to an intellectual belief system, as they became in European culture. The retention of these beliefs in the power of the old ways after contact, sometimes in defiance of, and sometimes integrated with, Western beliefs, is a form of cultural resistance and identity.

Many non-Western cultures, confronted with the European intruder, perceived the actions of the newcomer in terms of their own construct of supernatural or occult power. For example, in the Americas, Native American Indians identified literate Europeans (mostly missionaries) as shamans, and their books as tools for manipulating natural/supernatural forces. Literacy was a powerful form of knowledge to acquire, and therefore was classified with other powerful forms of knowledge in their culture held by their shamans (magic and magicians to the Europeans). Thus literacy, and the Christian religion wound around it, was incorporated into the native belief system (Axtell, 1988). Indeed, in the same way that the taking of photographs was feared by some groups as a type of sorcery for capturing their souls, so too the writing down of history and ethnographic observations (by Westerners) is sometimes perceived as a kind of sorcery: the power of shaping and defining cultural identity past and present.

Just as the magic–religion divide is indistinct, so too the magic–science line is easily crossed. Westerners cast doubt on the ancient Polynesians' ability to navigate the Pacific to reach new islands such as the Hawaii chain without the technological tools used by Europeans to accomplish such tasks. New research and recreated voyages, however, confirm the knowledge of wind, sky, and water, and the skills of navigation contained in the remnants of Polynesian chants and other oral traditions, usually categorised as magico-religious rituals. This ambiguous line between magic and science is also visible in modern globalised medical practices that incorporate both Western medical techniques and traditional medicine from non-Western societies. The Chinese practice of acupuncture, once viewed in the United States as magic or pseudo-science, is now being mainstreamed into American medicine in lieu of drugs. In Java, a doctor claims the ability to produce heat and electricity from his own body for healing purposes and he documents this ability using the Western scientific mode of proof (experimentation, repeatability), performing spontaneous combustion on video *(Ring of Fire).* Such global syncretism is increasing rather than decreasing, blurring the distinctions between magic and science as defined in European intellectual history.

Traditionally, then, magic has been defined in opposition to science or religion in Western intellectual development. However, such practices in non-Western cultures that appear to resemble this category or are similar to practices Westerners label magic, may not in those cultures have been defined as a class in opposition to some concept of religion or science. The realisation that "one man's religion (or science) is another man's magic" is leading to redefinitions of these terms and the development of meanings and categories unique to each cultural context.

Karen Louise Jolly

REFERENCES

Axtell, James. *After Columbus: Essays in the Ethnohistory of Colonial North America.* New York: Oxford. 1988.

Flint, Valerie I.J. *The Rise of Magic in Early Medieval Europe.* Princeton: Princeton University Press. 1991.

Hanlon, David. "Sorcery, 'Savage Memories,' and the Edge of Commensurability for History in the Pacific." In *Pacific Islands History: Journeys and Transformations.* Ed. Brij Lal. Canberra: The Journal of Pacific History. 1992.

Herbert, Christopher. *Culture and Anomie: Ethnographic Imagination in the Nineteenth Century.* Chicago: University of Chicago Press. 1991.

Hsu, Francis L.K. *Exorcising the Trouble Makers: Magic, Science and Culture.* Westport, Connecticut: Greenwood Press. 1983.

Kendall, Laurel. *Shamans, Housewives, and Other Restless Spirits: Women in Korean Ritual Life.* Honolulu: University of Hawaii Press. 1985.

Kieckhefer, Richard. *Magic in the Middle Ages.* Cambridge: Cambridge University Press. 1990.

Luck, Georg. *Arcana Mundi: Magic and the Occult in the Greek and Roman Worlds.* Baltimore: Johns Hopkins University Press. 1985.

Malinowski, Bronislaw. *Magic, Science and Other Essays.* Boston: Beacon Press. 1948.

Neusner, Jacob, Ernest Frerichs, and Paul Virgil McCracken Flesher, eds. *Religion, Science, and Magic: In Concert and In Conflict.* New York: Oxford University Press. 1989.

Seigel, Lee. *Net of Magic: Wonders and Deceptions in India.* Chicago: University of Chicago Press. 1991.

Tambiah, Stanley. *Magic, Science, Religion, and the Scope of Rationality.* Cambridge: Cambridge University Press. 1990.

Thomas, Keith. *Religion and the Decline of Magic.* New York: Charles Scribner's Sons. 1971.

Magic Squares in Indian Mathematics

The oldest datable magic square in India occurs in Varāhamihira's encyclopaedic work on divination, *Bṛhatsaṃhitā* (ca. CE 550). He utilised a modified magic square of order four in order to prescribe combinations and quantities of ingredients of perfume. It consists of two sets of the natural numbers 1 to 8, and its constant sum (p) is 18. It is, so to speak, pan-diagonal, that is, not only the two main diagonals but also all "broken" diagonals have the same constant sum. Utpala, the commentator (CE 967), also points out many other quadruplets that have the same sum.

One of the four candidates for Varāhamihira's original square (Figures 1 and 2), with a rotation of 90°, coincides with the famous Islamic square (Figure 3), which al-Bīrūnī and al-Zinjānī frequently used as a basic pattern for talismans.

Varāhamihira called his square *kacchaputa* (the carapace of a turtle?), which reminds one of the title of a book on magic, *Kakṣaputa* (date unknown). The book contains a method for constructing a magic square of order four (Figure 4) when a constant sum (p) is given. It also contains a square having the sum 100, which is attributed to Nāgārjuna (Figure 5).

In his medical work, *Siddhayoga* (ca. CE 900), Vṛnda prescribed a magic square of order three (Figure 6) to be employed by a woman in labour in order to have an easy delivery. Its sum is thirty. This is the first datable instance of a magic square of order three in India, although there is a legend that a Garga, who may or may not be the author of the *Gargasaṃhitā* (ca. first century BCE or CE), recommended magic squares of order three in order to pacify the *navagraha* (nine planets).

The famous Jaina magic square (Figure 7), which is incised on the entrance of a Jaina temple, Jinanātha, in Khajuraho, is assignable to the twelfth or the thirteenth century on a palaeographical basis. It is pan-diagonal. Several Jaina hymns that teach how to make magic squares have been handed down, but their dates are uncertain.

As far as is known, Ṭhakkura Pherū, a Jaina scholar, is the first in India who treated magic squares in a mathematical work. His *Gaṇitasāra* (ca. CE 1315) contains a small section on magic squares that consists of nine verses. He gives a square of order four, and alludes to its rearrangement (Figure 8); classifies magic squares (*jaṃta* = Sanskrit *yantra*) into three (odd, even, and evenly odd) according to the order (n), i.e. the number of cells (*kuṭṭha* = Sanskrit *koṣṭha*) on a side of the square, gives a square of order six (Figure 9) and prescribes one method each for constructing even and odd squares.

The method for even squares (Figure 10) divides the square into component squares of order four, and puts the numbers into cells according to the pattern of a standard sqaure of order four. That for odd squares (Figure 11) first places in the central column the arithmetical progression whose first term and common difference are unity and $(n + 1)$ respectively; and then, starting from the numbers in the central column and proceeding by knight move, successively increases the number by n. The square thus obtained is the same as the one obtained by the so-called diagonal method (cf. Figure 21).

Nārāyaṇa wrote a comprehensive work on mathematics entitled *Gaṇita kaumudī* (CE 1356). Its last chapter, called *bhadra-gaṇita* (Mathematics of Magic Squares), comprises fifty-five verses for rules and seventeen verses for examples, and is devoted exclusively to magic squares and derivative magic figures (*upabhadra*) of various shapes.

2	3	5	8
5	8	2	3
4	1	7	6
7	6	4	1

Figure 1 *Varāhamihira's magic square (p=18).*

10	3	13	8
5	16	2	11
4	9	7	14
15	6	12	1

(a)

2	11	5	16
13	8	10	3
12	1	15	6
7	14	4	9

(b)

10	3	5	16
13	8	2	11
4	9	15	6
7	14	12	1

(c)

2	11	13	8
5	16	10	3
12	1	7	14
15	6	4	9

(d)

Figure 2 *Magic squares reconstructed from Varāhamihira's square (p=34).*

8	11	14	1
13	2	7	12
3	16	9	6
10	5	4	15

Figure 3 *The Islamic square of order four (p=34).*

$a-3$	1	$a-6$	8
$a-7$	9	$a-4$	2
6	$a-8$	3	$a-1$
4	$a-2$	7	$a-9$

(a) $a = \frac{p}{2}$ *(p: even)*

$b-3$	1	$a-6$	8
$a-7$	9	$b-4$	2
6	$a-8$	3	$b-1$
4	$b-2$	7	$a-9$

(b) $a = \frac{p+1}{2}, b = \frac{p-1}{2}$ *(p: odd)*

Figure 4 *Patterns for magic squares of order four given in the Kakṣapuṭa.*

30	16	18	36
10	44	22	24
32	14	20	34
28	26	40	6

Figure 5 *Nāgārjuna's magic square (p=100).*

16	6	8
2	10	18
12	14	4

Figure 6 *Vṛnda's magic square of order three (p=30).*

7	12	1	14
2	13	8	11
16	3	10	5
9	6	15	4

Figure 7 *Jaina magic square of Khajuraho (p=34).*

12	3	6	13
14	5	4	11
7	16	9	2
1	10	15	8

(a) Original square

1	7	12	14
10	16	3	5
8	2	13	11
15	9	6	4

(b) Rearranged

Figure 8 *Pherū's square of four and its rearrangement (p=34).*

1	32	34	33	5	6
30	8	28	27	11	7
24	23	15	16	14	19
13	20	21	22	17	18
12	26	9	10	29	25
31	2	4	3	35	36

Figure 9 *Pherū's square of order six* (p=111).

15	9	6	4
8	2	13	11
1	7	12	14
10	16	3	5

(a) Standard square

			4				8
	2				6		
1				5			
		3				7	
			12				16
	10				14		
9				13			
		11				15	

(b) First stage

63	33	30	4	59	37	26	8
32	2	61	35	28	6	57	39
1	31	36	62	5	27	40	58
34	64	3	29	38	60	7	25
55	41	22	12	51	45	18	16
24	10	53	43	20	14	49	47
9	23	44	54	13	19	48	50
42	56	11	21	46	52	15	17

(c) Magic square obtained (p = 260)

Figure 10 *Pherū construction method for 'even'-order magic squares.*

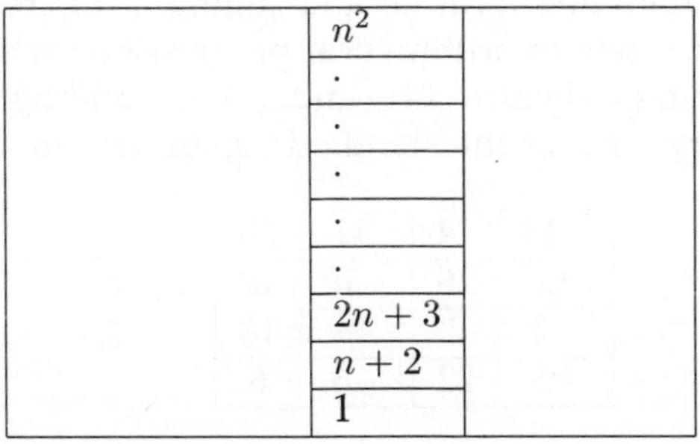

(a) Central column

37	48	59	70	81	2	13	24	35
34	38	49	60	71	73	3	14	25
26	28	39	50	61	72	74	4	15
16	27	29	40	51	62	64	75	5
6	17	19	30	41	52	63	65	76
77	7	18	20	31	42	53	55	66
67	78	8	10	21	32	43	54	56
57	68	79	9	11	22	33	44	46
47	58	69	80	1	12	23	34	45

(b) Square of order nine obtained (p =369)

Figure 11 *Pheru's construction method for 'odd'-order magic squares.*

The topics treated are: definitions of technical terms; determination of the mathematical progressions to be used in magic squares by means of *kuṭṭaka,* i.e. indeterminate equations of the first degree; how to make a square of order four by *turagagati* (horse move)(Figure 12); the number of pan-diagonal magic squares of order four, 384, including every variation made by rotation and inversion.

1	8	13	12
14	11	2	7
4	5	16	9
15	10	3	6

(a)

1	14	4	15
8	11	5	10
13	2	16	3
12	7	9	6

(b)

Figure 12 *Nārāyaṇa's square of four made by 'horse-move' (p =34).*

Then Nārāyaṇa gives three general methods for constructing a square having any optional order (n) and constant sum (p) when a standard square of the same order is known – (1) by means of an arithmetical progression having an appropriate first term (a) and common difference (d)(Figure 13), (2) by means of n sets of arithmetical progressions whose common differences are all unity (Figure 14), and (3) by adding an appropriate number (t) to every term of the standard square (Figure 15).

−5	9	19	17
21	15	−3	7
1	3	25	11
23	13	−1	5

(a) $a = -5, d = 2, p = 40$

−14	14	34	30
38	26	−10	10
−2	2	46	18
42	22	−6	6

$a = -14,\ d = 4,\ p = 64$

Figure 13 *Nārāyaṇa's square of four made by an arithmetical progression.*

7	15	22	20
23	19	8	14
10	12	25	17
24	18	9	13

(a) $p = 64$ Cf. Fig. 12a

16	14	7	30	23
24	17	10	8	31
32	25	18	11	4
5	28	26	19	12
13	6	29	22	20

(b) $p = 90$ Cf. Fig.21

Figure 14 *Nārāyaṇa's square of four made by* n *sets of arithmetical progressions.*

$\frac{35}{2}$	$\frac{49}{2}$	$\frac{59}{2}$	$\frac{57}{2}$
$\frac{61}{2}$	$\frac{55}{2}$	$\frac{37}{2}$	$\frac{47}{2}$
$\frac{41}{2}$	$\frac{43}{2}$	$\frac{65}{2}$	$\frac{51}{2}$
$\frac{63}{2}$	$\frac{53}{2}$	$\frac{39}{2}$	$\frac{45}{2}$

$t = \frac{33}{2}, \; p = 100$

Based on Fig. 12a

Figure 15 *Nārāyaṇa's square of four made by addition of a number* (t).

Nārāyaṇa next explains two methods each for constructing *sama-garbha* (even-womb) or evenly even, *viṣama* (odd-womb) or evenly odd, and odd squares when the sum is given. The two methods for the first kind are: (1) by folding two preliminary squares *karasaṃpuṭa-vat* (just like the folding of two hands) (Figure 16), and (2) by arranging numbers in the component squares of order four according to the pattern of a standard square (Figure 17). For the second kind they are: (1) by putting numbers zigzag in the square (Figure 18), and (2) by tranposing certain numbers in the natural square (Figure 19). These two methods are not completely mechanical, and require *mati* (intelligence). The methods for the third kind are: (1) by folding two preliminary squares just as in the case of the first kind (Figure 20), and (2) by starting from the central cell of any side of the square and proceeding diagonally (Figure 21).

2	3	2	3
1	4	1	4
3	2	3	2
4	1	4	1

(a) Prelim-A

5	0	10	15
10	15	5	0
5	0	10	15
10	15	5	0

(b) Prelim-B

17	13	2	8
1	9	16	14
18	12	3	7
4	6	19	11

(c) B over A
$p = 40$

8	2	13	17
14	16	9	1
7	3	12	18
11	19	6	4

(d) A over B
$p = 40$

Figure 16 *Nārāyaṇa's method for 'even-womb' squares (I): folding method.*

1				2				
		8					7	
4				3				
		5					6	

(a) First stage

1	32	49	48	2	31	50	47
56	41	8	25	55	42	7	26
16	17	64	33	15	18	63	34
57	40	9	24	58	39	10	23
4	29	52	45	3	30	51	46
53	44	5	28	54	43	6	27
13	20	61	36	14	19	62	35
60	37	12	21	59	38	11	22

(b) Magic square obtained ($p = 260$)

Figure 17 *Nārāyaṇa's method for 'even-womb' squares (II): according to the pattern of standard square (Fig. 12a).*

In the last section Nārāyaṇa gives a number of examples for two kinds of derivative magic figures, *saṃkīrṇa* (miscellaneous) and maṇḍala (circular). Both kinds are made by rearranging ordinary magic squares (Figures 22 and 23).

In his encyclopaedic work on Hindu Law, *Smṛtitattva* (ca. CE 1500), Laghunandana gives a method for making squares of order four having any optional sum that is determined according to the purpose (see Table 1).

Significant scholarly work has been done on the importance of magic squares both mathematically and philosophically. Cammann and Roṣu both discuss the significance of magic squares in Indian thought and compare Indian, Islamic, and Chinese magic squares. Kusuba provides an English translation with mathematical commentary, as well as an edition of the Sanskrit text, of Nārāyaṇa's work on magic squares.

1	195	194	193	5	6	190	7	9	10	186	185	184	14
169	27	171	25	173	23	175	22	20	178	18	180	16	182
168	167	166	32	33	34	162	35	37	38	39	157	156	155
141	142	143	53	52	51	147	50	48	47	46	152	153	154
140	139	138	60	61	62	134	63	65	66	67	129	128	127
113	114	115	81	80	79	119	78	76	75	74	124	125	126
112	111	92	88	89	90	98	110	93	94	95	101	100	106
85	86	105	109	108	107	99	87	104	103	102	96	97	91
57	58	59	137	136	135	64	133	132	131	130	68	69	70
56	55	54	144	145	146	49	148	149	150	151	45	44	43
29	30	31	165	164	163	36	161	160	159	158	40	41	42
28	170	26	172	24	174	21	176	177	19	179	17	181	15
196	2	3	4	192	191	8	189	188	187	11	12	13	183

$n = 14, p = 1379$

Figure 18 *Nārāyaṇa's method for 'odd-womb' squares (I): zigzag method.*

100	92	93	94	5	6	7	8	9	91
20	89	83	84	16	15	87	18	82	11
30	29	78	77	75	26	74	73	22	21
40	39	38	67	65	66	64	63	32	31
41	52	43	44	56	55	47	48	59	60
51	42	58	57	46	45	54	53	49	50
61	69	68	37	35	36	34	33	62	70
71	72	28	27	25	76	24	23	79	80
81	19	13	14	86	85	17	88	12	90
10	2	3	4	96	95	97	98	99	1

$n = 10, p = 505$

Figure 19 *Nārāyaṇa's method for 'odd-womb' squares (II): transposing method (conjectural reconstruction).*

4	5	1	2	3
5	1	2	3	4
1	2	3	4	5
2	3	4	5	1
3	4	5	1	2

(a) Prelim-A

20	25	5	10	15
25	5	10	15	20
5	10	15	20	25
10	15	20	25	5
15	20	25	5	10

(b) Prelim-B

19	15	6	27	23
25	16	12	8	29
26	22	18	14	10
7	28	24	20	11
13	9	30	21	17

(c) B over A: $p = 90$

Figure 20 *Nārāyaṇa's method for odd squares (I) : folding method*

22	21	13	5	46	38	30
31	23	15	14	6	47	39
40	32	24	16	8	7	48
49	41	33	25	17	9	1
2	43	42	34	26	18	10
11	3	44	36	35	27	19
20	12	4	45	37	29	28

$n = 7, p = 175$

Figure 21 *Nārāyaṇa's method for odd squares (II): diagonal method.*

1	24	37	36	2	23	38	35	3	22	39	34
42	31	6	19	41	32	5	20	40	33	4	21
12	13	48	25	11	14	47	26	10	15	46	27
43	30	7	18	44	29	8	17	45	28	9	16

(a) Preliminary magic oblong

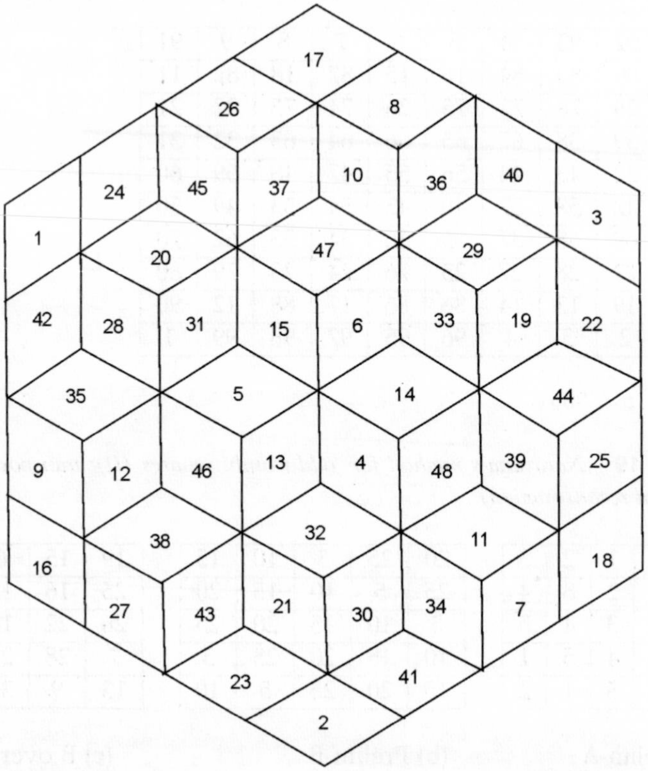

(b) Magic lotus: $p = 294$

Figure 22 *(a) Nārāyaṇa's magic lotus with six petals: preliminary magic oblong. (b) Nārāyaṇa's magic lotus with six petals (p=294)*

1	16	25	24	2	15	26	23
28	21	4	13	27	22	3	14
8	9	32	17	7	10	31	18
29	20	5	12	30	19	6	11

(a) Preliminary magic oblong

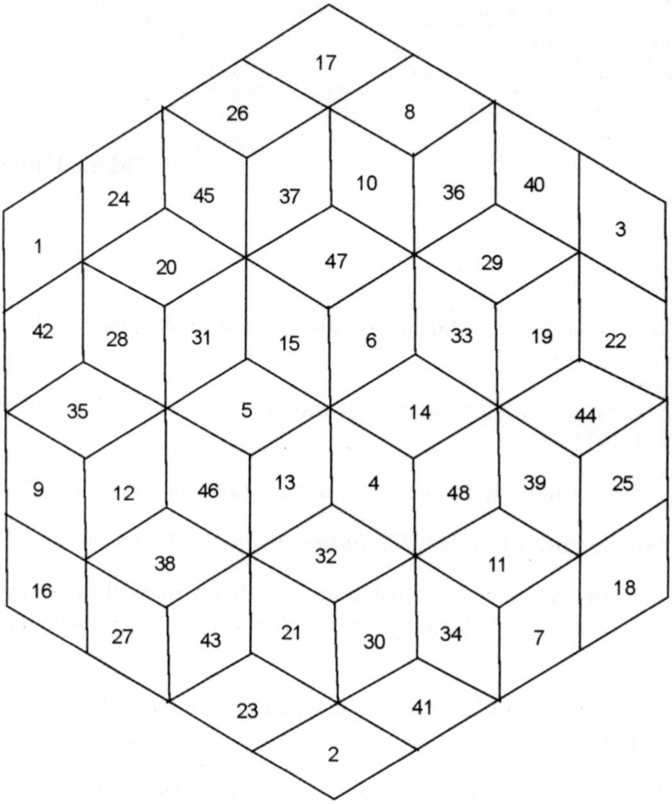

(b) Magic circle: $p = 360$

Figure 23 *(a) Nārāyaṇa's magic circle: preliminary magic oblong. (b) Nārāyaṇa's magic circle (p=360).*

1	8	$a-7$	$a-2$
$a-5$	$a-4$	3	6
7	2	$a-1$	$a-8$
$a-3$	$a-6$	5	4

$a = \frac{p}{2}$

Figure 24 *Pattern for magic square of four by Laghunandana.*

Table 1 *Purposes of magic squares of order four*

Sum (p)	Purpose
20	To neutralise poison
28	To protect crops from insects
32	To accelerate delivery
34	To protect travellers
50	To exorcise evil spirits
64	To protect warriors
72	For women having no children
84	To soothe crying children

Takao Hayashi

REFERENCES

Cammann, S. "Islamic and Indian Magic Squares." *History of Religions* 8: 181–209 and 271–299. 1968 and 1969.

Datta, B. and A.N. Singh. "Magic Squares in India." *Indian Journal of History of Science* 27:51–120. 1992.

Goonetilleke, W. "The American Puzzle." *The Indian Antiquary* 11: 83–83. 1882.

Grierson, G.A. "An American Puzzle." *The Indian Antiquary* 10: 89–90. 1881.

Hayashi, T. "Hōjinzan (A Japanese translation with mathematical commentary of Nārāyaṇa's 'Mathematics of Magic Squares')." *Epistēmē* (Tokyo: Asahi Press), Series II, 3: i–xxxiv. 1986.

Hayashi, T. "Varāhamihira's Pandiagonal Magic Square of the Order Four." *Historia Mathematica* 14: 159–166. 1987.

Kapadia, H.R. "A Note on Jaina Hymns and Magic Squares." *Indian Historical Quarterly* 10: 148–153. 1934.

Kusuba, T. *"Combinatorics and Magic Squares in India. A Study of Nārāyaṇa Paṇḍita's Ganitakaumudī, Chaps. 13–14."* Ph.D. Dissertation, Brown University. 1993.

Ojha, G.K. *Aṅkavidyā.* Delhi: Motilal Banarsidass. 1982.

Roṣu, A. "Études ayurvediques III:Les carrés magiques dans la médicine indienne." In *Studies on Indian Medical History.* Ed. G.J. Meulenbeld and D. Wujastyk. Groningen: Egbert Forsten. 1987. pp. 103–112

Roṣu, A. "Les carrés magiques indiens et l'histoire des idées en Asie." *Zeitschrift der Morgenländischen Gesellschaft* 139: 120–158. 1989.

Singh, P. "Total Number of Perfect Magic Squares: Nārāyaṇa's Rule," *Mathematics Education* 16A: 32–37. 1982.

Singh, P. "Nārāyaṇa's Treatment of Magic Squares." *Indian Journal of History of Science* 21:123–130. 1986.

Vijayaraghavan, T. "On Jaina Magic Squares." *Mathematics Student* 9: 97–102. 1941.

See also: Varāhamihira – Nārāyaṇa

Mahādeva

Mahādeva (ca. 1275–1350) composed the extensive set of planetary tables, *Mahādevī,* named after him and dated 28th March, 1316. He belonged to a family of astronomers and in his work *Grahasiddhi* describes himself as the son of the astrologer Paraśurāma, son of Padmanābha, son of Mādhava, son of Bhogadeva of the Gautamagotra, a follower of *Sāmaveda* and a performer of sacrifices. The planetary tables contained in the *Mahādevī,* prepared for facilitating the computation of the daily almanac, were extremely popular in the Gujarat and Rajasthan regions, and numerous manuscripts of the work have been located in these places. While the basic text was restricted to 43 verses, the author himself wrote a set of instructions for using the tables, called *Grahasiddhi,* which also were extensively used. The popularity of *Mahādevī* is also attested to by the several commentaries that were written on the work, including that of Nṛsiṃha (1528), Dhanarāja (1635) and Mādhava, and a host of anonymous commentators.

Mahādeva is a synonym of Śiva, one of the trinities of Hinduism, and so formed one of the words commonly used to name Hindus. There are several astronomers of medieval India who bore the name Mahādeva. Among these are: Mahādeva, younger brother of Viṭṭhala, from Gujarat, author of *Tithicakranirṇaya,* also called *Tithinirṇaya, Tithiratna,* and *Mahādevasiddhānta;* Mahādeva, author of *Jātakāpaddhati,* called also *Mahādevapaddhati* after his name; Mahādeva, son of Luṇiga and author of commentaries on the *Cintāmaṇiśraṇikā* of Daśabala and the *Jyotiṣaratnamālā* of Śrīpati; Mahādeva of the Kauṇḍinyagotra, son of Bopadeva, and author of *Kāmadhenu* called also *Tithikāmadhenu;* Mahādeva, son of Kahnaji Vaidya, author of *Muhūrtadīpaka* in 57 verses, written in 1640, *Praśnapradīpa* called also *Praśnaratna* (1647), *Bhāveśaphalapradīpa* (1647), *Kālanirṇayasiddhānta* (1652), and a commentary on his own *Muhūrtadīpaka* (1661); Mahādeva Pāṭhaka (1842–1899), son of Revāśaṅkara, and author of *Varṣadīpaka,* called also *Varṣadīpikā* (1861), *Jātakatattva,* written in 1872,

Pitṛmārgapradīpa in 57 verses (1874), *Varṣapaddhati*, a compilation (1874), and *Āśubodhajyotiṣa.*

K.V. Sarma

References

Dvivedi, Sudhakara. *Gaṇaka Taraṅgiṇī or Lives of Hindu Astronomers:* Ed. Padmakara Dvivedi. Benares: Jyotish Prakash Press. 1933.

Pingree, David. "Sanskrit Astronomical Tables in the United States." *Transactions of the American Philosophical Society* 58 (3): 1–77. 1968.

Pingree, David. *Census of the Exact Sciences in Sanskrit, Ser. A, Vol, 4.* Philadelphia: American Philosophical Society. 1981.

Mahāvīra

The Rāṣṭrakūṭa dynasty of the medieval period was founded in the Deccan, South India, by Dantidurga about the middle of the eighth century CE. A king of this dynasty named Amoghavarṣa ruled from CE 815 to 877. The long period of his rule is known for its material prosperity, political stability and academic fertility. He was rich, powerful, and peace-loving, and he patronised art and learning.

In the later part of Amoghavarṣa's reign there lived a great mathematician named Mahāvīrācārya. Mahāvīra was a Digambara Jaina and wrote an extensive Sanskrit treatise called *Gaṇitasāra saṅgraha* (Compendium of the Essence of Mathematics) about CE 850. It is devoted to elementary topics in arithmetic, algebra, geometry, mensuration, etc. The work is important because it is a collection summarising elementary mathematics of his time and providing a rich source of information on ancient Indian mathematics. It is written in the style of a textbook and was used as one for centuries in all of South India. Its importance is greater still because the *Pāṭīgaṇita* of Śrīdhara (ca. CE 750), written in the same style as the *Gaṇitasāra saṅgraha*, is not extant in full.

Mahāvīra shows sufficient originality not only in presenting older material lucidly, but also in introducing several new topics. A commentary called *Bālabodha* in Kannada was written by Daivajña Vallabha. A Sanskrit commentary was composed by Varadarāja. Dates for these two commentators are not known, nor are their works available in print. There were other translations in the eleventh century and in 1842.

The nine chapters of the *Gaṇitasāra saṅgraha* are as follows:

(1). Terminology (70 verses);

(2). Arithmetical operations (115 verses);

(3). Operations involving fractions (140 verses);

(4). Miscellaneous operations (72 verses);

(5). Rule of three (43 verses);

(6). Mixed operations ($337\frac{1}{2}$verses);

(7). Geometry and mensuration ($232\frac{1}{2}$verses);

(8). Excavations ($68\frac{1}{2}$verses);

(9). Shadows ($52\frac{1}{2}$verses);

The total number of verses, 1131, shows that the book is quite comprehensive. Another noteworthy feature is that the Jaina tradition of Indian mathematics is also preserved within the scope of the *Gaṇitasāra saṅgraha.*

The authorship of the astronomical work *Jyotiṣapaṭala* is also ascribed to Mahāvīra. Some manuscripts of this title are mentioned in the *Jinaratna-kośa* and the *New Catalogus Catalogorum,* but no author is mentioned. Another work attributed to him is *Chattisu,* but this is also a sort of elaboration of part of the *Gaṇitasāra saṅgraha* made by Mādhavacandra Traividya (about CE 1000). Whatever the case, the reputation of Mahāvīra relies solely on his *magnum opus.*

During ancient times, unit fractions were considered quite important. There are some interesting results in the *Gaṇitasāra saṅgraha* on this topic. One rule gives a practical method for expressing any given fraction as a sum of unit fractions. Let p/q be the given fraction (p being less than q). We add a suitable integer x to q such that $(q+x)$ become exactly divisible by p, say, r times. Then Mahāvīra's rule is.

$$\frac{p}{q} = \frac{1}{r} + \frac{x}{(r.q)}.$$

Mahāvīra made a very significant remark in connection with the square-root of a negative number. He said "A negative number is non-square by its nature, whence there is no (real) square-root from it."

This remark is the first clear recognition of the imaginary quantities in mathematics which had to wait for several more centuries for their formal definition.

Mahāvīra was one of the earliest Indian mathematicians to deal with the lowest common multiple which he calls *niruddha.* It was evolved to simplify operations with fractions.

Arithmetical and geometrical progressions were already handled earlier. An extensive treatment is available in the *Gaṇitasāra saṅgraha.* In the absence of modern theories of logarithms and equations, problems were solved by methods of trial and repetition.

Mahāvīra seemed to be expert in handling all sorts of equations reducible to quadratic forms and gave a variety of examples of them.

In mensurational problems in geometry, Mahāvīra usually gave two rules: one for rough and the other for better or accurate results. He dealt with all the usual plane figures. For π, he conformed to the Jaina values 3 (rough), and $\sqrt{10}$ (better). He was the first Indian to deal with mensuration related to an ellipse which he calls *āyata-vṛtta* (elongated circle), but his rules are approximate.

For an ellipse of semi major and minor axes a and b his "accurate" results are:

$$\text{Area} = b\sqrt{(4a^2 + 6b^2)};$$

$$\text{Perimeter} = \sqrt{(16a^2 + 24b^2)}.$$

For the exact rectification of the ellipse, one had to wait for about eight hundred years to acquire the powerful tool of calculus. In this situation Mahāvīra's first attempt is to be appreciated.

Regarding the volume of a sphere of radius r, Mahāvīra gave the formula:

$$v = \frac{9}{2}r^3,$$

which was the practical rule of Jaina tradition. He gave another rule for the purpose, but it gives a better result only with an amendment of the text. For the curved surface of a spherical segment, his rule has been newly interpreted to yield the formula

$$S = \pi r^2 \theta \sin\theta$$

where θ is the semi-angle subtended by a diameter of the base of the segment at the centre of the sphere. This peculiar formula gives quite a good result in all practical cases (i.e. for θ up to 60°). The modern exact formula is $2\,\pi r^2(1 - \cos\theta)$. For the volume of frustum-like solids, Mahāvīra gave a generalisation of Brahmagupta's rule based on the theory of averages.

In the end we mention Mahāvīra's extensive contribution to the formation of rational figures. He calls a triangle or quadrilateral *janya* (generated) when its sides, altitudes, and other important measures can be expressed in terms of rational numbers.

R.C. Gupta

References

Datta, B. "On Mahāvīra's Solution of Rational Triangles and Quadrilaterals." *Bulletin of the Calcutta Mathematical Society* 20:267–294. 1928–29.

Dube, Mahesh. "Poet and Mathematician Mahāvīrācārya" (in Hindi). *Arhat Vacana* 3(1): 1–26. 1991.

Gupta, R.C. "Mahāvīrācārya on the Perimeter and Area of an Ellipse." *Mathematics Education* 8(1B): 17–19. 1974.

Gupta, R.C. "Mahāvīrācārya Rule for the Surface-Area of a Spherical Segment." *Tulasī Praj?ñā* 1(2): 63–66. 1975.

Gupta, R.C. "Mahāvīrācārya's Rule for Volume of Frustum-like Solids." *Aligarh Journal of Oriental Studies* 3(1): 31–38. 1986.

Gupta, R.C. "The Mahāvīrā-Fibonacci Device to Reduce *p/q* to Unit Fractions." *HPM Newsletter* 29: 10–12. 1993.

Jain, Anupam, and S.C. Agrawal. *Mahāvīrācārya: A Critical Study* (in Hindi). Hastinapur (Meerut): Digambar Jain Cosmographical Research Institute. 1985.

Jain, L.C., ed. *The Gaṇitasāra saṅgraha* (with Hindi translation). Sholapur: Jain Sanskriti Sanraksaka Sangh. 1963.

Rangacharya, M., ed. and trans. *Gaṇitasāra saṅgraha of Mahāvīrācārya*. Madras: Government Press. 1912.

Sarasvati, T.A. "Mahāvīra's Treatment of Series." *Journal of the Ranchi University* I: 39–50. 1962.

See also: Śrīdhara

Mahendra Sūri

Mahendra Sūri, Jain astronomer, and pupil of Madana Sūri, was a protégé of the progressive minded Sultan Fīrūz Shāh Tuglaq, who ruled in Delhi from CE 1351 to 1388. The Sultan was one of the pioneers of the cultural exchange between Hindus and Muslims and was much interested in astronomy. His most important contribution in this field was the introduction of the astrolabe into India from the Islamic world. He induced Mahendra Sūri to study the astrolabe and familiarise Indian astronomers with the instrument through the Sanskrit language. This persuasion resulted in Mahendra Sūri's writing the *Yantrarāja* (King of Instruments), in 1370, the first work written in Sanskrit on the astrolabe.

The *Yantrarāja* sets out in five chapters the theory of the astrolabe, the construction of the instrument with its several planes and designs, and lines, circles, and other markings to be made on the planes while making the instrument and graduating it for making observations and recordings. It is to be noted that Mahendra Sūri first describes the ordinary astrolabe, which he calls *saumya-yantra* (northern instrument), wherein the astrolabe is projected from the South Pole, and then the *yāmya-yantra* (southern instrument), where the instrument is projected from the North Pole. He then introduces a *miśra* (mixed) instrument, which he calls *phaṇīndra-yantra* (the serpentine instrument), wherein the two types are combined.

The commentary on the *Yantrarāja* by the author's pupil, Malayendu Sūri (fl. 1377) explains the practical application of the instrument for taking readings. He also provides tables of the latitudes of about 75 cities in and outside India, and also one for 32 fixed stars. In this connection the commentator says that the Muslims have recorded more than 1022 stars, but that he selected only 32, those that were required for practical use in Indian astronomy and astrology. There is still another commentary on the work, which is more elaborate than that of Malayendu Sūri, written by Gopīrīja about CE 1540, which is still in manuscript form.

K.V. Sarma

References

Dvivedi, Sudhakara. *Gaṇaka Taraṅgiṇī* or *Lives of Hindu Astronomers*. Ed. Padmakara Dvivedi. Benares: Jyotish Prakash Press. 1933.

Pingree, David. *Census of the Exact Sciences in Sanskrit. Series A, vol.* 4. Philadelphia: American Philosophical Society. 1981.

Yantrarāja of Mahendra-guru with the commentary of Malayendu Sūri, along with the *Yantraśiromai* of Viśrāma. Ed. K.K. Raikwa. Bombay: Nirnaya Sagar Press. 1936.

Yantrarāja of Mahendra Sūri, with the commentary of Malayendu Sūri. Jodhpur: Rajasthan Prachyavidya Shodh Samsthan. 1936.

See also: Astronomy – Astronomical Instruments

Makaranda

Makaranda was a resident of Kāśī (or Benares, Varanasi). In CE 1478 he wrote an extensive astronomical manual with the title *Makaranda*. This was doubly significant, first because the title was reminiscent of his own name, and, second because it called the computed result obtained *makaranda* (honey), and gave the several astronomical terms names of parts of plants, such as *guccha* (flower cluster), *kanda* (bulb), *vallī* and *latā* (creeper), and the like. Makaranda based his work on the parameters of and practices prescribed by the modern *Sūryasiddhānta*, to which he added certain corrections to insure greater accuracy, and provided a number of astronomical tables for ease in computing the daily almanac.

Makaranda's tables, which are often long and extend to several centuries, involved much labour and ingenuity in their preparation. They cover such subjects as *tithi* (lunar day, 5 tables), *nakṣatras* (asterisms, 4 tables), *yogas* (complementary positions of the Sun and the Moon, 3 tables), *saṅkrāntis* (entry of the Sun into the zodiacal signs, 3 tables), mean motion of planets and their anomalies (11 tables), length of daylight on different days (1 table), weekdays (2 tables), and eclipses and allied matter (10 tables).

In order to render the work of the user easier, Makaranda provides, in certain cases, two sets of tables, one for single years and the other for groups of years, all from the epoch date of the commencement of Śaka year 1400 (CE 1478). Thus, when calculations are made for a date which is several years after the epoch, multiples of the group-years can be skipped, just taking note of the readings for the number of group-years skipped and applying the same to the relevant year of the current group. In the case of *tithis* the group is taken as 16 years; for *nakṣatras* and *yogas*, it is nearly 600 years, from CE 1478 to 2054, in 24-year periods. For the precession of the equinoxes, it is nearly 100 years, from CE 1758 to 1838, in 20-year periods, and for *saṇkrāntis* for 400 years, from CE 1478 to 1877, in 57-year periods.

Makaranda's tables were widely used in the entire northern belt of India, comprised of Gujarat, Rajasthan, Uttar Pradesh, Bihar, and Bengal, as attested by the profusion of manuscripts of the work found in this region. The work has also been commented on by several authors, from Ḍhuṇḍirāja (fl. 1590), through Puruṣottama Bhaṭṭa (fl. 1610), Divākara (fl. 1606), and Kṛpakara Miśra (fl. 1815), to Nīlāmbara Jhā (b. 1823). The *Siddhāntasudhā* of Paramānanda Ṭhakkura is based on the work of Makaranda.

K.V. Sarma

References

Dvivedi, Sudhākara. *Gaṇaka Taraṅgiṇī or Lives of Hindu Astronomers.* Ed. Padmakara Dvivedi. Benares: Jyotish Prakash Press. 1933.

Makaranda with the Ṭīkā of Gokulanātha, Divākara and Viśvanātha. Kasi: Benarsi Press. 1884.

Pingree, David. "Makaranda." *Transactions of the American Philosophical Society.* 58(3): 39–46. 1968.

Pingree, David. *Census of the Exact Sciences in Sanskrit, Series A, Vol. 4.* Philadelphia: American Philosophical Society. 1981.

See also: *Sūryasiddhānta* – Lunar Mansions – Astronomy

Maps and Mapmaking

In comparison to Europe, the Islamic world, and East Asia, the cartographic achievements of South Asia appear modest and until the 1980s were scarcely recognised by historians of science. In recent years, new archaeological evidence, surviving sacred and secular texts, and other evidence have led to a change in scholarly opinion. We now believe that some form of mapping was practiced in what is now India as early as the Mesolithic period, that surveying dates as far back as the Indus Civilisation (ca. 2500–1900 BCE), and that the construction of large-scale plans, cosmographic maps, and other cartographic works has occurred continuously at least since the late Vedic age (first millennium BCE). Because of the ravages of climate and vermin, surviving maps from prior to the eighteenth century are rare and are largely in stone, metal, or ceramic. However, cosmographies painted on cloth date back at least to the fifteenth century, while palm leaf architectural plans from seventeenth-century Orissa are believed to be copies of manuscripts originally prepared as early as the twelfth century. Few surviving maps on paper predate the eighteenth century.

Though not numerous, a number of map-like graffiti appear among the thousands of Stone Age Indian cave paintings; and at least one complex Mesolithic diagram is believed to be a representation of the cosmos. Other map-like grafitti continued to be produced by tribal Indians over most of the historic period. The principal reason for supposing that surveying was a feature of the Indus civilisation is that the regularity of its planned grid-pattern urban settlements could not easily be achieved without it. Moreover, excavations have uncovered several objects that appear to have been simple surveying instruments and measuring rods. The uniformity and modular dimensions of the bricks for so much of the architecture over the vast extent of the Indus civilisation are also noteworthy. During the ensuing period of the Vedic Aryans, the building of enormous sacrificial altars was an important religious activity. Texts known as *Śulbasūtras* set forth in great detail how these altars were to be constructed and called for drawing on the ground a plan prefiguring each altar. Again, a system of modular measures was employed. The ancient practice of building gigantic altars, once widespread, has died out over most of India, but survived into the latter half of the twentieth century in the Indian state of Kerala. Hindu temples were also built according to ancient detailed textual prescriptions, known as *Śilpaśāstras*, which also specified drawing on the ground, at full scale, the outline of the structure to be. That practice is still followed. Comparable, though simpler, rules applied, at least in theory, to the building of houses. The *Śilpaśāstras* also contained a variety of models for laying out towns and cities; relatively few present-day settlements in India suggest that such models were actually followed.

For the historic period, one of the earliest surviving artifacts that clearly embodies recognisable map symbolisations is an allegorical wall sculpture from Udayagiri in Madhya Pradesh, ca. CE 400, which shows the confluence of the Ganges and Yamuna Rivers in Madhyadesa (the sacred Central Region of India) over which the then Gupta Empire held sway. Other early datable works are sculpted bas relief cosmographies, some of which were quite elaborate. The earliest of these, depicting *Nandīśvardvīpa*, the eighth continent of the Jain cosmos, was carved in Śaka 1256 (CE 1199–1200). It may safely be assumed that cosmographic paintings adorned the walls and portals of many ancient temples and monasteries of India's main religious groups – Hindus, Jains and Buddhists – just as they still do in Jain religious edifices and in Buddhist establishments in other parts of the world. However, largely because of the ravages of time and partially because of the iconoclasm of Muslim invaders, virtually none of these survive from prior to the twelfth century, apart from fragmentary remains in the Buddhist caves at Ajanta.

Astronomy and its handmaiden astrology were well developed sciences in ancient India. The major texts provided detailed observations that enabled latter-day scholars to prepare elaborate celestial diagrams. Despite these, there is no unequivocal evidence that astronomical charts accompanied those early works or were otherwise drawn. Horoscopes were prepared to show the positions of major heavenly bodies at particular times (especially at times of birth), and iconic representations of those bodies as deities often appeared in sculpture and paintings, as did the signs of the zodiac. But none of these were formed into assemblages that one would readily designate as maps. During the Mughal period (1526–1857), however, planar astrolabes and celestial globes were manufactured in north-western India. One Muslim family practiced the trade in Lahore over a period of several generations. Though some of these works used Sanskrit, rather than Persian, as was the norm, in naming various heavenly bodies, the tradition in which all were made has been dubbed "Islamicate". A variety of related objects in the form of giant masonry instruments appeared in the astronomical observatories constructed during the period ca. 1722–39 by the Rajput king, Sawai Jai Singh, in his capital at Jaipur and in Delhi, Varanasi, Ujjain, and Mathura. Many of these are still usable.

At least five Hindu cosmographic globes are known, all based largely on *Purāṇic* texts from the mid-first millennium BCE to the mid-first millennium CE. The oldest of these is a brass globe, probably from Gujarat, dated Śaka 1493 (CE 1571). The largest (diameter ca. 45 cm) and most elaborate, and the only one to contain a substantial component of geographic information along with its mainly mythic elements, is a papier-maché globe, probably from eastern India, that appears to date from the mid-eighteenth century. One of the other globes is a painted wooden production, also thought to be from eastern India, probably from the mid-nineteenth century. Two are late nineteenth century bronze creations of unknown provenance. Unlike most Western globes, none of these was constructed with the use of gores, the triangular or moon-shaped pieces that form the surface of modern globes.

A number of planispheric world maps also survive. All but one of these, a crude Marathi map on paper, probably from the mid-eighteenth century, are essentially Islamicate productions, in which mythic elements (e.g. the Land of Gog and Magog) coexist with known geographic places. The largest and most ornate of the world maps is a richly illuminated eclectic painting on cloth, with text in Arabic, Hindustani, and Persian; it probably dates from the eighteenth century and may be of either Rajasthani or Deccani provenance. Additionally, a 32-sheet atlas of the "Inhabited Quarter" (i.e. the part of the world suitable for human life), forms part of a 1647 encyclopedic work by Sadiq Isfahani of Jaunpur in what is now Uttar Pradesh. The orientation on maps made by Muslims is typically toward the south.

Indigenous regional maps, mostly from the eighteenth and early nineteenth centuries, derive mainly from Rajasthan, Kashmir, Maharashtra, and Gujarat, probably in that order of frequency. These range in size from very large (several metres on a side) to page-size productions, the larger works being almost always painted on cloth. No clear schools of cartography emerge, though one can usually distinguish among regions of origin. Map symbols and orientation vary markedly, though some regional tendencies may be noted. No map has a consistent scale, though some contain textual notes on distances between places.

Route maps most commonly appear in strip form, occasionally as lengthy scrolls, and typically show the places and physical and man-made features encountered between two given points. Some route maps, largely relating to pilgrimages, which frequently take the form of circuits, are likely to be more complex. Surviving navigational charts are few in number, entirely from Gujarat, and in a tradition presumably derived from the Middle East. The oldest such known work is dated CE 1710.

The most common genre of maps are those that relate to relatively small localities, especially individual towns, as well as plans of specific forts, palaces, temples, gardens, and tombs. Such maps served many purposes: guides to pilgrims, aids for engineering projects, plans or documents for military activities, commemorations of historical events, text illustrations, interior adornments, and so forth. Locality maps typically combine a largely pictorial rendition of specific structures, drawn in either an oblique perspective or in frontal elevation, with an essentially planimetric rendition of the encompassing space. Hill features on such maps (as well as on regional maps) are characteristically shown in frontal perspective. Colours for rendering hills, water features, and vegetation are naturalistic and not very different from what one would encounter on Western maps. The largest known Indian map, depicting the former Rajput capital at Amber in remarkable house-by-house detail, measures 661 × 645 cm. (260 × 254 inches, or approximately 20 × 21 feet).

Although hundreds of Indian maps have now been studied and described, hundreds of additional recently discovered works await analysis; and it may be safely predicted in light of the interest aroused by recent research that many more maps will soon come to light.

Joseph E. Schwartzberg

REFERENCES

Arunachalam, B. "The Haven-Finding Art in Indian Navigational Traditions and Cartography." In *The Indian Ocean: Explorations in Exploration, Commerce, and Politics*. Ed. Satish Chandra. New Delhi: Sage Publications. 1987. pp. 191–221.

Bahura, Gopal Narayan and Singh, Chandramani. Catalogue of Historical Documents in Kapad Dwara, Jaipur; Part II: Maps and Plans. Jaipur: published with the permission of the Maharajah of Jaipur. 1990.

Caillat, Collette and Ravi Kumar. *The Jain Cosmology*. Basel: Ravi Kumar. 1981.

Digby, Simon. "The Bhugola of Ksema Karna: a Dated Sixteenth Century Piece of Indian Metalware." *AARP (Art and Archaeology Research Papers)* 4: 10–31. 1973.

Gole, Susan. *Indian Maps and Plans: From Earliest Times to the Advent of European Surveys*. New Delhi: Manohar. 1989.

Habib, Irfan. "Cartography in Mughal India." *Medieval India, a Miscellany* 4: 122–34, 1977; also published in *Indian Archives* 28: 88–105. 1979.

Savage-Smith, Emilie. *Islamicate Celestial Globes: Their History, Construction, and Use*. Washington, D.C.: Smithsonian Institution Press. 1985.

Schwartzberg, Joseph E. "South Asian Cartography." Part 2 of *Cartography in the Traditional Islamic and South Asian Societies*, Vol 2, Book 1 of *The History of Cartography*. Ed. J.B. Harley and David Woodward. Chicago: University of Chicago Press. 1992. pp. 295–509.

Tripathi, Maya Prasad. *Development of Geographic Knowledge in Ancient India*. Varanasi: Bharatiya Vidya Prakashan. 1969.

See also: Astrology – Astronomy

Mathematics

A widely held view is that Indian mathematics originated in the service of religion. Support for this view is sought in the complexity of motives behind the recording of the *Śulbasūtras*, the first written mathematical source dated around 800–500 BCE, dealing with the measurement and construction of sacrificial altars. This view ignores the skills in mensuration and practical arithmetic that existed in the Harappan (or Indus Valley) culture which dates back to 3000 BCE. Archaeological remains indicate a long established centralised system of weights and measures. A number of different plumb-bobs of uniform size and weights have been found that could be classified as decimal, i.e. if we take a plumb-bob weighing approximately 27.534 grams as a standard representing 1 unit, the other weights form a series with values of 0.05, 0.1, 0.2, 0.5, 2, 5, 10, 20, 50, 100, 200 and 500 units. Also, scales and instruments for measuring length have been discovered, including one from Mohenjodaro, one of the two largest urban centres, consisting of a fragment of shell 66.2 mm long, with nine carefully sawn, equally spaced parallel lines, on average 6.7056 mm apart. The accuracy of the graduation is remarkably high, with a mean error of only 0.075 mm.

A notable feature of the Harappan culture was its extensive use of kiln-fired bricks and the advanced level of its brick-making technology. While fifteen different sizes of Harappan bricks have been identified, the standard ratio of the three dimensions – the length, breadth and thickness – is always 4:2:1, considered even today as the optimal ratio for efficient bonding. A close correspondence exists between the standard unit of measurement (the "Indus inch" (33.5 mm) and brick sizes, in that the latter are integral multiples of the former.. (An Indus inch is exactly twice a Sumerian unit of length (*sushi*). 25 Indus inches make a Megalithic yard, a measure probably in use in North-west Europe around 2000 BCE. These links have led to the conjecture that a decimal scale of measurement originated somewhere in western Asia and then spread as far as Britain, Egypt, Mesopotamia and the Indus Valley.)

This relationship between brick-making technology and metrology was to reappear 1500 years later during the Vedic period in the construction of sacrificial altars of bricks. However, a most intriguing suggestion of B.V. Subbarayappa is that the Harappan numeration system contains certain similarities with the Kharoṣṭīh and the 'Asokan' variant of the *Brāhmi* numeration systems which emerged in India about two thousand years later. He notes the following similarities:

- There are identical symbols for the numbers one to four and for a hundred in all three numeration systems, and

- All three were ciphered systems employing a decimal base.

He suggests that deciphering the inscriptions on the large number of excavated seals and other artefacts require that they be recognised as numerical records rather than as literary passages. Given the failure so far to decipher the Harappan script, this approach is certainly worth further examination.

The earliest written evidence in India of a recognisable antecedent of our numeral system is found in an inscription from Gwalior dated 'Samvat 933' (CE 876) where the numbers 50 and 270 are given as ५० and २७०. Notice the close similarity with our notation for 270 showing in both an understanding of the place value principle as well as the use of zero. There is earlier evidence of the use of Indian system of numeration in South-east Asia in areas covered by present-day countries such as Malaysia, Cambodia and Indonesia, all of which were under the cultural influence of India. Also, as early as CE 662, a Syrian bishop, Severus Sebokt, comments on the Indians carrying out computations by means of nine signs by methods which "surpass description" (Joseph, 1993).

The spread of these numerals westwards is a fascinating story. The Arabs were the leading actors in this drama. Indian numerals probably arrived at Baghdad in 773 CE with the diplomatic mission from Sind to the court of Caliph al-Manṣūr. In about 820, al-Khwārizmī wrote his famous *Kitāb ḥisāb al-adad al-hindī* (Book of Addition and Subtraction According to the Hindu Calculation, also called just Arithmetic), the first Arab text to deal with the new numerals. The text contains a detailed exposition of both the representation of numbers and operations using Indian numerals. Al-Khwārizmī was at pains to point out the usefulness of a place value system incorporating zero, particularly for writing large numbers. Texts on Indian reckoning continued to be written, and by the end of the eleventh century this method of representation and computation was widespread from the borders of Central Asia to the southern reaches of the Islamic world in North Africa and Egypt.

In the transmission of Indian numerals to Europe, as with almost all knowledge from the Islamic world, Spain and (to a lesser extent) Sicily played the role of intermediaries, being the areas in Europe which had been under Muslim rule for many years. Documents from Spain and coins from Sicily show the spread and the slow evolution of the numerals, with a landmark for its spread being its appearance in an influential mathematical text of medieval Europe, *Liber Abaci* (Book of Computation), written by Fibonacci (1170–1250), who learnt to work with

Indian numerals during his extensive travels in North Africa, Egypt, Syria, and Sicily. The spread westwards continued slowly, displacing Roman numerals, and eventually, once the contest between the abacists (those in favour of the use of the abacus or some mechanical device for calculation) and the algorists (those who favoured the use of the new numerals) had been won by the latter, it was only a matter of time before the final triumph of the new numerals occurred, with bankers, traders, and merchants adopting the system for their daily calculations.

The beginnings of Indian algebra may be traced to the *Śulbasūtras* and later Bakhshālī Manuscript, for both contain simple examples involving the solution of linear, simultaneous and even indeterminate equations. An example of an indeterminate equation in two unknowns (x and y) is $3x + 4y = 50$, which has a number of positive whole number (or integer) solutions for (x, y). For example, $x = 14$, $y = 12$ satisfies the equation as do the solution sets (10, 5), (6, 8) and (2, 11).

But it was only from the time of Āryabhaṭa I (b. CE 476) that algebra grew into a distinct branch of mathematics. Brahmagupta (b. CE 598) called it *kuṭṭaka gaṇitā*, or simply *kuṭṭaka*, which later came to refer to a particular branch of algebra dealing with methods of solving indeterminate equations to which the Indians made significant contributions.

An important feature of early Indian algebra which distinguishes it from other mathematical traditions was the use of symbols such as the letters of the alphabet to denote unknown quantities. It is this very feature of algebra that one immediately associates with the subject today. The Indians were probably the first to make systematic use of this method of representing unknown quantities. A general term for the unknown was *yāvat tāvat*, shortened to the algebraic symbol *yā*. In Brahmagupta's work Sanskrit letters appear, which are the abbreviations of names of different colours, which he used to represent several unknown quantities. The letter *kā* stood for *kālaka*, meaning 'black', and the letter *nī* for *nīlaka* meaning 'blue'. With an efficient numeral system and the beginnings of symbolic algebra, the Indians solved determinate and indeterminate equations of first and second degrees and involving in certain cases more than one unknown. It is likely that a number of these methods reached the Islamic world before being transmitted further westwards by a similar process and often involving the same actors as the ones that we discussed earlier in the spread of Indian numerals.

The beginnings of a systematic study of trigonometry are found in the works of the Alexandrians, Hipparchus (ca. 150 BCE), Menelaus (ca. CE 100) and Ptolemy (ca. CE 150). However, from about the time of Āryabhaṭa I, the character of the subject changed to resemble its modern form. Later, it was transmitted to the Arabs who introduced further refinements. The knowledge then spread to Europe, where the first detailed account of trigonometry is contained in a book entitled *De triangulis omni modis* (On All Classes of Triangles), by Regiomontanus (1464).

In early Indian mathematics, trigonometry formed an integral part of astronomy. References to trigonometric concepts and relations are found in astronomical texts such as *Sūryasiddhañta* (ca. CE 400), Varāhamihira's *Pañcha Siddhāntạ* (ca.CE 500), Brahmagupta's *Brāhma Spuṭa Siddhānta* (CE 628) and the great work of Bhāskara II called *Siddhānta Śiromaṇi* (CE 1150). Infinite expansion of trigonometric functions, building on Bhāskara's work, formed the basis of the development of mathematical analysis – a precursor to modern calculus to be discussed later.

Basic to modern trigonometry is the sine function. It was introduced into the Islamic world from India, probably through the astronomical text, *Sūryasiddhānta,* brought to Baghdad during the eighth century. There were two types of trigonometry available then: one based on the geometry of chords and best exemplified in Ptolemy's *Almagest,* and the other based on the geometry of semi-chords which was an Indian invention. The Arabs chose the Indian version which prevailed in the development of the subject. It is quite likely that two other trigonometric functions – the cosine and versine functions – were also obtained from the Indians.

One of the most important problems of ancient astronomy was the accurate prediction of eclipses. In India, as in many other countries, the occasion of an eclipse had great religious significance, and rites and sacrifices were performed. It was a matter of considerable prestige for an astronomer to demonstrate his skills dramatically by predicting precisely when the eclipse would occur.

In order to find the precise time at which a lunar eclipse occurs, it is necessary first to determine the true instantaneous motion of the moon at a particular point in time. The concept of instantaneous motion and the method of measuring that quantity is found in the works of Āryabhaṭa I, Brahmagupta and Muñjāla (ca. CE 930). However, it was in Bhāskara II's attempt to work out the position angle of the ecliptic, a quantity required for predicting the time of an eclipse, that we have early notions of differential calculus. He mentions the concept of an "infinitesimal" unit of time, an awareness that when a variable attains the maximum value its differential vanishes, and also traces of the "mean value theorem" of differential calculus, the last of which was explicitly stated by Parameśvara (1360–1455) in his commentary on Bhāskara's *Līlāvatī.* Others from Kerala (South India) continued this work with Nīlakaṇṭha (1443–1543) deriving an expression for the differential of an inverse sine function and Acyuta Piṣāraṭi (ca. 1550–1621) giving the rule for finding the differential of the ratio of two cosine functions.

However, the main contribution of the Kerala school of mathematician-astronomers was in the study of infinite series expansions of trigonometric and circular functions and finite approximations for some of these functions. The motivation for this work was the necessity for accuracy in astronomical calculations. The Kerala discoveries include the Gregory and Leibniz series for the inverse tangent, the Liebniz power series for π, the

Newton power series for the sine and cosine, as well as certain remarkable rational approximations of trigonometric functions, including the well-known Taylor series approximations for the sine and cosine functions. And these results had been obtained about three hundred years earlier than the mathematicians after whom they are now named. Referring to the most notable mathematician of this group, Mādhava (ca. 1340–1425); Rajagopal and Rangachari (1978) wrote: "(It was Mādhava who) took the decisive step onwards from the finite procedures of ancient mathematics to treat their limit-passage to infinity, which is the kernel of modern classical analysis". The growing volume of research into medieval Indian mathematics, particularly from Kerala, has refuted a common perception that mathematics in India after Bhāskara II made "only spotty progress until modern times" (Eves, 1983).

George Gheverghese Joseph

References

Primary sources

Algebra with Arithmetic and Mensuration from the Sanscrit of Brahmegupta and Bhāscara. Ed. H.T. Colebrooke. London: Murray, 1817; reprinted Wiesbaden: Sandig, 1973.

Āryabhaṭiya of Āryabhaṭa. Ed. and trans. K.S. Shukla and K.V. Sarma. New Delhi, Indian National Science Academy. 1976.

Gaṇitāsāra-saṃgraha of Mahāvīrācārya. Ed. and trans. M. Rangacharya. Madras: Government Press. 1912.

Pañcasiddhāntikā of Varāhamihira. Ed. and trans. T.S.K. Sastry and K.V. Sarma. Madras: PPST Foundation. 1993.

Secondary sources

Bag, A.K. *Mathematics in Ancient and Medieval India*. Varanasi: Chaukhambha Orientalia. 1979.

Datta, B. and A.N. Singh. *History of Hindu Mathematics*, 2 vols. Bombay: Asia Publishing House. 1962.

Eves, H. *An Introduction to the History of Mathematics*, 5th ed. Philadelphia: Saunders. 1983.

Joseph, George G. *The Crest of the Peacock: Non-European Roots of Mathematics*. London: Penguin. 1993.

Rajagopal, C.T. and M.S. Rangachari. "On an Untapped Source of Medieval Keralese Mathematics." Archives for History of Exact Sciences, 18, 1978. 89–108.

Sarma, K.V. *A History of the Kerala School of Hindu Astronomy*. Hoshiarpur: Vishveshvaranand Institute. 1972.

Srinivasiyengar, C.N. *History of Indian Mathematics*. Calcutta: World Press. 1967.

Subbarayappa, B.V. *Numerical System of the Indus Valley Civilisation*. Mimeo. 1993.

See also: *Śulbasūtras* – Weights and Measures – Technology and Culture – Geometry – Arithmetic – *Bakhshālī* Manuscript – Eclipses – Brahmagupta – Āryabhaṭa – Bhāskara – Parameśvara – Nīlakaṇṭha Somayāji – Acyuta Piṣāraṭi – Mādhava

Medical Ethics

The origins of medicine in India stretch back to antiquity. The Harrapan city culture flourished in and around the Indus Valley ca. 2500 BCE; it is known for its elaborate bathhouses and drains and sewers built under the streets leading to soak pits. In the second millennium BCE, the north-western parts of India were host to a series of Indo-European immigrants and invaders from Central Asia. With them began the classical culture of India. Vedas, the sacred lore of the Indo-Europeans, celebrate the *Bheṣaj*, one knowledgeable in medicinal herbs. One of the four Vedas, the *Atharvaveda* contains many chants, mantras, and herbal preparations to ward off evil, enemies, and diseases. The priest-physicians prescribed preparations of plants and herbs, and prayers and fasts for their patients. The Indian medical tradition, *Āyurveda*, meaning the science of vitality and long life, is considered a limb of the *Atharvaveda*.

A more formal system of medicine evolved from around the time of the Buddha (ca. 500 BCE). It became organised in textual form in the first century CE, and reposes in a vast body of literature redacted and updated from that time to the present. There are six principal texts of the Āyurveda. The older three are the two compendia, *Carakasaṃhitā* and *Suśrutasaṃhitā*, named after the two legendary physicians, Caraka and Suśruta, and the *Aṣṭāṅgahṛdaya*, the eightfold essence attributed to an eighth century physician named Vāgbhaṭa. The younger three are the *Mādhavanidāna* (ninth century), *Śārṅgadharasaṃhitā* (thirteenth or early fourteenth century), and *Bhāvaprakāś* Bhāvamiśra (sixteenth century). The word *caraka* also means one who moves about, and may have referred to the itinerant Buddhist and Jain monks who played a pioneering role in the evolution of the Indian medical tradition. In the realm of King Aśoka (273–232 BCE), who embraced Buddhist ideals, Buddhist monasteries served as institutions, like hospitals and hospices, for the care of the sick and the dying.

The earliest medical writings known as the Bower manuscripts, discovered in a Buddhist Stupa in Kashgar (modern China), and translated by Rudolph Hoernle, are considered to have been written by Buddhist authors around CE 450. These texts contain medical treatises which describe the virtues of garlic in curing diseases and extending the life span, elixirs for a long life, ways of preparing medical mixtures, eye lotions, oils, enemas, aphrodisiacs, and procedures for the care of children. Early Indian medicine was carried to Tibet along with Buddhism and was best preserved there, as well as in China. Travellers to and from China, Greece, Persia, and Arabia contributed to the spread of Indian medicine outside India.

The basic assumptions of Indian medicine are rooted in the religious and philosophical traditions of India. Early developments exhibited great diversity in opinion and formulation in keeping with the diversity in Indian thought, tied to Hindu, Buddhist, or Jaina philosophies in various measures. Similarly the system allowed for significant geographic variation as knowledge spread through the subcontinent over a long period of time.

The medical ethics which are closely linked to these religious and philosophical perspectives *(darśanas)* reveal variable, shifting, and accommodating attitudes.

Āyurvedic constructs of the body and the self, central to the medical enterprise, grew in tandem with the faith traditions. The primary vehicles of ayurvedic pathophysiology are the *doṣas* (humours): *vāyu* or *vāta* (wind), *pitta* (bile), and *kapha* (phlegm), and the *dhātus* (body substances). The three humours represent movement, heat, and moisture respectively in the body. The primary body substance, *rasa,* organic sap, is derived from food, moves throughout the body, is stored in various reservoirs, and is finally excreted as waste products. In processes of sequential transformation, the *dhātus,* flesh, fat, bone, marrow, and semen, are derived, semen being the purest and most vital product of this process.

The Indian system of medicine views health as a state of balance of body substances, *dhātusāmya,* and illness as a state of disequilibrium. The body responds to many kinds of inputs: physical, as in food and drink, psychological, as in emotions of anger or jealousy, and social, as in affection, praise, or scorn. Each input is a potential source of a disease or a cure.

The theory of *guṇas* (lit. strands or qualities) introduces the notion of ethics as a material basis in the ayurvedic pathophysiology. Inherent and substantial, *sattva* (goodness), *rajas* (vitality or activity), and *tamas* (inertia) are qualities or traits found in all substances in various combinations. The balance determines the overall dispositions of persons, foods, activities, bodily substances, and so forth. *Sattva,* which is cool and light, produces calmness, purity, or virtue; *rajas,* which is hot and active, produces passion, happiness, or sorrow; and *tamas,* which is dark, heavy, and dull, produces sloth, stupidity, and evil. Contemplation, meditation, silence, devotion, and fasting promote goodness; love, battle, attachment, pleasure seeking, and emotionality enhance vitality. Sleep and idleness increase inertia. In a hierarchy of values, the *sattva* categories reign supreme and become less material, closer to the idea of *sat* (truth or essence), and often the same as the mind or self. The object of the therapeutic is to transform a person from lower to higher strands or qualities, which is accomplished through the prescription of foods and activities which build goodness. Thus the therapeutic and the ethical become coterminus.

In the Indian view life is not the opposite of death; birth is the opposite of death. Life begins when an embryo is formed out of the union of male and female germinal substances. Defining when human life begins was neither easy nor uniform and straightforward. Some texts maintained that life began with the aforesaid union, and others at the moment of quickening or the descent of the foetus into the pelvis; the latter was more frequently understood as a point of viability. Abnormal pregnancies, congenital deformities, multiple pregnancies, and infertility were explained in terms of defective germinal substances, unnatural coitus, failure in nourishment, or disturbances in humours in the mother or the foetus.

Among the religious obligations, having male progeny was imperative in order to secure a passage to the land of forefathers through the performance of funerary rites. In situations in which a woman failed to have a son, the man was to take another wife, or otherwise adopt a son. If the problem appeared to be male impotence or infertility, the husband's younger brother or another suitable man was to impregnate the wife (a custom called *niyoga*). Early medical texts elaborate on the ways of enhancing conception, and later texts discuss problems of contraception. Mythology also testifies to *in vitro* fertilization and embryo transfer.

The *Suśrutasaṃhitā* describes various forms of arrested foetal development or obstructed deliveries and describes ways of inducing labour and/or destroying the foetus, especially in the case of danger to the mother's life. A seventeenth century text also describes ways of inducing labour for purposes of abortion in cases of women in poor health, widows, and women of liberal morals.

In contemporary problems of medical ethics, no problem has caused as much furore as has amniocentesis. Preference for a male child, with an easily available technology to determine gender prenatally, has resulted in inordinate and indiscriminate use of abortions. Some states in India have enacted laws to restrict the scope of indications and use of amniocentesis.

There are three categories for the etiology of diseases in *Āyurveda.* External or invasive diseases are caused by foreign bodies, injuries, infestations, and possession by evil spirits. Internal diseases are disturbances of humours, in part caused by lapses in discretion, as in faulty or unseasonable diets, overexertion, sloth, sexual indulgence, or mental disturbances.

In either case, the final pathway for the pathology of a disease is an imbalance of humours. The third category contains the diseases which are the fruits of *karma*, the operative principle of Hindu ethics. A very simple explanation might be "every action has a reaction" or "as you sow, so shall you reap", but the logic extends beyond one life. In *karma* theory, when a person dies his self moves to the other world, enveloped in the part material and part ethereal covering which carries the traces of all actions performed, and comes to determine its condition in the next life. Thus some diseases are the fruits of actions from past lives. The unseen hand of *karma* is invoked to explain the not so easily explicable. Events like epidemics and disasters are a result of bad actions of a whole community or the actions of a king.

Mental illnesses also arise from these etiologies: possession states, disturbances in humours, and lapses in discretion. Some disease states are also seen as the workings of time, as in aging.

Physicians in ancient India did consider *karma* in etiology, but they agreed that the passivity that results from assumptions of predetermination made the whole medical enterprise meaningless. Human effort was always a factor in the workings of *karma*, and caring and healing must be actively pursued by the physician. There was also a recognition of incurable diseases, in the face of which human effort was futile. The physician was prudent if he avoided heroic efforts to prevent the inevitable, which not only led to loss of income but also loss of prestige. If the case was hopeless, the physician was to do no more than attend to the nutrition of the dying patient, and even that might be withdrawn at the request of the family.

A category of "willed death" was also recognised in the various religious traditions and was understood to be different from suicide. Suicide was regarded as an act of desperation and willed death an act of determination. It involved permission of the religious order and was resorted to only when the quality of remaining life was likely to be poor.

The ayurvedic physician, called a *vaidya*, was esteemed for his powers but also shunned because of his contact with impurities such as body products, suppurative lesions, and corpses, and his mingling with common people. Taboos around touching ultimately resulted in palpation falling into disuse.

The physician was enjoined to strive constantly to acquire new knowledge, advance through practical experience, and enter into learned dialogues with practitioners from other places. His education began as an apprentice, with the teacher and pupil choosing each other. A good teacher was free of conceit, greed, and envy, and a student had to be calm, friendly, and without physical defects. Later on the *vaidya* became a sub-caste or occupational division, and the profession passed from father to son.

The *Caraksaṃhitā* contains an extensive list of ethical directives in the form of an oath to be taken by one entering medical practice. Among these were injunctions never to abandon a patient even if that interfered with one's livelihood, to be modest in dress and conduct, gentle, worthy, and wholesome. A physician must not enter a patient's house without permission, and be mindful of the peculiar customs of a household. He was to avoid women who belonged to others and maintain confidentiality.

Quacks and charlatans were known by their pretense and arrogance, boastfulness and superficial knowledge. The fate of their patients was worse than death. The *Caraksaṃhitā* says that one can survive a thunderbolt but not the medicine prescribed by quacks.

Medical ethics was an integral part of ancient Indian medicine. The texts addressed ethical issues that arose at both ends of life, birth and death. Their approach was pragmatic and flexible, and the purpose of alleviating an illness was always considered in the context of geographic locale, time (the era and the stages of a patient's life), and the particularities of a person. The physician's conduct was also to be always above reproach both in his professional and personal conduct.

Prakash N. Desai

References

Primary sources

Brihadaranyaka Upaniṣad. Trans. F. Max Muller in the *Upaniṣads* part 2. New York: Dover. 1962.

Carakasaṃhitā. 2 vols. Trans. Priyavrat Sharma. Varanasi, India: Chaukhambha Orientalia. 1981–83.

Mahābhārata, vols 1–3. Trans. J.A.B. Van Buitenen. Chicago: University of Chicago Press. 1973.

Suśrutasaṃhitā. Trans. Vaidya Jadavji Trikamji Acharya and Narayana Ram Acharya. Varanasi, India: Chaukhambha Orientalia. 1980.

Secondary sources

Chandrashekar, Sripati. *Abortion in a Crowded World: The Problem of Abortion with Special Reference to India.* London: Allen and Unwin. 1974.

Chattopadhyaya, Debiprasad. *Science and Society in Ancient India.* Calcutta: Research India Publications. 1977.

Dasgupta, Surendranath. *A History of Indian Philosophy.* 5 vols London and New York: Cambridge University Press. 1922–1955.

Desai, Prakash. "Medical Ethics in India." *Journal of Medicine and Philosophy* 13: 231–255. 1988.

Desai, Prakash. *Health and Medicine in the Hindu Tradition.* New York: Crossroads. 1989.

Desai, Prakash. "Hinduism and Bioethics." In *Bioethics Yearbook*, vol. I, *Theological Developments in Bioethics*, 1988–1990. Eds. Baruch A. Brody et al. Dordrecht: Kluwer Academic Publishers. 1991. pp. 41–60.

Fujii. Masao. "Buddhism and Bioethics in India: A Tradition in Transition." In *Bioethics Yearbook*, vol. I, *Theological Developments in Bioethics*, 1988–1990. Eds. Baruch A. Brody et al. Dordrecht: Kluwer Academic Publishers. 1991. pp. 61–68.

Jolly, Julius. *Indian Medicine.* Trans. Chintamani Ganesh Kashikar. New Delhi: Munshiram Manoharlal. 1977.

Purushottama, Bilimoria. "The Jaina Ethic of Voluntary Death." *Bioethics* 6(4): 331–355. 1992.

Young, Katherine K. "Euthanasia: Traditional Hindu Views, and the Contemporary Debate." *In Hindu Ethics: Purity, Abortion. and Euthanasia.* Ed. Harold G. Coward, Julius J. Lipner, and Katherine K. Young. Albany: State University of New York Press. 1989. pp. 71–130.

See also: Medicine – Caraka – Suśruta

Medicine: Āyurveda

The science and art of Āyurveda are integral parts of the cultural heritage which is preserved, fostered, and promoted in India and its neighbouring countries for the preservation and promotion of positive health and the prevention and cure of disease. As a science, it is based on sound and rational principles of physiology, pathology, pharmacology, diagnostics, and therapeutics which have been critiqued, systematised, and generalised on the rigid principles of logic. Even today it is followed as a healing art, not only by rural and poor people but also by well-placed persons of learning in all walks of life, including affluent intellectuals who could otherwise easily afford to obtain the services of top-ranking physicians of modern Western medicine.

Literally meaning the "science of life", Āyurveda is often used in a narrow sense as a "system of medicine", which considerably dilutes and distorts its real scope and objective. Health according to Āyurveda is not only freedom from disease. According to Suśruta, one of the great early practitioners, it is a state of the individual where, in addition to harmony among the functional units (*doṣas*), digestive and metabolic mechanisms (*agnis*), structural elements (*dhātus*), and waste products (*malas*), a person should also be in an excellent state (*prasanna*) of the spirit (*ātman*), senses (*indriyas*), and mind (*manas*).

The history of Āyurveda is as old as that of the Vedas, as the former is considered to be one of the limbs (*aṅgas*) of the latter. It has eight specialised branches:

(1) *Kāya-cikitsā* (internal medicine),

(2) *Śalya-tantra* (surgery),

(3) *Śālākya-tantra* (treatment of diseases of the eyes, ears, nose, and throat),

(4) *Agada-tantra* (toxicology),

(5) *Bhūta-vidyā* (treatment of seizures by evil spirits and other mental disorders),

(6) *Bāla-tantra* (pediatrics),

(7) *Rasāyana-tantra* (geriatrics, including rejuvenation therapy), and

(8) *Vājīkaraṇa-tantra* (sexology and the use of aphrodisiacs).

Several classics were composed for each of these branches, and each was widely practiced.

Āyurveda is a holistic science of life. Each part of the physique is interrelated with other parts for their proper functioning. The functions of the body are closely related to the mind and soul (spirit) of the individual. To maintain health, the body should be free from disease, the mind should be happy, and the person should be spiritually elevated. To cure diseases, the physician examines and treats the whole body and the mind as well as the spirit. Therefore, for the preservation of positive health and prevention as well as cure of diseases, several codes of conduct and religious rituals are prescribed, along with medicines and food. Thus, Āyurveda advocates the psychosomatic interrelation of health and disease.

Several categories of germs and organisms are said in Āyurveda to cause diseases like *kuṣṭha* (obstinate skin diseases, including leprosy). But these organisms are considered causative factors of a secondary nature, the primary cause being the disturbance in the equilibrium of the *doṣas* (three factors regulating the functioning of the body). As seeds germinate only when sown over fertile land, and decay if the land is barren, similarly germs, however virulent they may be, will not be able to multiply and cause a disease if the constituents of the body are in a state of equilibrium. Āyurvedic therapies are prescribed not to destroy the invading germs (seeds), but to bring about equilibrium of the constituents by which these germs, finding the atmosphere hostile (barren field) for their survival, get destroyed.

In Āyurveda, drugs of vegetable origin, animal products, and metals, minerals, gems, and semi-precious stones are used for therapeutic purposes. These are used in their natural form and processed in order to obtain their whole extract or to make them non-toxic, palatable, and therapeutically more potent. Different parts of these drugs, like alkaloids, glucosides, and other active principles are not extracted for therapeutic use. According to Āyurveda, every drug has therapeutically useful parts that may produce toxic effects if used alone. The same drug, however, contains other parts that counteract these adverse effects. Therefore, the use of the whole drug is emphasised, and no isolated section or synthetic chemicals are used. Some ingredients used in recipes are no doubt toxic in their raw form. But these are processed and detoxified according to prescribed methods before being added to recipes. These recipes have been in constant and regular use for thousands of years, which testifies to the absence of any acute, sub-acute, or chronic toxicity and teratogenic effects. These recipes are equally useful for patients and healthy persons. In the former, while curing the disease, they produce immunity against future attacks, and therefore, instead of giving toxic side effects, they produce side benefits. In healthy individuals, they revitalise the body cells and stimulate the immune system as well as fortify the body's resistance against disease.

Considerable emphasis is laid upon proper diet in Āyurveda for the prevention and cure of disease. If a person consumes a proper diet, medicines are superfluous for him because he will not fall victim to diseases, and if he does, the ailment will get corrected easily. If, however, he does not follow a proper diet, he will not get cured in spite of the best medicines. Therefore, Āyurvedic works describe in detail the properties of various ingredients of diet and beverages.

According to Āyurveda, the individual (microcosm) is a replica of the universe (macrocosm). Virtually every phenomenon of the universe or *brahmāṇḍa* can be found to take place in the individual or *piṇḍa*, albeit in a subtle form. Therefore, every action of the individual in its turn has an impact on the environment as well as the universe. This intracosmic relationship which is the very foundation of Āyurveda is further elaborated in Tantra and Yoga.

The scope and utility of Āyurveda are not confined to any particular community or any political, geographical, economic, or sociocultural group. The principles of Āyurveda are equally useful for and applicable to all the people in different parts of the world. According to Caraka, an early physician (before 600 BCE), the "science of life" has no limitations. Therefore, humility and relentless industry should characterise every endeavour of the physician to acquire knowledge. The entire world consists of teachers for the wise, and only a fool finds enemies in it. Therefore, according to him, the knowledge conducive to health, longevity, fame, and excellence coming even from an unfamiliar source (lit. enemy) should be received, assimilated, and utilised with earnestness. This approach of Āyurveda is evident from the descriptions in the classics. Several plants, animal products, and minerals which are not available in India are described in detail with reference to their therapeutic properties. The properties of several ingredients of food and beverages which are normally not used in India, such as meat, especially beef, are also described. The fundamental principles of Āyurveda are also described. These principles are applicable in any part of the world, and its practice can be modified according to regional requirements.

Pañcamahābhūta Theory

The body and the universe, according to Āyurveda, are composed of five *mahābhūtas* (basic elements), namely, *pṛthvī, ap, tejas, vāyu,* and *ākāśa.* These terms are often mistranslated as earth, water, fire, and ether, which are misleading even for well-meaning critics of Āyurveda. These five categories of matter correspond to five senses, namely smell, taste, vision, touch, and hearing. The objective matters of the universe are subjectively comprehended by means of these five senses. These five states of matter (subjective series) and the five senses (objective series) are both the evolutive products of *prakṛti* (primordial-matter-stuff). These five *mahābhūtas* do not stand for the elements of the present day physical sciences as the mistranslation of these terms would have them, but the five classes of objects of our material universe correlated to the five senses by means of which one subjectively contacts the objective universe.

The body, as well as food and drugs, are all composed of these five *mahābhūtas.* It is disturbances in the equilibrium of these which cause disease. To correct the ailments so caused, appropriate *mahābhūtas* in the form of drugs and food must be used. Though the characteristic features of these *mahābhūtas* are described in detail in Āyurvedic texts, in practice it is extremely difficult to ascertain the precise requirements. Hence, their presence in the body is described in terms of *doṣas* (functional units of which there are three types), *dhātus* (tissue elements of which there are seven categories), and *malas* (waste products). Similarly, the mahābhautic composition of drugs, diet, and beverages is described in terms of *rasa* (taste), *guṇa* (attribute), *vīrya* (potency), *vipāka* (aftertaste), and *prabhāva* (specific action).

The *doṣas, vāyu, pitta,* and *kapha* are the three elementary functional units or principles on which the building and sustenance of the body depends. The individual remains healthy as long as these three elements remain in a state of equilibrium. But if this is disturbed beyond certain limits, the individual succumbs to disease and decay. The *doṣas* remain in two different forms in the body, namely, *sthúla* or gross, and *sūkṣma* or subtle. In their subtle state, they are *atīndriya* or beyond the normal cognition of the senses. Their normal and abnormal states are ascertained by the manifestation of their respective actions. These three *doṣas* control all the physical and psychic functions of an individual. Each one of these is further subdivided into five categories on the basis of their actions on different parts of the body.

The physical structure of the body, according to Āyurveda, is composed of seven categories of *dhātu*s or tissue elements:

- *rasa* (plasma, including chyle),
- *rakta* (blood, particularly the haemoglobin fraction of it),
- *māṃsa* (muscle tissue),
- *medas* (fat),
- *asthi* (bone tissue),
- *majjā* (bone marrow), and
- *śukra* (generative fluid, including sperm and ovum).

There are two categories of tissue elements, *poṣaka* (nutrients of tissues) and *poṣya* (stable tissues). The former moves through different channels to provide nourishment to the latter. The circulation of these nutrient tissue elements is controlled by *vāyu*, their metabolism including assimilation is controlled by *pitta*, and *kapha* helps in the maintenance of their cohesion and compactness. During the metabolic transformation of these tissue elements, a substance called *ojas* is produced as their essence. This *ojas* provides immunity or the power of resistance to attacks of disease. It is said that the diminution of this *ojas* (*ojaskṣaya*) is responsible for the disease called AIDS.

During the process of digestion and metabolism, several waste products are formed. Some of these serve a useful purpose, but ultimately they are eliminated from the body through faeces, urine, sweat, and so on. Their regular elimination from the body is essential for the maintenance of good health. In Āyurvedic therapeutics, attention is always paid to the elimination of these wastes.

Āyurveda classifies individuals broadly into seven categories called *prakṛtis*. These are defined by the predominance of *vāyu, pitta*, and *kapha*, or their combinations, or by the state of equilibrium of them all. This trait of the individual is developed at the time of birth and continues throughout one's life. Individuals are also classified into sixteen categories, on the basis of their psychic disposition. The former seven categories are called *deha prakṛti* or physical constitution, and the latter sixteen categories are called *mānasa prakṛti* or psychic temperament. Knowledge of these *prakṛtis* is essential for selecting the appropriate food, drink, and other regimens for different categories of healthy individuals, and likewise for the treatment of their ailments.

As has been explained earlier, the ingredients of drugs, diet, and drinks are composed of five *mahābhūtas*. In order to facilitate examining the exact nature of these *mahābhūtas*, easy practical methods are described in Āyurveda in terms of *rasas* (tastes), of which there are six, twenty *guṇas* (attributes), *vīryas* (potencies) of which there are broadly two, and three *vipākas* (aftertastes). Āyurvedic texts also describe the specific and general actions of drugs.

There are several other unique concepts of Āyurveda, understanding of which is essential before appreciating the merits of this health science. Āyurveda is a repository of therapeutically useful and time-tested recipes for curing several obstinate and otherwise incurable ailments. But its prescriptions for the preservation and promotion of positive health and prevention of disease is what is especially useful to relieve human suffering.

Bhagwan Dash

References

Agniveśa. *Carakasaṃhitā*. Bombay: Nirnayasager Press. 1941.

Bhela. *Bhelasaṃhitā*. Delhi: Central Council of Research in Indian Medicine. 1977.

Dash, Bhagwan. *Fundamentals of Āyurvedic Medicine*. Delhi: Konark Publishers. 1992.

Dash, Bagwan, and Lalitesh Kashyap. *Materia Medica of Āyurveda*. Delhi: Concept Publishing Co. 1980.

Dash, Bagwan, and Lalitesh Kashyap. *Basic Principles of Āyurveda*. Delhi: Concept Publishing Co. 1980.

Śarmā, Śiva. *Āyurvedic Medicine: Past and Present*. Calcutta: Dabur. 1975.

Suśruta. *Suśrutasaṃhitā*. Varanasi: Chaukhamba Oriyantalia. 1980.

Vāgbhaṭa. *Aṣṭāṅgasaṅgraha*. Pune: Athvale. 1980.

Vāgbhaṭa. *Aṣṭāṅgahṛdaya*. Varanasi: Kṛṣṇadāsa Academy. 1982.

Vṛddhajīvaka. *Kāśyapasaṃhitā* Varanasi: Chaukhamaba Sanskrit Series Office. 1953.

See also: Suśruta – Caraka

Medieval Science and Technology

This article deals not with the ancient period of Indian history, but with the time after the conquest of northern India by the Turko-Afghāns between CE 1192 and 1206. Its focus is on Medieval India.

India had already had commercial and cultural relations with Greece, Afghanistan, and Central Asia since prehistoric times. When the Muslims arrived in India a cross fertilisation of scientific and intellectual ideas began. Muslims migrated to India after CE 1258 because of the Mongol invasion of Iran, Iraq, and Central Asia; some of them were astronomers, mathematicians, and physicians.

"With the establishment of the Ghaznavid and Mughal rule in India the Greek or rather more advanced Ptolemaic astronomy in an Arabic version reached India and began to be studied and taught among the Muslim and Hindu astronomers who appreciated its merit." (Sen, 1971). Al-Bīrūnī (d. CE 1050) claims that he translated Euclid's *Elements* and Ptolemy's *Almagest* into Sanskrit, but these translations are not available. In any case, the Arabic versions of these two books were introduced by the Muslims, and towards the close of the twelfth century CE mathematical books in Arabic also began to trickle into India.

Some of the Muslim mathematicians and astronomers who settled in India knew these Arabic translations and translated them into Persian. Thus Greco-Arab astronomy and mathematics were introduced. Some of this material was incorporated into textbooks used in the educational institutions. Akbar (d. CE 1605) made those subjects compulsory.

Both mathematics and astronomy were used by the Hindus and Muslims for religious purposes. Mathematics was also studied for its practical utility, as in the construction of buildings and monuments. Later, geometry was further developed for this purpose.

The outstanding mathematician-astronomers of this period were Śrīdharācārya (ca. CE 991), Śrīpati (CE 1039–1056), Śatānanda (fl. CE 1099), and Bhāskarāchārya II (CE 1150).

Al-Bīrūnī studied the Indian sciences in the original Sanskrit texts and translated some of them into Arabic. His book *Kitāb al-Hind* (Book on India) gives valuable information about astronomical methods and Indian astronomers, some of whom were personally known to him. His *Ras›āil* (Treatise), the *al-Qānūn al-Mas‹ūdī* (Masudic Canon) contains useful source material for the history of science in India in the first half of the eleventh century.

A new development in astronomy was the introduction of Arabo-Persian-Greek *Zīj* literature (astronomical tables). 'Abu›l-Faḍl mentions eighty-six *Zīj*es in his *Ā›īn-i Akbarī*. There were several *Zīj*es prepared in pre-Mughal India.

Astrolabes were used in India after the Muslims arrived. In 1370, during the reign of Sultan Fīrōze Shāhā Tughluq, an astrolabe was constructed and named *Asṭurlāb-i Fīrōze Shāhī*. This testifies to the fact that Indians possessed adequate knowledge of applied technology during the fourteenth century.

Ḍiyā᾽ad-Dīn Baranī records the names of the astronomers and astrologers who flourished during this period, and adds that other sciences such as *ramal* (geomancy) and *al-Kīmyā* (alchemy) also flourished.

Another fact worth mentioning is that the first substantial contact between the Āyurvedic (Indian) and Ūnānī (Greco-Arab) systems of medicine began during this period, which resulted in mutual enrichment in therapeutics, materia medica, and pharmacology. There were seventy hospitals in Delhi alone under Sulṭān Muḥammad ibn Tughluq (CE 1297–1348) having 12,000 *Vaidyas* (Hindu physicians) and *Ṭabībs* (Arab physicians) paid by the state.

Medicinal herbs and plants were widely cultivated. Al-Bīrūnī's *Kitāb al-Saydanah fī᾽ṭ-Ṭibb* (Book on Pharmacy and Materia Medica) contains useful information for the pharmacographia Indica. Compiled in the middle of the eleventh century, it is an encyclopedia of simple drugs containing medicinal herbs and minerals used in the Ūnānī system, arranged in alphabetical order. There are names of hundreds of Indian medicinal herbs and plants.

The arrival of the Turko-Afghāns in India also brought some changes in the area of technology. In the field of metallurgy, iron, copper, brass, gold, silver, and other minerals were extracted from the mines in different parts of India by a simple technology with the help of a clay furnace in which wood and charcoal were burnt. As the army was important, iron was much in use for the manufacture of arms and armour. For the minting of coins, mostly gold, silver, lead, bronze, brass, and copper were used.

The premier industry in India was cotton textile; the dyeing, printing, and painting of clothes had been done since ancient times. In connection with the innovation in textile technology, the general use of the spinning wheel should be mentioned. There is a lack of positive evidence ascribing its origin to India. Moreover, the word for it is *charkha* which is Persian. In the absence of better evidence, literary ones from Indo-Persian works can be put forward to show that the spinning wheel in which the belt-drive technique was employed was in use at the end of the twelfth century. Another piece of equipment which was much used in this period was the bow-string (*Tant* and *Kamān*) for cleaning the cotton and separating the seeds (ginning). This technique is still used in India today.

Al-Bīrūnī gives evidence that around CE 1030 the materials used for writing in India were mainly black tablets, palm leaves, the bark of the Tūz tree called *Bhūrja*, and silk. The manufacture of white paper started in India in the thirteenth century. Amīr Khosrāw mentions paper (*Kāghaz*) several times in his work. He states that paper was made with cotton, linen, silk (*Qash, Ḥarīr*), and reed (*Kilk*). They were soaked in water, then pounded and turned into pulp, and dried. After that, they were cut with sharp scissors. He adds that this light paper was quite costly. As regards the preparation of ink, the earliest recipes for hair dye in India are found in the *Navanītaka* (ca. second century CE). From this Nityanath Siddha (CE 1200) derived his recipes for ink as recorded in his *Rasaratnākara*. The ingredients used were metallic or herbal substances such as lamp-black charcoal, gum, burnt almond husk, or gold and silver powder.

In the field of irrigation, the large Perumamilla tank bears an inscription dated CE 1291. There is a difference of opinion about the device for water raising used in ancient India called *Araghaṭa* or *Ghati-Yantra* which are Sanskrit words. Several scholars have argued that it was the same as the Persian wheel or *sāqiyah* and it was not introduced into India by the Turko-Afghans. But these two words are not clearly mentioned in any early source. It was actually the simple *noria*, a wheel carrying pots and buckets on its rim which did not involve gearing. One view is that the Persian wheel was introduced by the Turko-Afghans in India in the thirteenth century. It was used to raise well water with the help of a chain of buckets driven by animal power using pin-drum gearing.

The planks of the hulls of Indian ships were sewn together by coir and generally not nailed. G.F. Ḥourānī says that al-Maʾsūdī states "...in ships of the Indian Ocean iron nails do not last because the sea-water corrodes the iron and the nails grow soft and weak in the sea and therefore the people on its shores have taken to threading cords of fibre instead and these are coated with grease and tar". Ibn Baṭṭūṭa, who arrived in India in September 1333, gave a detailed account of the technique used in the ships' manufacture.

Indian builders were employed by the Muslim rulers for the construction of monuments, and several expert masons and craftsmen were also brought from Persia and Afghanistan, introducing the Arabo-Persian architectural and decorative traditions. They used the lettering of the Quranic verses as well as leaves, flowers, buds, and geometrical designs (*arabesque*). This brought about great changes in Indian architectural style. The imported style consisted of domes, portals, minarets (*mīnār*), and pendentive and squinch arches. In so far as building materials are concerned, stones and bricks were used, and lime-mortar was employed perhaps for the first time in India.

So many articles made of glass have been discovered, especially at Kopia in Uttar Pradesh, and the Taxila and Satavahana sites, that it cannot be argued that all of them were imported goods. Most of the articles are beads, bangles, bowls, slags, small vessels, tiles, and flasks. But very little information is available concerning the technique of glass-making, including the raw materials, furnaces, or tools used.

In the field of military technology, it is generally asserted that the stirrup was used in ancient India, and some evidence shows that the flat iron stirrup was introduced by Turkish conquerors in the early thirteenth century. Nailed horseshoes were used after the Turko-Afghan conquest.

To manufacture a sword the two ingots were covered with soft earth in order to prevent carbon loss; then both the ingots were forge-welded by hammering when the steel was softened. The process of hardening steel by quenching the red hot metal into water or oil is still employed today. The first hammering was for preparing a blank, while the second was to give it shape. The blunt sword was sharpened on an abrasive wheel and polished later with oxymel.

The question arises whether artillery and gunpowder were used before Babar's time (CE 1530) or not. This is a controversial issue. Some historians state that mechanical artillery was in use in India as early as the first quarter of the thirteenth century; others are of the view that it was not used until CE 1365–1366 (Makhdoomee, 1936). Naptha or Greek fire was used by the Turko-Afghan invaders, but this also may mean gunpowder.

One of the difficulties in writing this history is that the source materials, especially the manuscripts in Sanskrit, Arabic, and Persian, have not been edited and published. Some of them are uncatalogued so that they are unknown even to those who do research on the history of science in India. Therefore, it is evident that a thorough assessment of India's contributions to scientific and technological development is not possible now, and the history of science and technology in late ancient and early medieval India presented here is neither thorough nor complete.

M.S. Khan

References

al-Bīrūnī. *Kitāb al-Hind* (Alberuni's India). Trans. E.C. Sachau. New Delhi: Oriental Reprint. 1983.

Bag, A.K. "Al-Bīrūnī on Indian Arithmetic." *Indian Journal of the History of Science* 10: 174–186. 1975.

Chowdhury, Mamata. "The Technique of Preparing Writing Materials in Early India with Special Reference to al-Bīrūnī's Observations." *Islamic Cultures* 48(1): 33–38. 1974.

Chowdhury, Mamata. "The Technique of Glass Making in India." Paper presented at *National Seminar on Technology and Science in India during 1400–1800* AD. New Delhi: Indian National Science Academy. 1978.

Gode, P.K. "Recipes for Hair Dye in the *Navanītaka* and Their Close Affinity with the Recipe for Ink Manufacture after AD 1000." *Studies in Indian Cultural History* 1: 101–110. 1961.

Ḥourānī, George Fadlo. *Arab Seafaring in the Indian Ocean in Ancient and Early Medieval Times*. Princeton: Princeton University Press. 1951.

Khan, M.S. *Al-Biruni and Indian Sciences*. forthcoming.

Khan, M.S. "Arabic and Persian Source Materials for the History of Science in Medieval India." *Islamic Culture* April–July. 1988. pp. 113–139.

Khan, M.S. "The Teaching of Mathematics and Astronomy in the Educational Institutions of Medieval India." *Muslim Education Quarterly* 6. 1989.

Khan Ghori, S.A. "Development of *Zīj* Literature." In *History of Astronomy in India*. Ed. S.N. Sen and K.S. Shukla. New Delhi: Indian National Science Academy. 1985. pp. 21–48.

Makhdoomee, M.A. "Mechanical Artillery in Medieval India." *Journal of Indian History* 11: 189–195. 1936.

Mukherjee, B.N. "Technology of Indian Coinage, Ancient and Medieval Period." In *Technology in Ancient and Medieval India*. Ed. Aniruddha Roy and S.K. Bagchi. Delhi: Sundeep Prakashan. 1986. pp. 47–70.

Ray, Priyadaranjin. *History of Chemistry in Ancient and Medieval India*. Calcutta: Indian Chemical Society. 1956.

Ray, Priyadaranjan. "Medicine as it Evolved in Ancient and Medieval India." *Indian Journal of the History of Science* 6:86–100. 1970.

Sen, S.N., D.M. Bose and B.V. Subbarayappa. *A Concise History of Science in India*. New Delhi: Indian National Science Academy. 1971.

White, Lynn. "Tibet, India and Malaya as Sources of Western Medieval Technology". *American Historical Review* 65(3): 517. 1960.

See also: al-Bīrūnī – Astronomy – *Zīj* – Medicine in India: Āyurveda – Agriculture – Technology and Culture – Metallurgy – Textiles – Irrigation – Navigation – Military Technology

Metallurgy: Bronzes of South India

The discovery and use of metals revolutionised the history of mankind and accelerated human evolution. The first metal to be discovered and used was copper in its native form, found along riverbeds. The period when man used both stone and copper is the 'Chalcolithic' period (5000 BCE), *chalkas* means copper and *lithos* means stone. All other ages, such as the Brass and Bronze Age that followed, were based on the foundations laid by the Chalcolithic Age. The present article describes the evolution of the Bronze Age with an emphasis on the marvels of South India. South

Indian bronzes are best appreciated if we know the global background of the growth of the Bronze Age.

Genesis of Bronze Age

Native copper, i.e., copper metal collected from riverbanks, is thought to have been one of the first metals used for practical applications, around 5000 BCE. The word copper came from *cuprum*, the Roman name for Cyprian metal. Later, around 4500 BCE, extraction of copper from its ore began in civilisations in the Middle East. The Chaldeans used objects of beaten copper about 4500 BCE. Copper weapons and ornaments from about the same time have been found in the ruins of Susa, an ancient civilisation located in the area of the nation that is now Iran. Clear early evidence for smelting copper comes from the Middle East from about the fourth to third millennium BCE onwards, from parts of Israel, Jordan and Egypt. Early copper artifacts of about the sixth millennium BCE are also reported from the pre-Indus Valley sites of Baluchistan in the north-western part of the Indian subcontinent close to the Iranian border. There is fairly extensive evidence for the ancient mining of copper ores from the Khetri region of Rajasthan in North-western India dating to about the 3rd–2nd millennium BCE. The craftsmen of Ganeshwar in Rajasthan were making copper items from about 3000 BCE, from a time even before the cities of the Harappan civilisation emerged. The copper artefacts from Ganeshwar found their way to Harappan sites and to many other places in North and Central India.

Brass, referred to in ancient literature as *pittala* or *ārakuṭa,* was not known in the Chalcolithic Age of India, but came into usage only in the Iron Age (1000 BCE to 300 CE), during the second urbanisation. By the sixth or seventh century, it became popular, along with bronze.

The most important metallurgical development was the discovery that adding tin to copper produced a far superior metal, eventually known as bronze (called "Vengalam" in Tamil– a South Indian Language). Thus the civilisation transitioned from the Copper Age to the Bronze. The Age of Bronze constitutes an important epoch in the history of culture. At first, bronze was only used decoratively, because many civilisations did not have the copper ore to produce large quantities of it. Gradually, as copper ore became more plentiful, bronze replaced stone as the material of choice for weapons, tools, and the like. Civilisations began their bronze ages at different times, with some as early as 2500 BCE and others not until 1000 CE. The use of bronze started a little later (2100 BCE) in the Harappan culture of India; it had already started in Iran (3900 BCE), Iraq (2800 BCE), Egypt (2600 BCE) and Turkey (2500 BCE). India was overtaken by the Iron Age before it achieved a fully-developed Bronze Age. On the technological side, nowhere in the world was icon making in bronze mastered as it was in India. Tin mines are known from ancient Turkey dating to the 3rd millennium BCE. The Sumerians were the first civilization to use bronze in commerce. Bronze artifacts dated 3600 BCE have been found in Thailand. The Shang dynasty's capital of Anyang in Northern China had a bronze-casting industry in 1400 BCE. Amongst the earliest bronze castings in the world are the famous cat from ancient Egypt, the very impressive ceremonial vessels of China and the Greek bronze figurines.

Indian artisans and craftsmen have long been masters at extracting and shaping metals and alloys, as proven by archaeological finds from the 2nd–3rd millennia BCE. A few traces of a bronze industry from before 2000 BCE are found in India. However, the great use of bronze in India came during the Indus Valley civilisation, when it became a medium for religious sculpture. In the Harappan culture, metal workers knew the technique of using copper, bronze and lead, apart from silver and gold. Lost wax casting was not only known but was practised with great zeal. The statue of a dancing girl from Mohenjodaro is a classical example of the excellence achieved in bronze casting using the lost wax process (Figure 1).

Figure 1

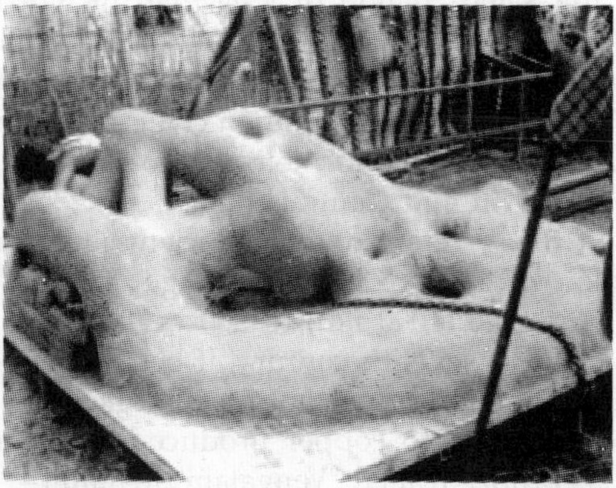

Figure 2

The bronze sculptures of *Gandhāra* in the 1st century CE mark the beginning of the history of Indian bronze. The recognition of the role played by copper in the cultural development of Indian civilisation is reflected in the label Copper–Bronze Age to a particular period of Indian history.

SOUTH INDIAN BRONZES

The South Indian bronzes are some of the finest creations of Indian visual art. Known for their aesthetic beauty, iconometry and casting quality, these bronzes are excellent examples of the fusion of technology with cultural traditions. These bronzes are considered global icons cherished by national, international and private antiquarians around the world and found in practically all the best museums in the world.

One of the most striking aspects of the ancient South Indian bronzes is their authentic representation of human and animal anatomy. Other items are also rendered in great detail. The reason for this detail was the process by which they were made, namely the *cire perdue* or lost wax process. In ancient Hindu scriptures, this process was referred to as *madūcchiṣṭhavidhānam, madūcchiṣṭha* meaning beeswax.

The lost wax process is divided into four distinct stages: 1) creation of the wax model; 2) forming the mould (with clay) around the wax model; 3) casting of the mould and 4) engraving and finishing the piece. The subject is first modelled in wax, and then coated with clay. The wax is then melted out, leaving a mould behind into which liquid metal is poured to cast a solid image. Thus, the wax model which serves as the core is lost or drained out before the actual casting takes place. After the casting of an image, its mould is destroyed, with the result that no two bronze icons are alike even if they are made by the same metalsmith or craftsman (also called *sthapati*, the local Tamil name given to a master craftsman who is an expert in either stone sculpting or bronze casting).

A pictorial illustration of the process is given in Figures 1–7. The process depicted is for producing solid castings. In the case of hollow castings there is one additional step. The *sthapati* models the object in clay first. The clay core model is then coated with a layer of wax. All the subsequent steps, such as forming the mould in clay over the wax model etc., are the same as outlined above.

Figure 3

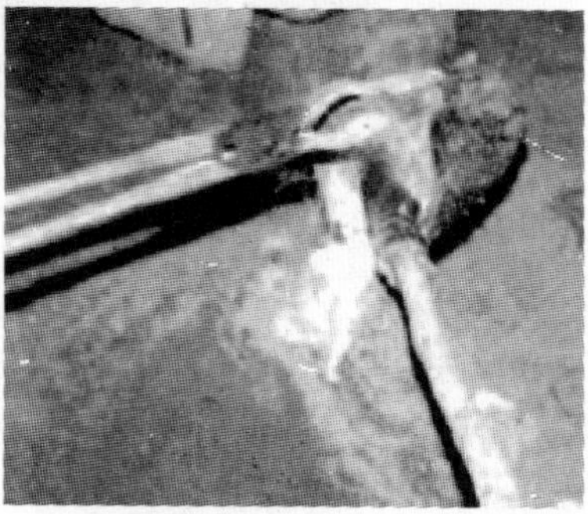

Figure 4

Figure 5

Figure 6

The art and science of casting bronze icons is a skill passed on from generation to generation and is still practised as a traditional village vocation. Some of the rare bronzes at the Government Museum in Chennai, India are shown in Figure 8.

Figure 7

Figure 8

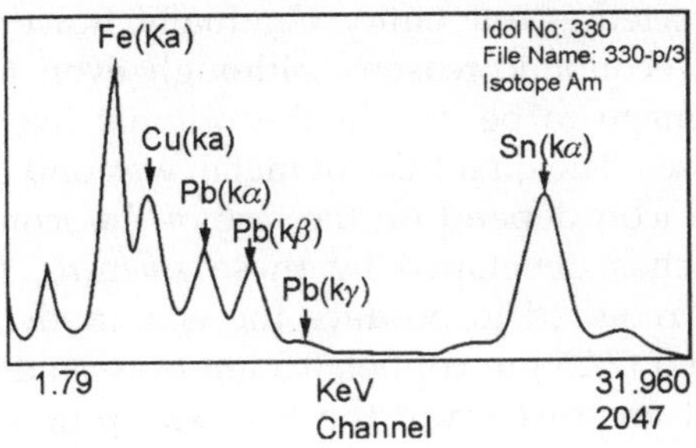

Figure 9

The excellence of the bronze icons that are cast by this method has been compared with those produced in Europe. W.S. Hadaway (1913) observed,

> The *cire perdue* process which is commonly, it might be said almost universally, used for either simple or intricate work, in India, usually produces, when manipulated in Western countries, a spongy and unsound casting almost impossible to work upon successfully and finish properly... In the West, when this process is used, the object is to obtain a casting, which requires as little

> finishing as possible, but it is always at the expense of the soundness of the whole mass, for to obtain the delicacy of the original wax model so fine, an earth must be used that it allows of no general ventilation of the mould... This finishing which the European hopes to avoid (but always at the expense of the soundness of the casting) by the *cire perdue* process is, in India, taken as a matter of course

From this account we understand that the European casters laid more emphasis on not having to polish the icon after casting. In the process, the cast icon itself was not sound enough.

An interesting aspect about the icons is that most of them were cast only in bronze. There could be many reasons for this, the most important being (a) castability, (b) workability after finishing the casting, (c) polishability, (d) melting point of the alloy (requiring a large quantity of molten alloy) and (e) availability of raw materials, ore, etc.

From the *sthapati's* point of view, castability, workability, polishability and melting point would have been the overriding factors. The shine one gets in bronze is generally not available in brass and other alloys that were prevalent in those times. This also should have been a major reason for choosing the alloying additions for obtaining desirable results. The other important reason is the scriptural command of the *Āgamas,* that prescribe metal to be used for casting idols for worship. Later, icons were also cast using *pañcaloha.* This term is derived from Sanskrit, in which *pañca* means "five" and *loha* means "metal". Formerly, these five consisted of the following metals which were considered to be auspicious – copper, silver, gold, brass and lead. Brass is referred to as metal, as copper and zinc were co-extracted as one entity. Gradually, however, gold and silver were deleted for economic reasons, although even today a client may commission an image to be cast in the original five metals for special devotional purposes. The quantities of metal, wax and other raw materials required for each icon depend on the size of the icon. According to the empirical approaches developed by these *sthapatis,* for every gram of various materials required to produce the wax model, 8 grams (gms) of alloy were required (7.25 gm copper, 0.5 gm brass and 0.25 gm lead). The bronze icons that are cast using the lost wax process can be classified broadly into two categories: those that are used (to be used) for worship and/or *pūjā* and those that are cast mainly for ornamental purposes. The composition of the latter category is roughly copper 82%, brass 15% and lead 3%, whereas in the former category, gold and silver are added in proportions that are dictated by the *Vedas* and the *Āgama Śāstras.* Table 1 gives the chemical compositions of typical icons (ancient to modern). In North India, icons were cast using *aṣṭaloha* – *aṣṭa* in Sanskrit means eight and is comprised of gold, silver, copper, iron, lead, mercury, zinc and tin.

Table 1 *Elemental composition of some South Indian icons*

Elements	9th century	13th century	16th century	21st century
Copper (wt%)	98.60	94.50	89.40	82.96
Tin (wt%)	0.88	1.10	3.40	3.58
Lead (wt%)	0.34	1.00	2.88	9.7
Zinc (ppm)	15.00	40.00	2646.00	2.5 (%)
Iron (wt%)	0.13	0.05	0.52	
Antimony (ppm)	476.00	1747.00	3095.00	
Silver (ppm)	541.00	596.00	823.00	
Arsenic (ppm)	1295.00	1785.00	1394.00	
Nickel (ppm)	436.00	415.00	636.00	
Bismuth (ppm)	85.00	164.00	351.00	
Cobalt (ppm)	45.30	135.00	114.00	
Manganese (ppm)	5.80	<0.80	1.80	1.26 (%)
Total	100.20	97.10	97.10	100

Archaeo-metallurgical Studies in South Indian Bronzes

The casting quality of the bronzes was quite excellent. A number of archaeo-metallurgical investigations using a variety of non-destructive test (NDT) techniques have been undertaken worldwide on such bronzes. NDT techniques, as the name implies, have been developed to determine quality and integrity without causing any harm or alteration to the objects under investigation. Table 2 lists some of the major NDT techniques used for investigations on objects of cultural heritage. Use of NDT techniques for such archaeo-metallurgical investigations also helps in the following ways.

Table 2 *Major NDT techniques used for authentication of art objects*

Sl. No.	Techniques	Capabilities
1.	Precision photography	Dimensions, measurements and record of intricate details
2.	Holography	3-D characterisation
3.	Moire fringes	Contour mapping
4.	Laser excited emission spectrometry	Chemical assay
5.	Mass spectrometry	Chemical assay
6.	Activation analysis	Trace analysis of elements
7.	X-ray energy spectrometry	Chemical composition
8.	In-situ metallography	Microstructural characterisation
9.	Radiography/Tomography	Internal defects, joints, repair regions
10.	Physical parameters: thermal emf, electrical conductivity, hardness	Characterisation of material and thermo-mechanical treatment

Understanding the style, period, structure and metallurgy of these icons.

Studies by Srinivasan (1999) have revealed that lead isotope and trace element analysis can be a powerful fingerprinting method for exploring classifications in metal artefacts and can be used as pointers to the periods in which such bronzes were cast.

Assessing the condition of the bronzes help in restoration and conservation.

Most bronzes have a well-developed patina which adds to their aesthetic beauty. Understanding the nature of the patina and the chemical composition is quite important from a conservation point of view. X-ray fluorescence techniques have been used to characterise the patina and also the surface composition of the bronzes. Figure 9 is a typical XRF spectrum from the surface of an ancient South Indian bronze. XRF investigations on more than 18 bronzes by Baldev Raj et al. (2000) have indicated that the major constituent in all these is copper with the additive elements being tin, lead and iron.

Authenticating and fingerprinting the bronzes.

These bronzes, famous for their beauty and excellent craftsmanship, are prone to theft. These investigations help in scientific and comprehensive fingerprinting of icons needed for documentation and authentic identification when they are retrieved. Icons are basically cast structures. During the process of casting, discontinuities or flaws are likely to be created for a variety of reasons. Radiography is the best technique to detect internal features and flaws in the castings. Depending on the casting process, the location of the defects if formed would be different in different icons. Thus, these defects can serve as authentic fingerprints of an icon. Duplication of these defects is not possible, as the probability of the defect occurring in the same location and with the same geometric contours and area is extremely low. Apart from serving as fingerprints, the radiographs also serve as indicators of the quality of the castings. Figure 10 shows a typical digitised radiographic image of a portion of the prabhāvalī of a dancing Śiva cast by the traditional lost wax process. We see only minor inclusions. Analysis of the radiographs of more than 200 rare South Indian bronzes by Baldev Raj et al. (2000) clearly indicates that the main body of the icons have excellent casting quality with porosities being the only major type of defect, especially in the thinner portions.

In the western world, objects of cultural heritage mainly comprise stone sculptures, paintings, coins, silverware and pottery. There are fewer bronze objects compared to what is available in India. Some of the scientific investigations using NDT include: radiography of the famous Liberty Bell to characterise the crack that had occurred in it before the restoration of the bell could be undertaken, authentication of Roman vessels and wares by radiography and digital radiography and computed tomography of ancient Chinese cast bronze at the Cincinnati Art Museum in Ohio.

Bronzes were also cast using the lost wax process by artisans in North India, especially at Bastar in Chattisgarh. The essential differences between the Bastar and the South Indian bronzes were the raw materials and the methodology. While South Indian bronzes had rigorous rules based on heuristics, Bastar bronzes were cast based totally on the imagination of the metalsmith and did not follow proportions as far as human anatomy was concerned. Cooper (1996) has written an excellent treatise on Bastar bronzes.

South Indian bronzes have been a point of attraction from ancient times and will continue to be so in the years to come. What gives a unique signature to these bronzes is the fact that each casting process is non-repeatable, since both the wax model and the clay shell are lost forever after each icon is cast. It is because of this that it is impossible to find two identical icons cast in this process. Each icon thus becomes a designer icon. Hence the lost wax method is also called the master technique. The making of images in this process is indeed laborious. But each item is characterised by a rare individuality and provides the possibility of continuous improvement, aesthetics and technology – a hallmark achieved by ancients as a matter of course.

Baldev Raj, B. Venkatraman and D. M. Vijayalakshmi

References

Raj, Baldev, C.Rajagopalan and C.V.Sundaram. *Where Gods Come Alive: A Monograph on the Bronze Icons of South India*. New Delhi: Vigyan Prasar. 2000.

Hadaway, W.S., E. Thurston and Velayuda Asari. *Illustrations of Metal Work in Brass and Copper Mostly in South India*. Madras: Madras Government Press. 1913.

Srinivasan, Sharada. *Archaeometry* 41 (1): 91–116.

Cooper, Ilay, John Gillow, and Barry Dawson. *Arts and Crafts of India*. London: Thames and Hudson. 1996.

Metallurgy: Iron and Steel

The ability to make iron from its ore marked a milestone in the history of civilisation. On the one hand, it required a degree of sophistication in technology, because it involved working at the limits of temperatures that could be achieved with simple implements. On the other hand, it provided a material with wide ranging properties which offered a variety of possibilities for tool making. Until recent times therefore, the progress of a nation was often judged by the per capita consumption of iron and its alloy, steel.

Iron and steel are of great importance for the following reasons.

- Iron is the fourth most abundant element. Among metals only aluminum is more abundant. Five per cent of the earth's crust is made of iron. Iron ores are also widely distributed. In India large and small deposits are distributed throughout the country, outside the Gangetic plains.
- Temperatures needed for iron smelting are easily attainable in a vigorously worked charcoal fire. They are, however, higher than those required for smelting some of the non-ferrous materials like copper, zinc, and tin, and therefore iron smelting needs great expertise.
- Iron, when mixed with the inexpensive alloy element, carbon, becomes steel. This alloying converts it into a material with a large possible range of properties from low strength, highly ductile, low carbon steels to high strength, but less ductile, high carbon steel. At higher concentrations of carbon, iron forms an alloy with excellent casting properties which melts at a low temperature. This alloy is called cast iron.
- Iron can be easily welded by forging. Before fusion welding came into existence, this property was important.

The origin of the knowledge of iron and steel technology in India is still a matter of debate. Recent archaeological evidence, however, is making it increasingly apparent that iron making in India is of an independent origin and may even predate the iron age in many other civilisations. Chakrabarti and to some extent Hegde have presented extensive evidence to support this view.

The iron age probably started in India not later than the second half of the second millennium BCE. Archaeological evidence pertaining to this period has come up in various places across the country, indicating that the origin might have been earlier. Though literary evidence also supports this hypothesis (e.g. the reference to a metal called *ayas* in the *Ṛgveda*), it is not conclusive.

It is undisputed that well before the beginning of the Christian era, artisans here had achieved a level of excellence, so that Indian steel was famous for its quality throughout the old world. There was extensive export trade even at that time. In the first millennium CE, the industry probably grew to large proportions. The Delhi pillar near the Kutub Minar, dated to the fourth to fifth centuries CE, is the classic example cited as evidence of the level of the iron industry in India.

The pillar, which is about twenty-five feet high and weighs several tons, has not rusted. Large iron objects of this age are found throughout the country (Figure 1) — for example, there is the broken pillar at

Dhar, Madhya Pradesh in Central India (thirteenth century CE) which originally measured about fifteen meters high, and the ten metre pillar at Kodachadri Hills in Karnataka State in South India. Dharampal estimates that by the eighteenth century the iron and steel industry in India was very extensive indeed. There are several reports of British observers in Dharampal's compilation wherein the Indian product has been rated "better than the steel produced anywhere else in the World".

Figure 1 *Iron pillar at Mehrauli*

In the traditional iron making process, iron ore is reduced in a low shaft furnace built of clay. Iron ore and charcoal form the top, and air is blown by bellows from the bottom. Charcoal provides the fuel as well as the reducing agent for smelting. Temperatures between 1000°C and 1475°C are generated in the reaction zone. The reduction takes place as a gas–solid reaction. The reduced iron collects with some iron oxide to form liquid slag. The slag is tapped periodically through slag holes provided near the bottom of the furnace. At the end of a blowing cycle of three to six hours, the sponge iron bloom weighing 3–10 kg is extracted from the bottom of the furnace. It is then hammered to squeeze out any entrapped slag and to consolidate the iron.

A process of making special steels of exceptionally good quality by melting them into crucibles (Figure 2) is probably unique to India and its vicinity. The special steels were traded extensively with other civilisations, especially for making swords and daggers and other implements where exceptional hardness and toughness were necessary. These swords became known as "Damascus swords" in Europe since they were obtained from the Middle East, and Damascus was the major trading centre. Only later was it known that the Arabs bought the steel from India and fashioned it into various implements. The *wootz* steel from which these blades were made thus became famous. A very large industry of *wootz* making is known to have existed in the southern parts of the country until the beginning of the nineteenth century. This industry slowly died down during the nineteenth century. The process is now extinct.

Figure 2 *Fired clay crucibles.*

British and European interest in Indian iron and steel has a long history. Several travellers' accounts such as one by Francis Buchanan provided detailed information on the extensive iron and steel industry around the seventeenth and eighteenth centuries. Crucible steel from India was also noted and described. In fact, technical papers and studies on various aspects of Indian iron and steel making continued until the turn of this century.

Starting from the turn of the eighteenth century, several efforts began to be made in England and later on in France, Italy, and much later in Russia to try to reproduce the Indian process of making steel. For a number of years various British scientists such as Pearson, as well as Stoddart and his student Faraday, tried to make *wootz* by the crucible process. Their experiments continued in England and France well into the 1850s and 1860s. They declined in importance in the 1860s, perhaps because of the general decline of interest in swords with the emergence of the gun. Also, by this time, the modern metallurgical processes for large-scale production of steel, using the Bessemer converter etc., had started to develop. In Russia these experiments continued well up to the turn of the twentieth century.

Around the turn of the century, the indigenous process had ceased to have any larger interest and the few observations and studies regarding it were by anthropologists and others concerned with exotic 'arts'. Ananda Coomara Swamy recorded the steel making process in Sri Lanka around 1908. In 1963 the National Metallurgical Laboratory at Jamshedpur hosted a symposium on the Delhi Iron Pillar. An interesting fact that then emerged was that the iron made by tribals in the early 1960s appeared to be comparable to the iron of the ancient Delhi Pillar from the point of view of corrosion resistance.

Beginning from the late 1950s a series of studies has been undertaken in England to characterise the various features of the bloomery process for producing wrought iron. The models taken up for reconstruction were from Africa or the earlier pre-medieval European designs. From the mid-1970s onwards there has also been a fresh spurt of interest in studying the traditional Indian process of making steel. Reacting to the discovery of a type of ultra-high carbon steel that was showing super plastic behaviour, the historian Cyril Stanley Smith felt that in these new alloys one might have only "rediscovered" some properties of traditional Indian steel. The crucible process of reduction is perhaps much more complex than was initially thought and needs to be understood and characterised afresh.

A.V. Balasubramanian

References

Bronson, Bennet. "The Making and Selling of Wootz: A Crucible Steel of India." *Archeomaterials* 1(1): 13–51. 1986.

Buchanan, Francis. *A Journey From Madras Through the Countries of Mysore, Canara, and Malabar.* 3 vols. Madras: Higgin Botham and Company. 1870.

Chakrabarti, D.K. *The Early Use of Iron in India.* New Delhi: Oxford University Press. 1992.

Dharampal. *Indian Science and Technology in the Eighteenth Century.* Hyderabad: Academy of Gandhian Studies. 1971. reprinted 1983.

Hadfield, Robert. "On Sinhalese Iron and Steel of Ancient Origin." *Proceedings of the Royal Society (London)* Series A. 86:94–100. 1912.

Hegde, K.T.M. *An Introduction to Ancient Indian Metallurgy.* Bangalore: Geological Survey of India. 1991.

Holland, Thomas H. "On the Iron Ores and Iron-Industries of the Salem District." *Records of the Geological Survey of India* 25 (3): 135–159. 1892.

Lahiri, A.K., T. Banerjee, and B.R. Nijhawan. "Some Observations on Corrosion-Resistance of Ancient Delhi Iron Pillar and Present-Time Adivasi Iron Made By Primitive Methods." *National Metallurgical Laboratory Technical Journal* 5 (1): 46–54. 1963.

Lahiri, A.K., T.Banjeree, and B.R. Nijihawan. "Some Observations on Ancient Iron." *National Metallurgical Laboratory Technical Journal* 8: 32–33. 1967.

Pearson, George. "Experiments and Observations to Investigate the Nature of a Kind of Steel, Manufactured at Bombay and There called Wootz." *Philosophical Transactions* 85: 580–593. 1795.

Sherbey, Oleg D. "Damascus Steel Rediscovered?" *Transactions of the Iron and Steel Institute of Japan* 19: 381–390. 1979.

Yater, Wallace M. "The Legendary Steel of Damascus Part III." *The Anvil's Ring,* 1983–1984. pp. 1–17.

Metallurgy: Zinc and !ts Alloys: Ancient Smelting Technology

The United Nations University says that "Traditional knowledge – technological, social, organisational or cultural – is a result of the great human experiment of survival and development".

One such experiment has to do with metals. This journey dates to around 5000 BCE, with the dawn of the Chalcolithic age. Copper was the first metal to be discovered and exploited by mankind, because of its abundant availability in its native form. This was followed by the Bronze Age (2100 BCE) and Iron Age (ca. 1200 BCE). Though zinc has not been one of the dominant metals like copper or iron, it played a crucial role in the ancient brass industry, even before the discovery of zinc. The present article describes the development of zinc and its alloys in its global perspective, with an emphasis on the contributions from the ancient Indians – who discovered the metal and invented its smelting technology.

It is impossible to conceive of modern life without zinc, the bluish-grey metal and the fourth-most used element after iron, aluminium and copper. Zinc – as "steel protector" – consumes more than 50% of today's market. Its second-most important application is that it is the major alloying element in "gold-like" brass, which probably paved the way for the discovery of zinc by natural evolution since the Chalcolithic age. Zinc is also used in die castings, batteries, purifying water, in power electric vehicles and as compounds by the rubber, chemical, paint, agricultural, pharmaceutical and cosmetics industries. Zinc is also an essential micronutrient in organisms.

Zinc exists naturally in air, water and soil. Zinc is the twenty-third most abundant element in the earth's crust (150 ppm). The major types of zinc ore deposits are the volcanic and sediment-hosted sulphides, Mississippi Valley-type carbonates, intrusion-related and broken hill-type ore deposits. Typically zinc ores contain only 3 to 10 % zinc, making it necessary, before smelting, to concentrate it up to about 55%. Zinc is completely recyclable without loss of its physical or chemical properties. Today, zinc is mined in more than fifty countries, with China, Canada, Russia and Australia being the largest producers.

Historic Evolution of Zinc

The historic evolution of zinc and its alloys can broadly be classified into two time zones–prior to the discovery of metallic zinc (200–300 BCE) and after the invention of the downward distillation process to extract zinc. The first era was marked by the prevalent use of zinc ore to make shiny, artistic brass. Neither brass nor bronze was known in the Chalcolithic period of India at all. India experienced the Iron Age (1000 BCE to 300 CE) before reaching a fully developed Bronze Age, despite the famous *cire perdue* (lost wax) technique of Harappan metalsmiths (2100 BCE) who mastered icon making in bronze. The famous 'dancing girl' (Figure 1) is an example.

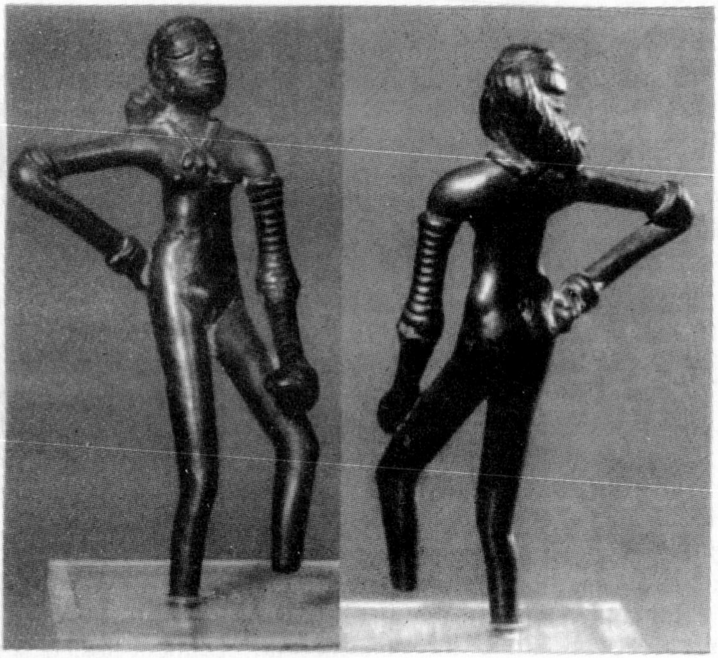

Figure 1 *Dancing girl: A bronze figurine excavated from Mohenjodaro*

The first era in the history of zinc and its alloys marks a period of using zinc ore, sphalerite its sulphide – for many medicinal and industrial purposes, without identifying metallic zinc as being responsible for the associated advantages. The earliest use of zinc is the gold-like brass – a copper–zinc alloy. Evidence suggests the use of copper alloys in the Sumerian civilisation of Mesopotamia around 3000 BCE and in China near the Hwang-Ho River. Archaeological remains of the Harappan (2000 BCE) civilisation from Lothal, Nautiyal Atranjikhera (in the Ganga Valley) and Rojdi in the Rajasthan–Gujarat area contained only 0.1 to 1% zinc. Fusion of zinc ore with copper increases its strength. With increasing amounts of zinc, the alloy developed a dazzling golden glitter which made it useful for statuary, temple roofs and cookery items. The Romans later identified the use of zinc ore for medicinal purposes. At the time of Augustus (20 BCE to 14 CE) Romans were known to use zinc ore for healing wounds and sore eyes.

Brass was made using a cementation process, in which finely divided copper particles were mixed with roasted zinc ore (oxide) and charcoal (reducing agent) and heated to 1000°C in sealed crucibles. This resulted in poor quality brass. High quality brass had to wait for the discovery of zinc. The temperature interval for reduction of zinc is a crucial stage in the above process. Below 950°C, zinc is not produced; above 1080°C, copper melted and flowed down with very little intake of zinc. The production of zinc was difficult because the boiling point of the metal is lower than the minimum temperature required for the conversion of its oxide to metal. This was mainly responsible for the delayed discovery of zinc. A proper reducing atmosphere is also essential for the extraction of zinc; otherwise the zinc vapour formed gets oxidised to form 'philosopher's wool'. The ancient Persians' attempt to reduce zinc oxide in an open furnace was not successful. Later, it turned out that the downward distillation technique used for zinc extraction was, indeed, an advanced technique and its evolution probably heavily depended on the Ayurvedic and alchemical expertise of ancient Indian artisans.

Genesis of Smelting Technology of Zinc

The second era of the discovery of zinc and the invention of its smelting technology began roughly around the third to fourth centuries BCE and gradually evolved to as late as the twelfth century CE. The production of zinc could be achieved only by proper design of various components like the furnaces, retorts and the arrangement for condensing vapours. Sphalerite, zinc sulphide, with impure dolomite, was mined, crushed, concentrated and roasted to get oxides. Small pellets containing the oxides with carbonaceous material and salt were used as smelting charges. The charge was loaded into small, thin-walled clay retorts. In order to retain the charge in the retorts while inverting, cylindrical reeds were inserted into charged retorts through funnel-like condenser tubes. When heated, the reeds burnt away, providing clear flow channels for the zinc vapours to flow out of the retorts. The cross-section of an ancient zinc distillation furnace, as found in Zawar of Rajasthan is shown in Figure 2. There is ample evidence, both archaeo-metallurgical and literary, to show that India developed a procedure to extract zinc from its ore in large amounts. Many other metals like gold, silver, tin and even deliberate production of iron, were well mastered by the ancient Indian artisans.

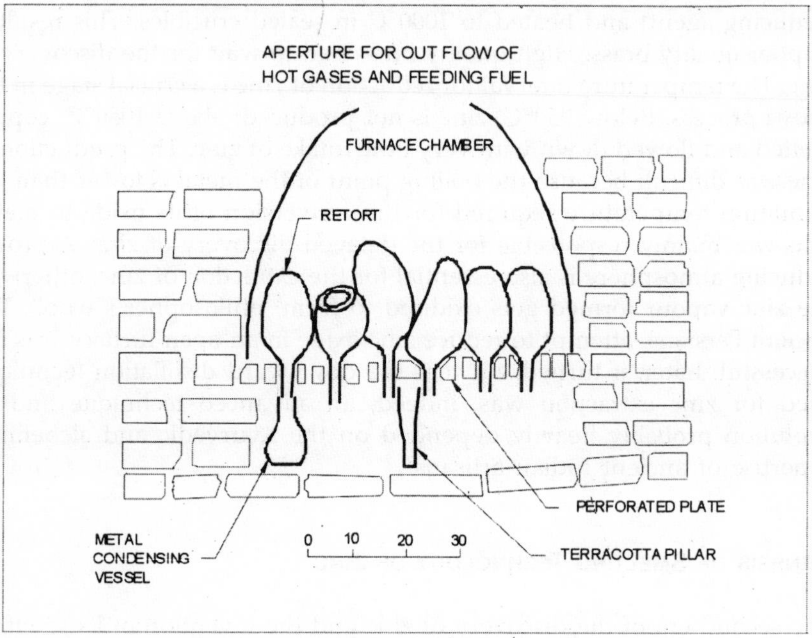

Figure 2 *Schematic of a smelting furnace for extraction of zinc in ancient India*

The archaeo-metallurgical evidence confirms the prevalence of advanced smelting practices of zinc in ancient India. The time of this technology has been traced to the third and fourth centuries BCE, based on earliest C^{14} dating of mines in Rampura–Agucha (40 km south of Ajmer), Rajpura–Dariba, and most crucially, the Zawar systems. Great heaps of small retorts at Zawar, 40 km south of Udaipur in Rajasthan, tubular retorts, and 130,000 tonnes of residues bear testimony to extensive zinc production from the twelfth to sixteenth centuries. A crude estimation from the extracts suggests extraction of equivalent of 1,000,000 tonnes of metallic zinc and its oxide.

There is evidence in a number of places for the use of zinc as a metal, without which good quality, high zinc containing brass could not have been made. At Prakasha, a Chalcolithic site in the Deccan plateau, copper objects with 18–26% zinc have been found, belonging to the second millennium BCE. Some Iron Age sites in the Ganga valley, such as Artranjikhera, contained copper alloys with 6–16% zinc. Taxila, 20 miles north of Rawalpindi, contained copper with 34% zinc and dated from the third to the first century BCE. A large number of artefacts and sites with copper alloys having high zinc content in the pre-Christian era were identified in places like Dwaraka, Ayodhya, Manikyala (second century CE) and Rajghat. There is evidence for the popular use of high zinc containing brass much later in history, as in the Mauryan, Sunga,

Kushana and Gupta eras. However, persistent efforts over centuries on a large scale can be traced only to Rajasthan, starting around 200–300 BCE.

The use of brass and other alloys for religious and domestic purposes reveals the tradition of arts and crafts in ancient India. Harappan culture is well known for its exquisite stone and metal sculptures. Another medieval craft which developed based on zinc technology is called Bidriware, prevalent even today. The fine craftsmanship in Bidriware, with its sleek and smooth dark-coloured metal work with intricate designs on its glossy surface, is famous. An example of Bidriware is shown in Figure 3. Basically, Bidri is a zinc–copper alloy, the opposite extreme of brass. Its surface is temporarily darkened using copper sulphate for engraving delicate designs into the metal. An inlay of either silver or brass or gold is used. The decoration is burnished with certain chemicals, which darken the body, producing a black patina without damaging the inlay. Historical evidence shows that a large amount of Bidriware was made for presentations to dignitaries, such as Alauddin Bahamani II in 1434 CE, Russian emperors (1470–74 CE) and the Prince of Wales in 1875. In fact, the story goes that Alauddin Bahamani II was so impressed that he invited his craftsmen in Bijapur to settle in Bidar itself.

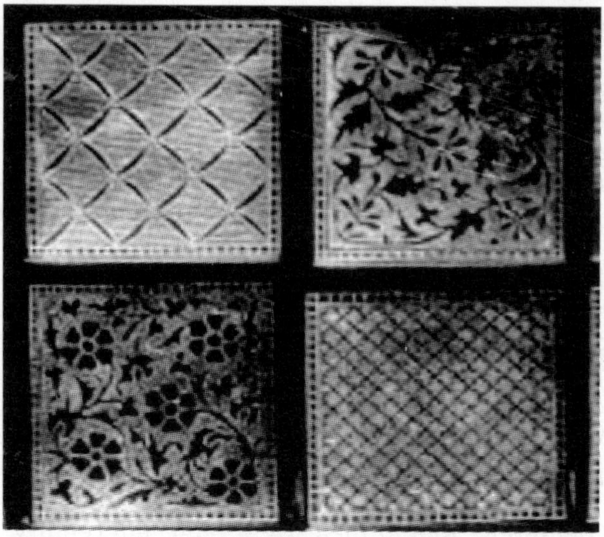

Figure 3 *An artistic design of 'Bidri boxes' from Bihar using the Bidri technology.*

Evidence from Ancient Literature

Most of the evidence from ancient literature for the production of metallic zinc in India originated from a much later period in history, somewhere between the twelfth and fourteenth centuries CE. The *Arthaśāstra* is the earliest literary evidence illustrating the production of metallic zinc on a regular basis in India. In the *Arthaśāstra,* brass is referred to as *arkuta* and liquid ore. There is also reference to burning *rasa* (metal) to produce an eye salve or zinc. The epics of India refer to brass as *kamsya* while Grihyasutra refers to it as *arakutah* or *riti* or *pitale.* The more popular name for brass was *riti* or *ritika* – a derivative of the word *herita* for yellow – a synonym in Vedic literature. Many records from the pre-Christian era of Manu, Yajnavakya and Patanjali refer to brass as *ritika.* The book *Rasārṇava* described the production of metallic zinc in 1200 CE. The *Rasaratasamuccaya,* a fourteenth century alchemical text, described a wide range of practices. It also described a type of special furnace for distillation, used by many alchemists such as Nagarjuna, who is considered the best chemist of the *Śātavāhana* period of the fourth century CE. However, *Rasavidya* (alchemy) was a closely guarded secret. The *Ain-i-Akbari,* a medieval text, refers to *ruh-i-tutiya* i.e., zinc found at Zawar. The word *ruh-i-tutiya* is probably derived from *tuttha;* this word was used in the *Arthaśāstra* to refer to silver ore.

Apart from these demonstrations of India's zinc smelting technology, there is some evidence showing ancient zinc in other countries. Excavations in the agora in Athens yielded a nearly pure zinc deposit dating back to the third or second century BCE. These observations suggest that its smelting was probably known both in India and Greece since the middle of the first millennium BCE. This is surprising, since there is no other evidence, neither archaeo-metallurgical nor literary, to confirm that Greeks were producing zinc at that time. Perhaps the Greeks obtained zinc from India. There are also occasional references to zinc coins of 200 BCE to 200 CE and brass objects of the Neolithic age in China. The routes of these are not clear. There is some literary evidence showing that the Chinese and Indians shared their alchemical knowledge. It has been found that zinc technology migrated to China through Buddhist scholars, where it developed into an industry to meet the needs of manufacturing brass. The Chinese apparently learned about zinc smelting around 1600 CE. This is supported by the absence of references to zinc in an encyclopaedia written in the latter half of the sixteenth century. The procedure of zinc manufacture is described in detail in a book by Dian Gong Gai (Tien-Kong- Kai) in the early seventeenth century. Zinc smelting in China seems to have begun in the Jiajing period (1552–1566 CE) of the Ming dynasty, based on excavations at Guizhou. It was exported to Europe by the end of the seventeenth century CE under the name *totamu* or *tutenag.* The word *tutenag* is *tutenagam* in the South Indian language Tamil, and in Persian *tutiya* is used for calamine and "tutty" is used in English for zinc oxide.

Industrialisation of Zinc Smelting Technology in Europe

The story of zinc in Europe started around the thirteenth century CE, again without recognising the individual property of zinc. It was found that, to colour copper to look like gold, glass had to be sprinkled on top of the crucible making copper. The glass acted as slag, preventing the escape of zinc vapours and increasing the zinc content of copper. Further refinement of this process was in progress throughout the sixteenth century. In an attempt to extract lead and silver using ores from Harz Mountain in Rammelsberg, a white metal was condensed and scraped off the walls of the furnace. This unknown metal used to glitter like gold, hence it was named "contrefey". It often contained zinc, which was not then recognised. However, the similarity of this new metal with those produced under similar circumstances in Silesia by the local people was identified. The local people in Silesia called it "zincum". Later in the sixteenth century, it was realised that "zincum" was a new metal and its properties were distinct from other metals known at that time. The knowledge of deliberate smelting of zinc in a retort was transferred when William Champion visited China in 1740 CE. He established the technology in Bristol, patented it, and commercialised the production process to the tune of 200 tons per year by 1758 CE.

By the turn of the eighteenth century, the zinc mines in Rajasthan and Gujarat had unfortunately to be closed down and the technology became extinct. This is attributed to a number of reasons, such as the devastation by the Maratha wars, the great famine of 1812 or the pollution due to clouds of cadmium dust found in residues of zinc smelting sites. However, none of the reasons can be verified. The persistent efforts of Indian metalsmiths, spanning many centuries, have made significant contributions towards the smelting technology of a difficult and enigmatic metal.

Baldev Raj, M. Vijayalakshmi and B.Venkatraman

References

Craddock, P.T. "The Early History of Zinc." *Endeavour*, New Series 11(4): 183–191. 1987.

Craddock, P.T., L.K.Gurjar and K.T.M.Hegde. "Zinc Production in Medieval India." *World Archeology* 15: 211–221. 1983.

Rao, R.P. and N.G.Goswami. *Metallurgy in India: A Retrospective*. New Delhi: India International Publisher. 2001.

Kumar, Biswas Arun. "The Primacy of India in Ancient Brass and Zinc Metallurgy." *Indian Journal of History of Science* 28 (4): 309. 1993.

Subbarayappa. B.V. *Chemistry and Chemical Tcchniques in India*. New Delhi: Project of History of Indian Science, Philosophy and Culture, Centre for Studies in Civilisations, Munshiram Manoharlal Publishers. 1999.

Meteorology

The very first stage of civilisation was marked by peoples' efforts to understand their surroundings and make use of their beneficial aspects. Their first action in this direction was to produce food, making use of the available water in the rivers and rainfall in the region. Though initially extreme phenomena like heavy rains, winds, cold and hot spells, droughts, and floods, appeared incomprehensible and hostile, early humans gradually sorted out their seasonal character and planned their agricultural operations accordingly. Thus began in a crude way the development of weather science all over the world.

In India, the development of this science commenced in the early Ṛgvedic period. That the heat of the Sun lifts the water to the atmosphere which after some time comes down as rain was recognised by the Vedic seers at a very early stage. In order to explain the occurrence of rain during a restricted period of about two months in their region in extreme North-west India, they imagined that water was absorbed by the Sun's rays in the vast ocean areas in the south during the winter season and the humid air carried northwards by the Sun's rays. When the Sun attains its extreme northward position and starts retracing its path, the humid air gets deflected near the foot of the Himalayas and brings rain from the east to their region. These moist easterlies replace the westerlies that were present in the region before the arrival of the monsoon rains. Whenever there was drought, they performed rituals to invoke the rain god. They believed that in Nature there is a feed of a substance called *soma* from above into the atmosphere, which aids the occurrence of rainfall. Therefore they fed into the ritual fire some substances like wild dates and some special types of grass which produced smoke and were believed to be effective in aiding rainfall.

The post-Vedic scholars developed the subject further, mainly working on the pregnancy concept of rainfall. They looked for symptoms in the winter season for the commencement of pregnancy and identified the characteristics of winter disturbances in their region as indicating the same. Working along these lines, they were able to observe weather very carefully during the pre-monsoon months and were able to define the course of events which go towards the nourishment of rain embryos and the delivery of good summer rainfall at the right time after 195 days. Any departure from the defined meteorological conditions during the growth period, such as too much rainfall, or snowfall, unfavourable winds, and temperature, was said to affect the quantity of rainfall delivered during the rainfall period. They also believed that hail would occur if the rain foetuses overstayed in the atmosphere. The Moon's position with respect to the Sun and the stars was believed to influence the formation of rain embryos. The Moon was conceived as a replica of *soma* in the heavens, and *soma* was capable of fertilising the atmosphere.

Based on such concepts and extensive observations, the post-Vedic scholars developed several rules of long-range rainfall forecasting. If they were at all successful, it was certainly due to their capacity to observe day-to-day weather and individual weather elements, like clouds, temperature conditions, wind, rain, lightning, and thunder. They were extremely clever in mentally working out correlations based on observed data. For short- and medium-range forecasting they framed many rules of thumb based on winds, clouds, temperature, lightning, thunder, moisture in the atmosphere, behaviour of people, animals, birds, snakes, worms, insects, trees, and plants, as well as visual impressions of the Sun, Moon, stars, and sky. They were so thorough with local weather that their capacity to forecast in the short- and medium-range was as high as that of any modern forecaster who does the same with sophisticated equipment and maps.

Measurement of rainfall in India dates back to the fourth century BCE. A standard rain gauge was constructed around the third century BCE, and this system of measurement was prevalent in North India for a very long time (third century BCE to CE sixth century).

Well before the birth of Christ, the Arab dhows sailed across the Indian Ocean for trade purposes. Hippalus, a Greek pilot of the first century, sailed across the Arabian Sea for the first time. A handbook for merchants called *Periplus* was written by a Greek around CE 50. Subsequently, Arab geographers wrote many books giving details of Indian ocean voyages. Sidi Alis' *Mohit*, written around CE 1554, not only gives a map of the Indian Ocean area but also mentions the occurrence of monsoons at fifty distinct places.

With the arrival of more voyagers from the west in the Indian Ocean, a steady effort for systematically observing the wind, weather, and weather systems of the Indian Ocean commenced. In his first voyage from Melinda to Calicut in 1499, Vasco da Gama made use of the monsoon winds and reached his destination in just three weeks. William Dampier published many observations of Indian ocean weather and weather systems in his travel accounts. He was a sixteenth century buccaneer who lived and worked with some of the rowdiest pirates in history. But he was also an astute observer of nature in general and weather in particular. In his *Discourse on Winds and Breezes, Storms and Currents*, he deals with general wind systems throughout the world and their seasonal changes, which include the South-east trades of the South Indian Ocean and Northeast and South-west monsoons of the North Indian Ocean. During the seventeenth and eighteenth centuries, the military and trade activities of the European powers in the Indian Ocean waters increased.

Matthew Maury in his *Physical Geography of the Seas* (1874) explained the formation of monsoon winds as resulting from the heat of the plains and deserts of the Asian region. The following ideas about the mechanism of the south-west monsoon and its rainfall were generally agreed upon by the meteorologists of the nineteenth century.

The plains of North India get very hot during the summer, and the air over that region ascends and becomes light. As a result, air over the sea areas where the pressure is high both in the neighbourhood of the Equator and south of it, moves towards the region of low pressure of the land. The south-east tradewinds, while moving northwards and crossing the equator, become south-west winds, owing to the rotation of the earth. Again, these south-west winds do not blow directly into the region of low pressure, but go around it in an anti-clockwise direction. If one stands with one's back to the wind, the pressure to the left is lower than to the right in the northern hemisphere. In the southern hemisphere, the realtion is reversed. The copious precipitation of the west coast is due to the high mountains which run along the coast. The higher the mountains, the heavier the precipitation. The monsoon is sustained by the latent heat released during the precipitation, which adds more heat to the atmosphere, and therefore further rarefaction takes place. Strong winds blow into the region of heavy rainfall, since air from the neighbouring regions rushes to occupy the space created by ascending air.

Meanwhile, more knowledge was added to the science of cyclones in the Indian Ocean. Henry Piddington made a monumental contribution to the science of storms. He was the first to coin the term "cyclone", which gained world usage later. In a series of papers he gave detailed accounts of many Indian Ocean cyclones. His best-seller at that time was the *Horn Book of Storms for the India and China Seas*, which was followed by another book called *Horn Book for the Law of Storms*, in which he explained the use of transparent horn cards provided in his book for finding out the centre of cyclones.

Many Indian meteorologists, led by Desai, Rao, Koteswaram and Majumdar, worked on various aspects of the formation of cyclones. They investigated the role of the upper tropospheric flow patterns in the intensification, movement, and dissipation of tropical disturbances in the Indian Ocean. The availability of aircraft winds and satellite pictures enabled the meteorologists, such as Raman and Srinivasan, to study the low-level convergence and associated winds around the calm eye region of the cyclone, upper-level divergence, and the relation of the direction of movement to the upper-level winds. They also studied the influence of sea surface temperature on the formation of the cyclone.

As regards the south-west monsoon, the upper air observations of wind and temperature and also the newly formulated dynamical concepts enabled the meteorologists to understand many synoptic aspects of the monsoon. Many meteorologists studied the role of the easterly jet stream and the Tibetan high, the northward shift of the westerly jet stream, the advance of the intertropical convergence zone to Northern India, and the extension of equatorial westerlies. Koteswaram and Flohn (1960) made important contributions in this field.

Today meteorology, in India is a highly developed subject, both from the research and service point of view. The country has produced many skilled meteorologists whose expertise is on par with that of meteorologists of developed countries.

A.S. Ramanathan

REFERENCES

Blanford, H.F. *Climates and Weather of India, Ceylon and Burma*. New York: Macmillan Company. 1889.

Capper, J. *Observations on the Winds and Monsoons*. London: Debrett. 1801.

Dampier, William. *Voyages and Descriptions*. London: J. Knapton. 1699.

Das, P.K. *The Monsoons*. New Delhi: National Book Trust India. 1988.

Maury, Matthew. *Physical Geography of the Seas*. New York: Harper. 1861.

Piddington, H. *Horn Book of Storms for the India and China Seas*. Calcutta: Bishop's College Press. 1844.

Piddington, H. *Horn Book for the Law of Storms*. New York: Wiley. 1848.

Ramage, L.S. *Monsoon Meteorology*. New York: Academy Press. 1971.

Ramanathan, A.S. "Weather Science in Ancient India, I–VIII." *Indian Journal of History of Science* 21(1): 7–21, 1986; 22(1): 1–14, 1987; 22(3): 175–197 and 198–204, 1987; 22(4): 277–285. 1987.

Simpson, G.C. "The South West Monsoon." *Quarterly Journal of the Royal Meteorological Society of London*. XI–XII:199. 151–172. 1921.

Walker, G.T. "On the Meteorological Evidence for Supposed Changes of Climate in India." *Memoires of the Indian Meteorological Department*. 21: 1–22. 1910.

Walker, G.T. "Correlation in Seasonal Variation of Weather." *Memoires of the Indian Meteorological Department*. 24: 275–332. 1924.

See also : Navigation

Military Technology

The term military technology is broad, and, as a subject restricted to non-Western cultures, potentially laden with analytical complexity. In fact, the contraints of a survey make it necessary to view the more technical innovations of the larger cultures rather than the myriad variations on pointed weapons fashioned by essentially all peoples. Stimulated by environment and nature, the gamut of world cultures have used artistic and functional inventiveness in weaponry. Non-Western ancient military technology provided significant origins for Western military technology as well.

The first most significant line in the military technology progression was metallurgy of copper in the transitional period between the Neolithic and true Bronze Ages, approximately between 4500 BCE (perhaps 5000 BCE) and 3500 BCE in the Near East arc from Mesopotamia to Egypt. Copper's cold malleability enabled the earliest metalworkers to beat, rather than fire it from the ore as with harder metals. In doing so, they could fashion a metal version of basic wood, bone, and stone pointed weapons: arrow tips, spears, and particularly, swords. This was followed by smelting (melting metal to separate out impurities) and founding (melting the purer metal for casting and molding). By about 3000 BCE the general use of copper and experimentation with its alloys (bronze with tin and brass with zinc) ushered in the Bronze Age to south-western Asia over five centuries before general use. This was an essentially Near Eastern phenomenon, probably disseminated to India, Anatolia, and surrounding areas after this.

As far as we know today, the first great civilisation of humanity was that of Sumer in southern Lower Mesopotamia after 4000 BCE. Among so many accomplishments handed down to subsequent Mesopotamian civilisations and the west, one especially important one was worked copper alloys and probably bronze swords. Mesopotamian cast copper mace heads, the first technical use of metal, date from 2500 BCE. About the same time Sumerian smiths were casting socketed axe heads. In the north, Semitic peoples to be called Akkadians, from which the Assyrian and Babylonian cultures developed, assimilated Sumerian technology. Before 2000 BCE non-Semitic, Indo-European invaders from Central Asia began various waves of infiltration from Asia through Asia Minor into Mesopotamia and on to India. All would leave their military mark. Among these were the Hittites from the north-west and later the Hurrians from the north-east and the Caucasuses – the one moving into Lower, the latter into Upper Mesopotamia.

The Hittites overran most of Asia Minor (Anatolia) after 2000 BCE and about 1500 BCE invaded Babylonia long enough to raze Babylon. Anatolia, a high plateau fringed by mountain ranges, was rich in mineral resources, among these gold and silver, but most importantly iron. The

Hittites probably ushered in the early Iron Age by their use of this much harder metal in their weapons. In the general extent of south-western Asia the Iron Age did not arrive until about 1000 BCE, although a few Mesopotamian objects of perhaps smelted iron have been dated before 2200 BCE, and some Egyptian work has been conjectured as even older. Iron was much superior to bronze in edged and projectile weapons and required higher temperature metallurgical processes of smelting iron ore and founding the crude metal. Although it was thought in some quarters of the last century that Egypt was the cradle of iron work, development of its metallurgy may have been contemporary with that of Asia Minor, considering abundant Egyptian iron resources. The use of iron also brought more effective defensive hardware, i.e. in armour and in horse trappings and the chariot.

The horse and the two-wheeled chariot were Hittite innovations to Western Asian warfare. The horse brought mobility to tactical manoeuvering on the battlefield for the specialised soldiers called cavalry. The chariot was introduced during the eighteenth century BCE and likely by the Indo-European Aryans (Indo-Iranian, also metalworkers — perhaps early iron weapon users) who invaded Iran from the north-east at that time and influenced the Hittites and evidently held sway over the Hurrians. The chariot provided a further tactical edge, allowing a soldier or two soldiers to act in concert in inflicting multiple casualties at one time. As specialised warriors, the charioteers introduced military class rule to Near Eastern civilisations. With the added innovations of scythe-like blades on its wheels, the chariot also added the mass fear psychological factor to warfare. The Hittites took Northern Syria in their clash with Egypt about 1400 BCE, the latter having adopted the horse and chariot after their temporary defeat by the chariot tactics of the Hyksos, Amorite peoples who invaded Palestine about 1700 BCE. These latter also contributed large fortification technology to the general mud wall military architecture pool of the Near East which started with the high curtain walls of ancient Jericho (8000 BCE), the first example of specialist military architecture.

By the middle of the fourteenth century BCE the Assyrians were able to take the military ascendancy in Upper Mesopotamia and eventually all of Mesopotamia by the late eleventh century BCE, to become a great empire. By the eighth century BCE the Assyrian army had reached an apex of coherency, a blueprint for the Persian army. Made up of both professional and militia soldiers, the Assyrian army equipped all troops with finely tempered iron weapons. They employed cavalry and chariots, archers (using the composite recurved or reinforced bow, found in the Middle East to 3000 BCE), and slingers, who used the simplest, oldest missile weapon. Adding to their siege tactics, sappers (essentially meaning diggers at that time) were used in approaches to mud-walled defenses, as were battering rams and wheeled-platforms, equipped with shielding defenses against arrows, for rolling against such walls. By the sixth century BCE the Persians had become heir to the Assyrian Empire and to the diverse military technology of the Near East. It remained dominant for two hundred years until the informal transition of east to west finally came face to face with the challenge of Greece under Alexander the Great in the middle of the fourth century BCE.

In the Far East, Chinese civilisation as far back at 2000 BCE was characterized by a value placed on functional technology. The integration of the wall into Chinese cultural architecture was given a profound military expression in the Great Wall, which was started in 214 BCE by the first emperor Shi Huangdi as a linking of earlier rampart walls. It was meant to keep out the north/north-western invaders who would plague China for centuries. The crenellated, brick-faced wall still stands, stretching some 4000 miles and 30 feet (9 metres) high, with regular spaced square watchtowers 40 feet (12 metres) high with a 9–12 foot (3–4 metre) passageway through them. Along with their own cultural variations on basic weapons, the Chinese designed light hunting crossbows by the fifth century BCE and were using them in combat by the second century BCE.

The use of iron metallurgy continued to be the prime advance in military technology. Iron ore is plentiful all over the world. Variations of alloying iron with carbon in smelting processes, which included the introduction of air blasting to fan the fire (the forge) to high temperatures, meant that steel (iron with a small proportion of carbon) and its hardening were probably fairly contemporaneous with iron working (from 1000 BCE). Although dating is indeterminate, the great deposits of iron in Central Africa and the proximity of the Egyptian influence point to limited iron and steel forging. Indian weapons of iron were prevalent by 500 BCE. In fact, tempered steel was produced fairly early in India. Bars, rods, and plates of raw steel were exported throughout the Near and Middle East. Indian steel was used in the founding of blades of "watered steel" (the process of folding malleable steel over and over then beating it out). These light, high tensile strength curved (Damascus) blades enabled the effective long sweeping offensive draw cut, used by both the infantry and cavalry of Western and Central Asia down through the last century.

The development of the relatively simple smelting methods of steel and steel weaponry was disseminated eastward to southeast Asia via Indian colonisation. Iron weaponry and working began independently in China about 500 BCE and smelting of crude steel was fairly contemporaneous (about 400 BCE). By the Middle Ages the effectiveness of the Mongolian steel sabre, influenced by Middle Eastern contacts, was supplanting the straight Chinese sword. Japanese iron weaponry, with Chinese influence, began about 200 BCE, although the earliest relics date between the second and eighth centuries CE. The best of the distinctive long, slightly curved samurai steel blades date from the twelfth century, and progress to the fine temper-lined watered blades of later centuries. All these areas applied iron and steel technology to military accouterments and armour. The work of the Near and Middle East, China, and Japan was particularly artistic as the Middle Ages progressed.

Although the steel sword would remain the principal weapon of the great non-Western cultures, the destructive potential of gunpowder technology into the High Middle Ages was to affect the larger non-Western cultures as it did the West. The use of incendiaries was already ancient, most noticeably in China. The so-called "Greek fire", the generic term for a variety of mixtures based on naphtha (a petroleum distillate) added to sulphur, pitch, turpentine, tars, and oils (in modern interpretation, probably a suspension of metallic sodium, lithium, or potassium in a petroleum base), was perhaps in crude use by the fifth century BCE. It is noted as being used by the Boeotian Greeks at the siege of Delium in 424 BCE during the Peloponnesian War.

The historiographic origin of gunpowder, that is black powder, is still controversial. The gunpowder recipe itself is of uncertain origin, but its basic constituents are now generally first attributed to ninth century Chinese alchemists. It might also be the independent product of Islamic lands, most likely Moorish Spain by the mid-twelfth century. From there it perhaps moved to India where there may have been independent knowledge and use of the chemical ingredients from the late eleventh century. It was known in Northern Europe by the early thirteenth century. The argument for an intermediary disseminator, the Eurasian Steppe lands, the European/Asain crossroad, to Islam and Europe, particularly by the thirteenth century Mongols is also plausible.

Explosive application of gunpowder in a weapon has also been controversial. Some theorists of Chinese primacy (Chinese toy rocket experiments for fireworks evolved early) date bamboo-tube hand guns or cannons and rockets for arrows and spears from CE 900–950. Various types of incendiary arrows, slings, and javelins, as well as incendiary and exploding bombs, grenades, and fire-balls are also attributed to the Chinese by the eleventh century. The historical point of military effectiveness of such devices remains uncertain. Widespread military use in China did not appear until the Song-Jin dynastic wars of the twelfth and thirteenth centuries. By the thirteenth century bomb technology with iron casings and large size was used by Chinese and Mongolian antagonists in land siege warfare. There is also evidence of time delay fusing using flintstone abraded against steel wheels to set off multiple mines in fourteenth century China. Rockets were introduced to Europeans during the Mongol western invasions at the Battle of Legnica in 1241.

Gunpowder weapons applications appeared about the middle of the thirteenth century in Muslim North Africa and Moorish Spain as crude iron and iron-reinforced wooden bucket mortars for flinging stones in fortifications warfare. Also, Moors were using effective rockets on Spanish soil by 1249. Thereafter some evidence shows that the evolution of mortars, cannons, and finally handheld firearms progressed with most tactical efficiency in Europe, although some historians date Chinese cannons of significant size and metal composition from as early as the tenth century. Non-Western applications were innovative in their own right. By the middle of the fourteenth century, cannons mounted on walls or on mobile carriages and cradles had replaced most of the traditional engines of war in both Europe and the Near East. And eastern projectiles ranged from stone balls to huge arrows with sheet-metal fins.

The growing threat to Eastern Europe, the Adriatic, and the Aegean by the ascendancy of the Ottoman Turks through the fourteenth century was furthered by their pursuing the use of artillery to challenge the weakening Byzantium Empire. A parallel was the thirteenth century Mongol challenge and conquest of China, with cavalry, siege tactics, and gunpowder technology. By the fifteenth century the Turks were casting – sometimes with the guidance of European renegades – huge bronze mortars and cannons, such as those used in the final siege and fall of Constantinople in 1453.

The Turks also turned to the Western matchlock arquebus, the first gunpowder longarm, which was the single most important transitional pivot from medieval to modern warfare. The Ottoman domination over the Arab world influenced firearm dissemination to Arabia and North Africa where, unlike the more angular stock of Turkish and Persian guns, styles reflected Arab and Kabyle preferences.

The influence of the West on the Asian Pacific was initially felt in trade and subsequently in acquaintance with western gunpowder technology. Perhaps the most interesting case involved the Japanese, who quickly adapted the matchlock arquebus which the ubiquitous Portuguese traders, already established in China, brought in 1542. The Japanese matchlock was an austere but highly stylish weapon, smaller in size and calibre than western matchlocks with a spring design firing mechanism, which soon joined the traditional feudal weapon array and went on to change the tactical manoeuvering of the civil warfare of the sixteenth century.

The Korean civilisation provides an interesting development in Asian and world naval warfare at this point in military technological history. Located on a strategic peninsula, the Korean people endured centuries of piratical incursions from the Japanese islands on one hand and politically complex dynastic invasions' from the Chinese mainland on the other. A sophisticated native culture, including science and technology (particularly, shipbuilding), was able to grow from the tenth century. Until 1592 peace and cultural advances continued. Then the Japanese general Toyotomi Hideyoshi unified Japan, calling for the invasion of China through Korea, which refused his passage. Although they had cannons, the Koreans did not have the matchlock longarms of the 200,000 Japanese invaders. The ensuing incursion was successful until 1593 when the Korean admiral Yisunsin invented what must be the first ironclad ship, evidently thin iron plating over a high, flattened oval-shaped ship of sixteen oars with circumference cannon ports. Burn- and board-proof, a fleet of these "tortoise boats" was sent against and defeated a Japanese armada in Chinhai Bay. This triumph provided the impetus to drive the Japanese out.

The Chinese perpetuated their own hand cannon, large wall artillery technology, and shipboard cannon well into the nineteenth century. They adapted the Portuguese style of longarm lock but designed their own pistol grip-like stock. Both features influenced the far away Malaysian peninsula gun style which itself influenced the intermediate region of the Gulf of Tonkin. On the under side of Asia, Indian matchlocks showed significant regional variations from both the Portuguese and Arabic initial introduction. Three basic subcontinent Indian matchlocks were joined by very stylised weapons from Ceylon (Sri Lanka). By the early seventeenth century the Ceylonese exceeded the Portuguese in the manufacture of musket size matchlocks, one type with a unique bifurcated scroll butt. The Burmese side of the Malaysian peninsula essentially used Indian matchlocks with local decorations.

Although more isolated non-Western peoples continued to use the matchlock (indeed, the Japanese did until the early nineteenth century), most succumbed to trade and import and adapted to the progression of firearms manufacturing and, just as significantly, ordinance technology in keeping with the single-minded exigencies of superiority in warfare. These latter factors inevitably and irrevocably set the new course of non-Western military technology as a dependent reflection of the West, a reflection all the more thought provoking in the modem shadows of nuclear and chemical weaponry.

William J. Mcpeak

References

Bhakari, S.K. *Indian Warfare.* New Delhi: Munshiram Manoharlad. 1981.

Bottomley, I. *Arms and Armour of the Samurai: the History of Weaponry in Ancient Japan.* New York: Crescent Books. 1988.

Creswell, K.A.C. *A Bibliography of Arms and Armour in Islam.* London: Royal Asiatic Society. 1956.

Held, Robert. *The Age of Firearms.* 2nd ed. Northfield, Illinois.: Gun Digest Co. 1970.

Needham, Joseph. *Science and Civilisation in China.* vol. 5: *Chemistry and Chemical Technology,* pt. 7: *Military Technology: The Gunpowder Epic.* Cambridge: Cambridge University Press. 1986.

Oman, C.W.C. *The Wars of the Sixteenth Century.* Reprint of 1937 ed. New York: E.P. Dutton. 1979.

Robinson, Charles A. Jr. *Ancient History From Prehistoric Times to the Death of Justinian.* New York: Macmillan. 1951.

Stone, George C. *A Glossary of the Construction, Decoration and Use of Arms and Armor in all Countries and in all Times.* Reprint of 1934 ed. New York: Brussel. 1966.

See also: Metallurgy – Navigation

Munīśvara

Munīśvara (b. 1603), son of Raṅganātha, was born into a family of reputed astronomers of several generations, who had migrated from their original home on the banks of river Godāvarī in the south to Varanasi in the north of India. Munīśvara's paternal uncle, Kṛṣṇa Daivajña, was patronized by the Mughal emperor Jehangir, who ruled from Delhi (1605–28). Elevating references by Munīśvara to Shahjehan, who succeeded Jehangir as emperor in 1628, and casting the horoscope of the time of Shahjehan's coronation are pointers to the continued royal patronage enjoyed by Munīśvara's family. In his commentary on the *Līlāvatī*, Munīśvara states that another name of his was Viśveśvara.

Munīśvara was a prolific writer, on both mathematics and astronomy, and wrote both original works and commentaries. The *Siddhāntasārvabhauma*, written in 1646, is his major work on astronomy. In twelve chapters, of which nine chapters constituted Part I, the work dealt with the subjects of a normal textbook. In Part II, the work dealt with the armillary sphere, astronomical instruments, and astronomical queries. He also composed a commentary on the work called *Āśayaprakāśinī*, which is dated 1650. On mathematics, Munīśvara has two works: *Pāṭīsāra* and *Gaṇitaprakāśa*. He was an admirer of Bhāskara II. His commentaries on Bhāskara's *Siddhāntaśiromaṇi*, entitled *Marīcī*, and on *Līlāvatī*, entitled *Nisṛṣṭārthadūtī*, are justly famous for their exhaustiveness, lucidity, and citations from earlier authors. He also commented on the *Pratodayantra* or *Cābukayantra*, a short work on an astronomical instrument used for the ascertainment of the time of the day, by Gaṇeśa Daivajña.

Munīśvara had professional detractors whose views differed from his. One was Raṅganātha, author of the manual *Siddhāntacūdamani* (CE 1640), who, in a short work called *Bhaṅgīvibhaṅgī*, criticized Munīśvara's *Bhaṅgī* (Winding) method of computing true planets. This work was refuted by Munīśvara in his *Bhaṅgīvibhaṅgī-khaṇḍana.* Another was Ekanātha, an astronomer of Maharashtra origin, settled in Varanasi, who seems to have passed strictures on Munīśvara's exposition of three verses on declension (*krānti*) in Bhāskara's *Siddhāntaśiromaṇi*. Munīśvara refuted Ekanātha's criticism and established his views in a short work entitled *Ekanāthamukhabhañjana* (A Slap in the Face of Ekanātha).

Though Munīśvara accepted Islamic trigonometry as an aid to studies in astronomy, he severely contradicted the theory of precession advocated by Kamalākara, against which Raṅganātha wrote a work entitled *Loha-gola-khaṇḍana*, which Munīśvara's cousin Gadādhara refuted in his *Lohagolasamarthana* (Refutation of the Loha-gola).

Characteristics that cannot be missed in Munīśvara's writings are the lucidity, chaste language, and the elegant style in which they are couched.

K. V. Sarma

References

Primary sources

Siddhāntasārvabhauma of Munīśvara. Ed. Mitha Lal Ojha. Varanasi: Sampurnanand Sanskrit University, 1978.

Siddhāntaśiromaṇi of Bhāskara with the Commentary Marīcī of Munīśvara. Ed. Muralidhara Jha. Benares: E.J. Lazarus and Co., 1917.

Siddhāntaśiromaṇi of Bhāskara with the Commentary Marīcī of Munīśvara. Ed. Dattatreya Apte. Poona: Anandasrama Sanskrit Series, 1943.

Siddhāntaśiromaṇi of Bhāskara with the Commentary Marīcī of Munīśvara. Ed. Kedardatta Joshi. Varanasi: Banaras Hindu University, 1964.

Secondary sources

Dikshit, S.B. *Bhāratiya Jyotish Śāstra (History of Indian Astronomy)* Trans. R.V. Vaidya. Pt. II: *History of Astronomy During the Siddhantic and Modern Periods*. Calcutta: Positional Astronomy Centre, India Meterological Department, 1981.

Dvivedi, Sudhakara. *Gaṇaka Taraṅgiṇī or Lives of Hindu Astronomers.* Ed. Padmakara Dvivedi. Benares: Jyotish Prakash Press, 1933.

Pingree, David. *Census of the Exact Sciences in Sanskrit, Series A, vol. 4.* Philadelphia: American Philosophical Society, 1981.

See also: Mathematics in India – Astronomy in India – Precession of the Equinoxes – Kamalākara – Bhāskara II

N

Nārāyaṇa Paṇḍita

Nārāyaṇa, the son of Nṛsiṃha (or Narasiṃha), was one of the major authorities on Indian mathematics after Bhāskara II. We do not know when or where he was born. He wrote two Sanskrit mathematical texts: the *Gaṇitakaumudī* on *pāṭi* (arithmetic) in 1356 (which is confirmed by the final verses of the book) and the *Bījagaṇitāvataṃsa* on *bīja* (algebra). Nārāyaṇa Paṇḍita was confused with another Nārāyaṇa, a commentator on the *Līlāvatī*.

The two books consist of rules (*sūtras*), examples (*udāharaṇas*) and commentary (*vāsanā*) thereon. It is in the *vāsanā* on the *Gaṇitakaumudī* but not in the *mūla* that a reference to the *Bījagaṇitāvataṃsa* is found. The *Gaṇitakaumudī* was published by P. Dvivedi in two volumes based on a single manuscript which had belonged to his late father. The numberings are not accurate. It consists of *paribhāṣā* (metrology units), *parikarma* (basic operations) and fourteen *vyavahāras*: the traditional eight *vyavahāras, kuṭṭaka, vargaprakṛti* (indeterminate equations), calculations for fractions, rule for fractionising, the net of numbers (combinatorics), and magic squares. A critical edition of the last two *vyavahāras* with an English translation and his own commentary was published by T. Kusuba in 1993. Nārāyaṇa's method of finding factors of a number given in the eleventh *vyavahāra* is equivalent to the one by the French mathematician Pierre de Fermat. The twelfth *vyavahāra* includes rules to express the number one as the sum of a number of unit fractions, which are similar to those given by Mahāvīra. The rules in the thirteenth chapter are modelled on those of the *Līlāvatī*, but are further advanced and can be compared to the rules for combinatorics in metrics and music. The *Gaṇitakaumudī* is the first Sanskrit mathematical text so far available that deals with magic squares.

Only the first portion of the *Bījagaṇitāvataṃsa*, based on a single and incomplete manuscript at Benares, has been published. The rules for *kuṭṭaka* and *vargaprakṛti* in the extant portion are similar to those in the *Gaṇitakaumudī* as well as those of Bhāskara II.

Takanori Kusuba

References

Primary sources

Gaṇitakaumudī. Ed. P. Dvivedi Benares. 1936–1942.

Bījagaṇitāvataṃsa. Ed. K.S. Shukla. Supplement to *Rtam* 1. pt. 2. 1969–1970.

Secondary sources

Cammann, Schuyler. "Islamic and Indian Magic Squares." *History of Religions* 7: 181–209 and 271–299. 1968 and 1969.

Datta, Bibhutibhusan. "Nārāyaṇa's Method for Finding Approximate Value of a Surd." *Bulletin of the Calcutta Mathematical Society* 23: 187–194. 1931.

Datta, Bibhutibhusan. "The Algebra of Nārāyaṇa." *Isis* 19: 472–485. 1933.

Datta, Bibhutibhusan, and A.N. Singh. *History of Hindu Mathematics*. 2 vols. Bombay: Asia Publishing House. 1935–38.

Kusuba, Takanori. *Combinatorics and Magic Squares in India*. Ph.D. Dissertation. Brown University. 1993.

Singh, Paramanand. "The So-called Fibonacci Numbers in Ancient and Medieval India." *Historia Mathematica* 12: 229–244. 1985.

Navigation in the Indian Ocean and Red Sea

Navigators in the Indian Ocean used the monsoons since before our era. The Chinese knowledge of monsoons was documented first, but the Indians and Middle Easterners benefited from them as well. The Greeks learned about sailing with monsoons between the Red Sea and India no later than the expedition of Nearchus (326–325 BCE). A Roman port was established at Adulis to trade with India under Ptolemy III Euergetes (247–221 BCE). The Greek *Periplus of the Erythrean Sea* (first century CE) attests to the Arab domination of routes between Arabia, East Africa, and India. With the rise of the Persian Sassanid Empire, Yemen, a crossroads of sea trade, became subject to rival interests of Persians, Byzantines, and Ethiopians. Persians seemed to control the navigation in the western part of the ocean until shortly before the rise of Islam. The revival and expansion of oceanic trade under the Abbasid caliphate (750–1258) must have involved not only Arabs and Persians but also coastal populations converted to Islam later, but the sources are not specific on this point. Participation of Indian Muslims as well as non-Muslims in navigation and piracy is recorded by Ibn Baṭṭūṭa in the fourteenth century. Islamic navigation in the Indian Ocean and the Red Sea and the Persian Gulf is often referred to as Arab navigation largely because the known sailing instructions and literary works describing methods of navigation are in Arabic. Early statements by some European scholars to the effect that the Arabs did not like or know the sea ignore the fact that Islam arose among northern Arabs at the time when south Arabians had accumulated many centuries' worth of sailing experience.

The Red Sea was the scene of early contacts between Muslims and Africa, especially Egypt and Ethiopia. It continued to play a role of conduit between the Mediterranean and the Indian Ocean and eventually carried heavy annual pilgrim traffic to Jedda and al-Jār (the port of Medina during much of the Middle Ages). Mocha was the main port in the south, Aqaba in the north-east; Qulzum (a major naval base) and Qusayr were prominent on the Egyptian coast, and ʿAidhāb on the African coast opposite Jedda. Outside the Bab el-Mandeb the ships stopped at Aden on the Arabian side or Zeila on the African side. Pirates found refuge on Dahlak and the smaller islands; once out in the Gulf of Aden, Socotran piracy was a threat.

During the Crusades the Red Sea became a scene of European attacks on Muslim shipping. Rulers of Egypt always tried to gain control of both coasts of the sea as well as Yemen. The Turkish conquest of Egypt and Yemen in the early sixteenth century made the Ottomans masters of the Red Sea and allowed them access to the Indian Ocean, where they tried to take over shipping routes leading to India, the Persian Gulf, and Africa. However, they were forced to yield to superior Portuguese force and later suffered naval intrusions into the Red Sea by both European and Indian (Gujarati) vessels. A major concern for Turkish authorities at Mocha and on the Ethiopian side (*eyalet of Habasha*) was the security and provisioning of the pilgrims to Mecca. Bombay, Goa, and Surat served as major ports for the Red Sea India trade, and Massawa had a colony of Indian merchants (*banyans*) from the late sixteenth century.

Sailing from the north was relatively easy, although passing the tip of the Sinai Peninsula was feared because the winds from the Gulf of Aqaba and Gulf of Suez met there. Ships carried from the ocean to the Red Sea by the monsoon had to sail against northerly winds once past the strait of Bab el-Mandeb. Jedda was the terminus of oceanic routes; transit further north had to use smaller boats. The only extant sailing instructions for the Red Sea cover the distance from Jedda to Aden (by Ahmad ibn Mājid, ca. 1500). Latitude measurements taken by the stars could use the Polaris *Jāh* because of the northerly location. "Triangular instruments", not described otherwise, and quadrants are mentioned. Finding one's location was not difficult in confined waters, but navigation was dangerous because of numerous coral reefs, contrary winds, and currents. The journey from the north to the south end took thirty days (sailing by day only and coasting), but the ships of Saladin's navy could reach the speed of 4–5 knots. The north wind of the Red Sea *shamāl* reached the southern part only from May to September, coinciding with the short period when the prevailing wind in the Gulf of Aden was westerly, thus propelling ships into the Indian Ocean. The south-easterly wind of the Red Sea *azyab* reached half way up.

Navigation in the Indian Ocean was dominated by the monsoon (from Arabic *mawsim* , "season"), a wind system that reverses direction seasonally. Both halves of the Indian ocean are subject to monsoon regime. The south-west monsoon *kaws* begins in March on the East African coast, slowly spreading eastwards. It reaches its maximum strength in June and blows across the ocean until October, bringing the heaviest rains to India in June and July and causing heavy swells which made landing difficult and even closed the ports. The north-east monsoon *azyab* originates from the Indian mainland in early October, reaching Zanzibar by late November. It makes it easy to sail almost directly from Malacca to Jedda as the wind continues into the Gulf of Aden. Between the monsoon periods, voyages were made in other directions, using variable winds and breezes. March to May are changeover months in the north-west corner of the ocean. In the Gulf of Aden the predominant non-monsoon wind is easterly. Travel from India to Africa had to be begun by early February. From Aden and Yemen one needed to leave in mid-October, and from north-east Arabia by late January, but one could not sail to Socotra during the same season. From Socotra to south-east Arabia one sailed in March–April, while India could be reached also by departing in May and during the August–September season *dāmānī*. Travel down the African coast was recommended from mid-November to April. From Bengal one had to leave westward by January; leaving from Malacca, Java, and Sumatra in February or March one could still reach Ceylon. Travel from Gujarat to Bengal and Indonesia began in April or late summer. October brought cyclones to the Bay of Bengal. The eastbound roundtrip journey from the Persian Gulf across the ocean took eighteen months. China-bound ships started from the Gulf in September or October, reached Kalah Bar in January and passed through the Strait of Malacca in time to use the southern monsoon in the Sea of China. Return to Malacca took place with the north-east monsoon between October and December; then ships could cross the Bay of Bengal in January and reach Arabia in February or March.

Navigation between the Middle East and China is confirmed by reports of an Arab embassy to China in the seventh century, a Persian settlement in the island of Hainan in 748, and a mixed Arab–Persian colony at Canton in the eighth–ninth centuries. In the seventh century Persian ships took twenty to thirty days to reach Sumatra from China, and one month from Ceylon to Palenbang. In the ninth century "Chinese ships" or "China ships" (that is, ships sailing to China) are reported in the Persian Gulf. Smaller boats brought goods from Basra and other ports to Siraf where they were reloaded on the large China boats. Ceylon (*Sarandīb*) was also visited and described by Arabic authors. Sea travel from the Persian Gulf to East African islands is mentioned by Arabic sources in the ninth century and described as routine in the tenth by the historian and traveler al-Mas'ṣdī, although the country of Sofala (southern Tanzania and northern Mozambique) was then still poorly known. The Persian Gulf ports of Siraf and Hormuz as well as Oman were dominant on that route at the time; Aden emerged to prominence somewhat later. Some of the tales of Sindbad the Sailor (of Basra) originated in stories of Indian Ocean sailor and merchant adventures collected in the book *The Marvels of India* by Buzurg ibn Shāhriyār (ca. 950). Much of Marco Polo's return journey in the late thirteenth century must have taken place in Muslim boats and followed routes mentioned in this book. In the fourteenth century Ibn Baṭṭūṭa travelled by ship across the Red Sea and later to Africa, visiting Mogadishu, Mombasa, and Kilwa. On other occasions he sailed along the west coast of India and possibly to China. In India he encountered merchants from Cairo and North-west Africa and fell victim to Indian pirates. He reported the presence of Chinese junks at Ceylon and planned to travel on one himself. The famous Chinese voyages led by the Ming court official Zheng He constitute the last known attempt by China to break into the Indian Ocean network. Two of these, in 1417–19 and 1421–22, reached Africa, visiting Malindi and the Horn of Africa. Even the arrival of the Portuguese caused only a disruption and reorganisation of shipping. Lodovico Varthema (ca. 1510) and the early Portuguese sources note the continuing international presence at western Indian ports: Egyptians and "Moorish" merchants and ships from Hormuz, Arabia, Abyssinia, Kilwa, Malindi, Mombasa, and Mogadishu at Cambay, on the Malabar coast and the islands (Maldives and Laccadives). Early naval battles between the Portuguese and combined Muslim navies (e.g. at Diu in 1512) resulted in capture and destruction of numerous Muslim vessels. (In the sixteenth century only the Chinese had ships able to withstand attacks of the Portuguese galleons). However, the Portuguese soon realised that they would be unable to stop native shipping, and turned their efforts to diverting trade to ports which they controlled, carriage in their own ships, and taxation of all others. By the time the Dutch and the English arrived, an accommodation had been reached. However, intra-European competition, added to Christian–Muslim rivalry, contributed to pre-existing pirate activity, especially in the Persian Gulf and on the Gujarat and Malabar coasts. The maritime Muslim trade revived somewhat in the seventeenth–eighteenth centuries and declined again in the nineteenth century, at least in part due to increased British control over the routes and ports of the western Indian Ocean. Another factor was the growing dominance of European companies in the long-distance East–West trade and their penetration of local trade.

The sources for traditional Arab navigation date mostly from the late fifteenth through the sixteenth centuries, while our knowledge of ships and shipbuilding in the region is modern or contemporary. There are vague, scattered references to earlier sailing guides *rāhnāmaj*, devices and ships. Naval law is best known from the Malacca code of Shāh Mahmūd (1488–1530). A sixteenth-century Persian source lists twelve categories of crew members with job descriptions. The best information on navigation proper and sailing routes comes from the works of Aḥmad ibn Mājid of Julfar in Oman (d. ca. 1504) whose recognised expertise made him into a patron saint of Muslim sailors. A learned practitioner, he composed navigation manuals and sailing instructions, largely in verse, to ease memorisation. From him we learn the names of several earlier pilots, dating back to the tenth–twelfth centuries. To these Ibn Mājid added the names of his own father and grandfather; apparently, the profession of pilot was hereditary but not highly regarded. It has been asserted by some scholars that Ibn Mājid was the pilot who guided Vasco da Gama from Malindi to Calicut, but this has been contested by others. Although he spent his life on the Indian Ocean, it appears that he was aware of the different methods of navigation in the Mediterranean. Other extant works on Islamic navigation belong to Sulaymān al-Mahrī (ca. 1511), a native of Shihr who wrote several practical and theoretical treatises, and the Ottoman writer Sidi Ali çelebi who compiled a Turkish summary of the former two authors' work while moored in Gujarat in 1554 after a Portuguese attack on the Turkish Indian Ocean fleet originally commanded by the portolan-maker Pirī Reis.

Contemporary Arabic names of ships: *baghala, ganja, sanbūq*, and *jihāzi*, apply to vessels with square, transom sterns showing European influence. The older type is represented by vessels now called *būm* and *zārūq* – double-edged, coming to a point both at bow and stern. The name *sanbūq* was formerly applied to the small craft of the Red Sea; *jalbah* was the sewn boat typical of the Indian Ocean region. First mentioned in the *Periplus of the Erythrean Sea*, sewn boats were carvel-built, with planks edge-to-edge, and stitched with ropes of palm fibre. The timber was teak or coconut wood. They were leaky and frail but had an advantage over clinker-built boats with overlapping planks when striking a coral reef. A common legend explained that nails were not used because of the dangerous power of a magnetic rock somewhere in the middle of the ocean which could attract the nails. However, iron as well as stone anchors were used. The generic names for "ship" were *markab* and *safīna*. Indian pirates had *bārijas*; smaller boats *zawraq, qārib* and *dūnij* are also mentioned in medieval texts, but no particular shape is indicated. *Dau* is a generic name for lateen-rigged vessels; it is not used by the Arabs. The basis of classification was the form of the hull. The ships were usually one-masted, often without deck, with a cargo capacity of up to 200 tons. Erecting the mast and the rigging, and even making the sail was the responsibility of the owner *nakhōda* and the crew rather than the shipwright. The lateen sail associated with Arab ships probably evolved from a square sail on the Indian ocean; its use in the Mediterranean is first noted in the ninth century. African and Indian vessels continued using square sails of coconut matting into the twentieth century. The lateen is a tall, triangular fore-and-aft sail with the fore angle cut off to form a luff. It allows sailing into the wind by going on the tack, although Arab mariners preferred not to sail closer to the wind than 90°. It is possible that a second sail (topsail) was sometimes used. No reefing was done in strong wind, but the yard could be lowered. Two side rudders were originally used for steering, although by the thirteenth century the stern rudder was known. Sailing speeds averaging 1–3 knots were normal but could reach six knots under favourable wind.

The most important person on board ship was the *mu'allim* who served both as captain *rubbān* and pilot. He was hired for the voyage and allowed to carry merchandise as part of his pay. The shipmaster *nakhōda* was a merchant; ships were often owned by shareholders. We know Persian terms for eleven ranks of crew members; of these, the most important were the bo'sun *tandīl*, the ship's mate *sarhang*, the steersman *sukkan-gir*, and the look-out *panjarī*. Sailors were called *khallāsī* or *khārwah*. The captain was responsible not only for navigation but for the safety of passengers and goods as well. Among the necessities he carried were a nautical directory *rahnāmaj*, a measuring instrument *qiyās*, a bussole *huqqa* or *dīra*, lodestone *hajar*, lot *buld*, and lantern *fānūs*. The pilot's principal science consisted of knowing the coasts, winds and seasons, and his ship. Before departure, the Muslim prayer *Fātiha* was recited and an invocation was made to Khidr, the mythical patron saint of mariners.

By the sixteenth century Arabic sailing manuals listed over thirty different routes. Navigational books *dafātīr* and charts *suwar* carried on board are mentioned in the tenth century; Ibn Mājid calls his "chart" *qunbās*, but no charts have come to light. G. R. Tibbets argues that the Arab pilot plotted his course in his head and did not need a chart; besides, proper charts could not be made because the Arabs had no way of correctly determining longitude at sea. The winds and geography of the Indian ocean allowed the pilot to be guided roughly by the latitude of his destination (determined by the Pole star altitude). Once that was reached, he sailed down the latitude toward his goal. Another way was to keep to a recommended bearing until land was in sight and then make corrections. Extant Arab maps do not allow practical application to navigation. A Chinese chart based on the Zheng He expedition shows the routes from China to Hormuz, the Red Sea and Africa, but no measurements can be taken from it. Charts from Muslim Indian (Gujarati) nautical manuals *roz nāmah* of the seventeenth–eighteenth centuries show some European influence and use stellar compass bearings and Arab units of time–distance. A possibility of Chinese influence has been suggested as well. Considering that Arab information is already found on early sixteenth-century Portuguese maps, it is clear that the sharing of information among mariners created a truly international maritime culture drawing on indigenous and regional traditions and innovations.

The Arab system of nautical orientation evolved on the Indian ocean in the intertropical region, but probably north of the Equator; it may have been representative of all Indian Ocean sailing. It is based on a 32-rhumb *khann* sidereal rose *dīra* divided into eastern and western halves separated by the Polaris *Jāh* in the north and the South Pole *al-Qutb* in the south. The east and west divisions approximate the rising and setting of certain bright stars and constellations (Ursa Minor, Ursa Major, Cassiopeia, Capella, Vega, Arcturus, Pleiades, Altair, Orion, Sirius, Scorpio, Antares, Centaur, Canopus, Achernar). This system may have been in place by the ninth century. The bearings *majrā* were set by the actual stars, visible in the clear skies, not by the mathematically correct rhumbs. The compass was not unknown but rarely used or even carried. Star altitude *qiyās* was measured in units called *isbaʿ* (finger), supposed to correspond to the arc covered by the little finger of an outstreched hand. Its degree value measured 1/2 of the distance from the Polaris to the true pole, and thus varied with precession. In 1394 one *isbaʿ* equalled 1°56′ but in 1550, 1°33′. The full circle of 360° corresponded alternatively to 210 or 224 *isbaʿ*. *Isbaʿ* also measured 1/24 of a cubit. For longitude estimates, one *isbaʿ* equalled eight *zām*, each *zām* corresponding to the distance covered in three hours of sailing. A variety of other measurements, including something approximating triangulation, were calculated in these units. The altitude of the Polaris was supposed to be taken at its inferior elevation.

The instruments used for measurements included *kamāl, lawh,* and *bilistī*. The *kamāl* was a rectangle of horn or wood with a string through the middle. It was held against the horizon in an outstretched hand, with the cord held in the teeth by the knot. Knots tied at certain intervals on the cord corresponded to the varying arcs covered by the rectangle. Variations of this instrument included knots tied at intervals corresponding to locations on a set route, a set of boards *lawh* corresponding to different arc values fixed on the cord, and the cord being held to the nose. The *bilistī* was a later version of the *kamāl,* with a rod replacing the cord, and four sliders of different sizes; most likely it post-dates the Portuguese arrival because its function is essentially that of the *balhestilha*. By the nineteenth century the system and the instruments had been largely driven out of use or forgotten, although the name and expertise of Aḥmad ibn Mājid were still respectifully remembered.

M.A. Tolmacheva

References

Arunachalam, B. "The Haven-Finding Art in Indian Navigational Traditions and Cartography." In *The Indian Ocean: Explorations in History, Commerce, and Politics*. Ed. Satish Chandra. New Delhi: Sage Publications. 1987. pp. 191–221.

Chandra, Satish, ed. *The Indian Ocean: Explorations in History, Commerce and Politics*. New Delhi and London: Sage Publications. 1987.

Chaudhuri, K.N. *Trade and Civilization in the Indian Ocean: An Economic History from the Rise of Islam to 1750*. Cambridge: Cambridge University Press. 1985.

Clark, Alfred. "Medieval Arab Navigation on the Indian Ocean: Latitude Determinations." *Journal of the American Oriental Society* 113 (3): 360–73. 1993.

Ferrand, Gabriel. *Instructions Nautiques et Routiers Arabes et Portugais*. 3 vols. Paris: P. Guetner. 1921–28.

Hall, Kenneth R. *Maritime Trade and State Development in Early Southeast Asia*. Honolulu: University of Hawaii Press. 1985.

Hourani, George Fadlo. *Arab Seafaring in the Indian Ocean in Ancient and Early Medieval Times*. Princeton, New Jersey: Princeton University Press. 1951.

Hsu, Mei-ling. "Chinese Marine Cartography: Sea Charts of Pre-Modern China." *Imago Mundi* 40: 96–112. 1988.

Schwartzberg, Joseph E. "Nautical Maps." In *The History of Cartography*, vol. 2, book 1 *Cartography in the Traditional Islamic and South Asian Societies*. Ed. J. B. Harley and David Woodward. Chicago: University of Chicago Press. 1992. pp. 494–503.

Serjeant, R. B. "Star-calendars and an Almanac from South-west Arabia." *Anthropos* 49: 478–502. 1954.

Severin, Timothy. *The Sindbad Voyage*. London: Hutchinson. 1982.

Sidi Çelebi. "Extracts from the Mohit, that is the Ocean, a Turkish Work on Navigation in the Indian Seas, translated by J. Hammer-Purgstall." *Journal of the Asiatic Society of Bengal*. 1834: 545–53, 1836: 441–68, 1839: 805–12, 1838: 767–80, 1839: 823–30.

Tibbets, Gerald R. *Arab Navigation in the Indian Ocean before the Coming of the Portuguese*. London: Luzac for the Royal Asiatic Society of Great Britain and Ireland. 1971.

Tibbets, Gerald R. "The Role of Charts in Islamic Navigation in the Indian Ocean." In *The History of Cartography*, Vol. 2, Book 1 *Cartography in the Traditional Islamic and South Asian Societies*. Eds. J. B. Harley and David Woodward. Chicago: University of Chicago Press. 1992. pp. 256–62.

Tolmacheva, Marina. "On the Arab System of Nautical Orientation." *Arabica* 27 (2): 180–192. 1980.

Varadarajan, Lotika. "Traditions of Indigenous Navigation in Gujarat." *South Asia: Journal of South Asian Studies*, n.s. 3 (1): 28–35. 1980.

Villiers, Alan J. *Monsoon Seas: the Story of the Indian Ocean*. London and New York: McGraw-Hill. 1952.

Villiers, Alan J. *Sons of Sindbad: an Account of Sailing with the Arabs*. London: Hodder and Stroughton. 1940.

See also: Precession of the Equinoxes

Number Theory

It is difficult to find 'number theory' in its proper sense in Indian mathematics. What I am going to describe below is how the Indians have treated kinds of numbers.

In the Vedas (ca. 1200–800 BCE), the oldest Hindu literature, a number of numerical expressions occur. Their favourite numbers were three and seven as well as a hundred and a thousand. The largest number contained in their common list of names for powers of ten is 10^{12} (called *parārdha*). Later (by the fourth century CE), those names came to be employed for denoting decimal places, and became the nucleus of the Hindu list of decimal names (eighteen in number), while the Buddhists and the Jainas developed longer lists, which contained numbers as large as 10^{53} (*tallakṣaṇa*) or more.

The Jainas even speculated about different kinds of uncountable and infinite numbers (*Aṇuogaddārāiṃ*, between the third and the fifth centuries CE). They divided the whole set of "numbers (*saṃkhyā*) concerning counting (*gaṇanā*)" into three subsets: (1) countable *(saṃkhyeya)*, (2) uncountable *(asaṃkhyeya)*, and (3) infinite *(ananta)* numbers; and further divided the last two sets into three each (Table 1). The entire system of countable–uncountable–infinite depends upon the smallest number (*a*) of the "restrictively uncountable" set, which in turn is defined by means of white mustard seeds. The text is not very clear on this last point, but its intention was probably the same as what is meant by "the aleph zero" in modern mathematics, the smallest transfinite cardinal number.

Vedic stanzas also contain various series of numbers such as an integer series up to 200, an odd series (1,3,5, ...) up to 99 (accompanied by 100), an even series (2,4,6, ...) up to 100, and series made from multiples of four, five, ten, and twenty, up to 100, etc.

The Vedas tell us that only the gods Indra and Viṣṇu could divide a thousand equally into three, but it is a matter of argument how they did it. Natural fractions ($\frac{1}{2}$, $\frac{1}{4}$, $\frac{1}{8}$ and $\frac{1}{6}$) also occur in the Vedas.

The *Śulbasūtras* (ca. 600 BCE and later), compendia of geometric knowledge related to the construction of various altars for the Vedic ritual, clearly state the so-called Pythagorean theorem: "The diagonal rope of an oblong produces both [areas] which its side and length produce separately." They explicitly mention several Pythagorean triples also: (3,4,5), (5,12,13), (8,15,17), (7,24,25), and (12,35,37). Moreover, they give an algorithm for calculating the diagonal *d* of a square whose side is *a*:

$$d = a + \frac{a}{3} + \frac{a}{3} \cdot \frac{1}{4} \cdot \frac{1}{34}$$

and call this value "one that has a difference" (*saviśeṣa*), which probably means the difference between this approximate value and the true one ($\sqrt{2}a$). The latter was called "one that makes [a square equivalent to] two [unit squares]" *(dvi-karaṇī)*. There is, however, no indication that they recognised the incommensurability of the diagonal and the side. A root approximation formula of the same type is employed in the Bakhshālī Manuscript.

Later Indian mathematicians and astronomers, such as Varāhamihira (ca. CE 550) and Brahmagupta (CE 628), used the word *karaṇī* in the two contradictory (but mutually related) senses, the square root of a non-square number and the number whose square root should be obtained (or the square of any number), and easily performed the six arithmetical operations involving irrational numbers. Even the irrationality of *karaṇīs* may have been understood by Bhāskara of the seventh century, because he says, in his commentary on the *Āryabhaṭiya* (CE 629), "*Karaṇīs* have a size that cannot be stated exactly", although whether he proved it or not is not known.

Varāhamihira recognised zero as a number. In the *Bṛhatsaṃhitā* he added and subtracted zero in exactly the same way as he did other integers. Brahmagupta in the *Brāhmasphuṭasiddhānta* gave a complete set of rules for the six arithmetical operations involving zero as well as negative and irrational numbers. Thus by the seventh century CE the Indians acquired a very large domain of numbers including positive and negative integers (they accepted both the positive and the negative roots of a square number), fractions, irrational numbers, and zero, which enabled them to develop *bījagaṇita* ("seed mathematics") or algebra, the theory of equations.

In his *Āryabhaṭīya* (CE 499), Āryabhaṭa provided a solution (called *kuṭṭaka* or the pulveriser) to the linear indeterminate equation: $n = ax + r = by + s$, or $y = (ax + c)/b$. He "pulverised" (i.e. reduced) the coefficients *a* and *b* by means of their mutual divisions (the so-called Euclidean algorithm), and found a set of solutions to the reduced form by trial and error. Mahavira (ca. CE 850) removed the trial and error by carrying out the mutual divisions until the remainder became 1.

Brahmagupta treated indeterminate equations of the type:

Table 1 *Classification of numbers according to the Jainas*

numbers for counting	countable		lowest: 2 middle: 3, 4..., $a-2$ highest: $a-1$
	uncountable	restrictively uncountable	lowest: a (= N_0 ?) middle: $a+1$, $a+2$, ..., $b-2$ highest: $b-1$
		properly uncountable	lowest: $b = a^a$ middle: $b+1$, $b+2$, ..., $c-2$ highest: $c-1$
		uncountably uncountable	lowest: $c = (b^2)^{b^2}$ middle: $c+1$, $c+2$, ..., $d-2$ highest: $d-1$
	infinite	restrictively infinite	lowest: $d = c^c$ middle: $d+1$, $d+2$, ..., $e-2$ highest: $e-1$
		properly infinite	lowest: $e = d^d$ middle: $e+1$, $e+2$, ..., $f-2$ highest: $f-1$
		infinitely infinite	lowest: $f = (e^2)^{e^2}$ middle: $f+1$, $f+2$, ... highest: does not exist

$Px^2 + t = y^2$. He showed, among other things, that this equation (called *vargaprakṛti* or the "square nature") can be solved for $t = 1$ if it is solved for $t = \pm 4, \pm 2$, or -1. Jayadeva (the eleventh century or before) gave a rule for arriving at a solution for $t = \pm 4, \pm 2$, or -1 from any solution for any t. Bhāskara of the twelfth century called it the "cyclic" (*cakravāla*) method.

Several rules given by Mahāvīra in his *Gaṇitasārasaṃgraha* indicate that he recognised two roots of a quadratic equation with one unknown, but all of his examples for those rules have two positive roots. (He, however, admits the negative root of a square number when he gives his rules for the six arithmetical operations). He was interested in the partition of numbers, and gave many rules for partitioning unity into the sum of unit fractions, a fraction into the sum of several fractions, etc. He also treated various mathematical progressions.

Śrīpati, perhaps for the first time in India, gave several rules for factorisation in a chapter devoted to algebra in his astronomical work, *Siddhāntaśekhara* (ca. CE 1040). This topic, as well as the partition of numbers, mathematical progressions, combinatorics, and magic squares were highly developed by Nārāyaṇa in his *Gaṇitakaumudī* (CE 1356).

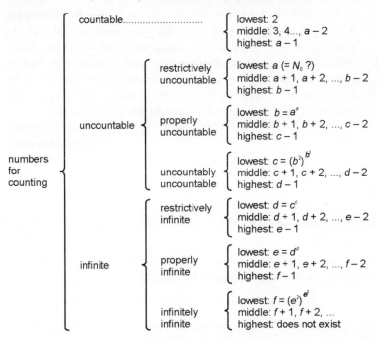

Bhāskara challenged various types of polynomial equations of the second and higher degrees (of special types) with the help of the "pulverizer" and the "square nature" in his work, *Bījagaṇita*, CE 1150.

Mādhava (fl. CE 1400) and his successors obtained, for the first time in the world, a number of power series for the circumference of a circle, sine, cosine, arctangent, etc. One of the most eminent scholars in his school was Nīlakaṇṭha. A great mathematician and reformer of Indian astronomy, he explicitly stated the incommensurability of the diameter and the circumference of a circle in his commentary on the *Āryabhaṭīya* (ca. CE 1540), although its proof is not found in his extant works.

Takao Hayashi

References

Datta, B. "The Jaina School of Mathematics." *Bulletin of the Calcutta Mathematical Society* 21: 115–145. 1929.

Datta, B. "Early Literary Evidences of the Use of the Zero in India." *American Mathematical Monthly* 33: 449–454. 1926 and 38: 566–572. 1931.

Datta, B. *The Science of the Śulba*. Calcutta: University of Calcutta. 1932.

Datta, B. "Vedic Mathematics."In *The Cultural Heritage of India*, vol. 3. Calcutta: Ramakrishna Centenary Committee. 1937. pp. 378–401.

Hayashi, T. Kusuba, T., and Yano, M. "The Correction of the Mādhava Series for the Circumference of a Circle." *Centaurus* 33: 149–174. 1990.

Michaels, A. *Beweisverfahren in der vedischen Sakralgeometrie*. Wiesbaden: Steiner. 1978.

Rajagopal, C.T. and M.S. Rangachari. "On an Untapped Source of Medieval Keralese Mathematics." *Archive for History of Exact Sciences* 18: 89–102. 1978.

Shukla, K.S. "Hindu Methods for Finding Factors or Divisors of a Number." *Gaṇita* 17: 109–117. 1966.

See also: Algebra – Arithmetic – *Śulbasūtras* – Geometry – Bakhshālī Manuscript – Varāhamihira – Brahmagupta – Nīlakaṇṭha – Mādhava – Magic Squares – Combinatorics – Nārāyaṇa – Śrīpati – Jayadeva – Mahāvīra – Bhāskara

O

Observatories

India has an ancient astronomical tradition. Information on its observatories is meagre, however. It is certain that a number of prominent astronomers, patronised by kings, carried out their own observations, which are mentioned in *karaṇas*, or practical manuals. The places of such observations, if operated for a reasonable period of time, technically could be called observatories. A court astronomer, Śaṅkaranārāyaṇa (fl. 869), mentions such a place with instruments in the capital city of King Ravi Varmā of Kerala. Astronomers of the Islamic school of astronomy, such as ʿAbd al-Rashīd al-Yāqūtī (fifteenth century) report an observatory in the city of Jājilī in India. The Emperor Humayun (d. 1556) is said to have had a personal observatory at Kotah, near Delhi, where he himself took observations.

An ambitious program of building observatories was undertaken by Sawai Jai Singh (Savā'ī Jaya Siṃha), an astronomer–statesman of India. Between 1724 and 1735, Jai Singh built observatories at Delhi, Jaipur (Figure 1), Mathura, Varanasi, and Ujjain. His observatories, except for that of Mathura, still exist today in varying degrees of preservation. Sawai Jai Singh's purpose in building observatories was to update the existing planetary tables. Toward this purpose, he designed and built instruments of stone and masonry. These instruments may be classified into three main categories based on their precision which varies anywhere from ±1′ to a degree. Table 1 presents an inventory of his masonry instruments according to their precision, with the low precision instruments listed first. Table 2 lists instruments added after Sawai Jai Singh's death.

Jai Singh constructed fifteen different types of masonry instruments for his observatories. Of these, the *Samrāṭ yantra, Ṣaṣṭhāṃśa, Dakṣiṇottara Bhitti, Jaya Prakāśa, Nāḍīvalaya,* and *Rāma yantras* are his most important instruments.

Figure 1 *Jaipur observatory of Sawai Jai Singh.*

Table 1 *Inventory of Jai Singh's masonry instruments*

Instrument	No.	Location
Dhruvadarśaka Paṭṭikā (North Star Indicator)	1	Jaipur
Nāḍīvalaya (Equinoctial dial)	5	Jaipur (2), Varanasi, Ujjain, Mathura
Palabhā (Horizontal sundial)	2	Jaipur, Ujjain
Agrā (Amplitude instrument)	5	Delhi, Ujjain, Mathura
Śaṅku (Horizontal dial)	1	Mathura
Jaya Prakāśa (Hemispherical instrument)	2	Delhi, Jaipur
Rāma yantra (Cylindrical instrument)	2	Delhi, Jaipur
Rāśi valaya (Ecliptic dial)	12	Jaipur
Śara yantra (Celestial latitude dial)	1	Jaipur
Digaṃśa (Azimuth circle)	3	Jaipur, Varanasi, Ujjain
Kapāla (Hemispherical dial)	2	Jaipur
Samrāṭ (Equinoctial sundial)	6	Delhi, Jaipur (2), Varanasi (2), Ujjain
Ṣaṣṭhāṃśa (60 degree meridian chamber)	5	Delhi, Jaipur (4)
Dakṣiṇottara Bhitti (Meridian dial)	6	Delhi, Jaipur, Varanasi (2), Ujjain, Mathura

Table 2 *Instruments added after Sawai Jai Singh's death*

Instrument	No	Location
1. Miśra yantra (Composite instrument)	1	Delhi
2. Śaṅku yantra (Vertical staff)	1	Ujjain
3. Horizontal scale (known as the seat of Jai Singh)	1	Jaipur

SAMRĀṬ YANTRA

The *Samrāṭ yantra* or the "Supreme Instrument" is Jai Singh's most important creation. The instrument (Figure 2) is basically an equinoctial sundial, which has been in use in one form or another for hundreds of years in different parts of the world.

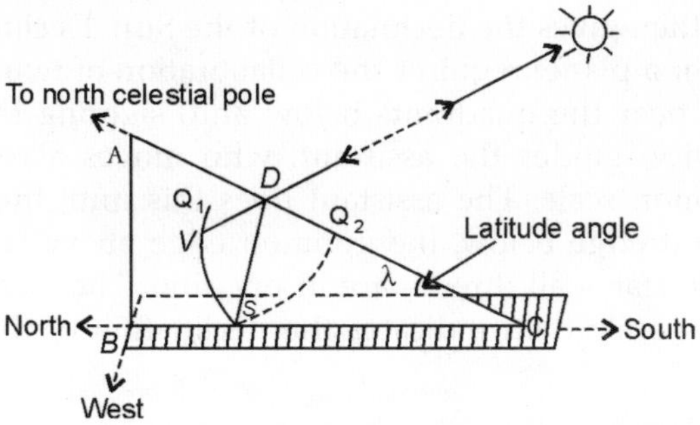

Figure 2 *Samrāṭ yantra: Principle and operation*

The instrument consists of a meridian wall ABC, in the shape of a right triangle, with its hypotenuse or the gnomon CA pointing toward the north celestial pole and its base BC horizontal along a north–south line. The angle ACB between the hypotenuse and the base equals the latitude λ of the place. Projecting upward from a point S near the base of the triangle are two quadrants SQ_1 and SQ_2 of radius DS. These quadrants are in a plane parallel to the equatorial plane. The centre of the two "quadrant arcs" lies at point D on the hypotenuse. The length and radius of the quadrants are such that, if put together, they would form a semicircle in the plane of the equator.

The quadrants are graduated into equal-length divisions of time-measuring units, such as *ghaṭikās* and *palas*, according to the Hindu system, or hours, minutes and seconds, according to the Western system. The upper two ends Q_1 and Q_2 of the quadrants indicate either the 15-*ghaṭikā* marks for the Hindu system, or the 6 a.m. and the 6 p.m. marks according to the Western system. The bottom-most point of both quadrants, on the other hand, indicates the zero *ghaṭikā* or 12 noon. The hypotenuse or the gnomon edge AC is graduated to read the angle of declination. The declination scale is a tangential scale in which the division lengths gradually increase according to the tangent of the declination.

The primary object of a Samrāṭ is to indicate the apparent solar time or local time of a place. On a clear day, as the Sun journeys from east to west, the shadow of the Samrāṭ gnomon sweeps the quadrant scales below from one end to the other. At a given moment, the time is indicated by the shadow's edge on a quadrant scale.

The time at night is measured by observing the hour angle of the star or its angular distance from the meridian. Because a Samrāṭ, like any other sundial, measures the local time or apparent solar time and not the "Standard Time" of a country, a correction has to be applied to its readings in order to obtain the standard time.

To measure the declination of the Sun with a Samrāṭ, the observer moves a rod over the gnomon surface AC up or down until the rod's shadow falls on a quadrant scale below. The location of the rod on the gnomon scale then gives the declination of the Sun. Declination measurement of a star or a planet requires the collaboration of two observers. One observer stays near the quadrants below and, sighting the star through a sighting device, guides the assistant, who moves a rod up or down along the gnomon scale. The assistant does this until the vantage point V on a quadrant edge below, the gnomon edge above where the rod is placed, and the star – all three – are in one line. The location of the rod on the gnomon scale then indicates the declination of the star.

Ṣaṣṭhāṁśa

A *Ṣaṣṭhāṁśa yantra* is a 60-degree arc built in the plane of meridian within a dark chamber. The instrument is used for measuring the declination, zenith distance, and the diameter of the Sun. As the Sun drifts across the meridian at noon, its pinhole image falling on the *Ṣaṣṭhāṁśa* scale below enables the observer to measure the zenith distance, declination, and the diameter of the Sun. The image formed by the pinhole on the scale below is usually quite sharp, such that at times even sunspots may be seen on it.

Dakṣinottara Bhitti Yantra

Dakṣinottara Bhitti yantra is a modified version of the meridian dial of the ancients. It consists of a graduated quadrant or a semicircle inscribed on a north–south wall. At the centre of the arc is a horizontal rod. The instrument is used for measuring the meridian altitude or the zenith distance of an object such as the Sun, the Moon, or a planet. According to Jagannātha Samrāṭ, this was the instrument with which Jai Singh determined the obliquity of the ecliptic (the band of the zodiac through which the Sun apparently moves in its yearly course), to be 23°28′ in 1729.

Jaya Prakāśa

The *Jaya Prakāśa* is a multipurpose instrument (Figure 3) consisting of hemispherical surfaces of concave shape and inscribed with a number of arcs. These arcs indicate the local time, and also measure various astronomical parameters, such as the coordinates of a celestial body and ascendants, or a sign on the meridian. *Jaya Prakāśa* represents the inverted image of two coordinate systems, namely, the azimuth–altitude and the equatorial, drawn on a concave surface. For the azimuth–altitude system, the rim of the concave bowl indicates the horizon. Cardinal points are marked on the horizon, and cross-wires are stretched between them. On a clear day, the shadow of the cross-wire falling on the concave surface below indicates the coordinates of the Sun. Time is read by the shadow's angular distance from the meridian along a diurnal circle.

Figure 3 *Jaya Prakāśa at Jaipur. The Nāḍīvalaya is in the background.*

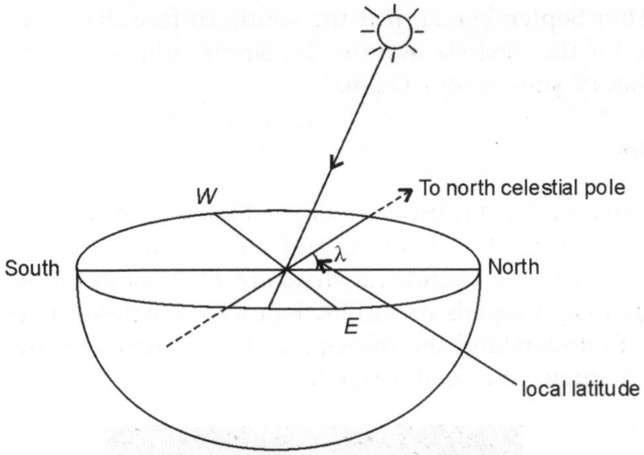

Figure 4 *The principle of Jaya Prakāśa and Kapāla yantra*

The instrument is built in two complementary halves (Figure 4), giving it the capacity for night observations. In the two halves the area between alternate hour circles is removed, and steps are provided in its place for the observer to move around freely for his readings. The space between identical hour circles of the two hemispheres is not removed, however. The sections left behind in the hemispheres complement each other. They do so in such a way that, if put together, they would form a complete hemispherical surface. For night observations the observer sights the object in the sky from the space between the sections. The observer obtains the object in the sky and the cross-wire in one line. The coordinates of the vantage points are then the coordinates of the object in the sky. Jai Singh built his *Jaya Prakāśa*s only at Delhi and Jaipur. These instruments survive in varying degrees of preservation. The instrument at Delhi has a diameter of 8.33 m and that at Jaipur, 5.4 m.

Nāḍīvalaya

A *Nāḍīvalaya* consists of two circular plates fixed permanently on a masonry stand of convenient height above ground level. The plates are oriented parallel to the equatorial plane, and iron styles of appropriate length pointing toward the poles are fixed at their centres. The instrument *Nāḍīvalaya* is, in fact, an equinoctial sundial built in two halves, indicating the apparent solar time of the place.

The *Nāḍīvalaya* is an effective tool for demonstrating the passage of the Sun across the celestial equator. On the vernal equinox and the autumnal equinox the rays of the Sun fall parallel to the two opposing faces of the plates and illuminate them both. However, at any other time, only one or the other face remains in the Sun. After the Sun has crossed the equator around March 21, its rays illuminate the northern face for six months. After September 21, it is the southern face that receives the rays of the Sun for the next six months. Jai Singh built *Nāḍīvalaya*s at each of his observatory sites except Delhi.

Rāma Yantra

The *Rāma yantra* is a cylindrical structure (Figure 5) in two complementary halves that measure the azimuth and altitude of a celestial object, for example the Sun. The cylindrical structure of *Rāma yantra* is open at the top, and its height equals its radius. Figure 6 illustrates its principle and operation. To understand the principle, let us assume that the instrument is built as a single unit as illustrated.

Figure 5 *Ramā yantras at Delhi*

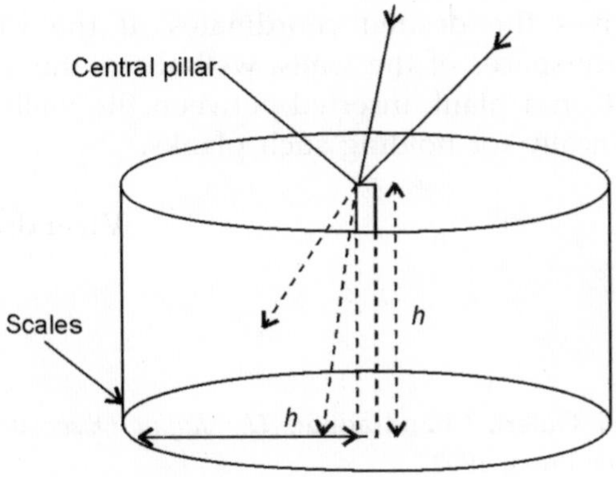

Figure 6 *The Principle of a Ramā yantra*

The cylinder, as in the figure, is open at the top and has a vertical pole or pillar of the same height as the surrounding walls at the centre. Both the interior walls and the floor of the structure are engraved with scales measuring the angles of azimuth and altitude. For measuring the azimuth, circular scales with their centres at the axis of the cylinder are drawn on the floor of the structure and on the inner surface of the cylindrical walls. The scales are divided into degrees and minutes. For measuring the altitude, a set of equally spaced radial lines is drawn on the floor.

These lines emanate from the central pillar and terminate at the base of the inner walls. Further, vertical lines are inscribed on the cylindrical wall, which begin at the wall's base and terminate at the top end. These lines may be viewed as the vertical extension of the radial lines drawn on the floor of the instrument.

In daytime the coordinates of the Sun are determined by observing the shadow of the pillar's top end on the scales, as shown in Figure 6. The coordinates of the Moon, when it is bright enough to cast a shadow, may also be read in a similar manner. However, if the Moon is not bright enough, or if one wishes to measure the coordinates of a star or planet that does not cast a shadow, a different procedure is followed. To accomplish this, the instrument is built in two complementary units.

The two complementary units of a *Rāma yantra* may be viewed as if obtained by dividing an intact cylindrical structure into radial and vertical sectors. The units are such that if put together, they would form a complete cylinder with an open roof. The procedure for measuring the coordinates at night with a *Rāma yantra* is similar to the one employed for the *Jaya Prakāśa.* The observer works within the empty spaces between the radial sectors or between the walls of the instrument. Sighting from a vacant place, he obtains the object in the sky, the top edge of the pillar, and the vantage point in one line. The vantage point after appropriate interpolation gives the desired coordinates. If the vantage point lies within the empty spaces of the walls, well above the floor, the observer may have to sit on a plank inserted between the walls. The walls have slots built specifically for holding such planks.

Virendra Nath Sharma

REFERENCES

Garrett, A ff. and Guleri, Chandradhar. *The Jaipur Observatory and Its Builder*. Allahabad: Pioneer Press. 1902.

Kaye, G.R. *The Astronomical Observatories of Jai Singh*. New Delhi: Archaeological Survey of India, reprint. 1982.

Sharma, Virendra Nath. *Sawai Jai Singh and His Astronomy*. New Delhi: Motilal Banarasidass. 1995.

Sarma, Sreeramula Rajeswara. "Yantraprakāra of Sawai Jai Singh." Supplement to *Studies in History of Medicine and Science*, vols. X and XI, New Delhi, 1986. 1987.

See also: Jai Singh – Gnomon – Time

P

Pakṣa

Pakṣa (Half) refers, in Indian astronomy, to half the lunar month. The lunar month, which is the interval between two successive new moons (*amāvāsyā*) or full moons (*paurṇamāsyā*), is a visible natural phenomenon, which occurs at regular intervals and, as such, has been used for reckoning time in all ancient civilisations, including that of India. For the people of India of the Vedic times, the two *pakṣas*, one *śukla* (bright) and the other *kṛṣṇa* (dark), provided an easy and most convenient instrument for reckoning time. From actual observation, it was found that a *pakṣa* was completed in fifteen lunar days (*tithis*) and these were named after their serial numbers, as *prathamā* (first), *dvitīyā* (second), *tṛtīya* (third), *caturthī* (fourth) etc., up to *caturdaśī* (fourteenth), and the fifteenth was *amāvāsyā* or *paurṇamāsyā*, according to whether it was the dark fortnight or the bright fortnight.

The Moon moves about thirteen degrees in its orbit, eastward, in a day of 24 hours, but since the Sun also moves in the same direction, one degree, during the same period of 24 hours, the Moon's resultant displacement, which constitutes the *tithi*, is about twelve degrees. A lunar month of thirty *tithis* would be around 29 and a half solar days, and in a solar year of 365 days, the Moon would have completed its twelve months in about 354 days, and would have moved further by eleven *tithis*. In order to correlate and bring together the solar and lunar years, the astronomical practice is to expunge and leave uncounted one lunar month every three years, when the extra eleven *tithis* would have accumulated to one month. This expunged month is called *adhika-māsa* or 'added' or 'intercalary' month.

Time reckoning by *tithi* and *pakṣa* is very much in vogue, even today, among Hindus, for fixing auspicious times for ritual, religious, and social functions, and for horoscopic astrology.

K.V. Sarma

Parameśvara

Among Indian astronomers hailing from Kerala, Parameśvara (ca. CE 1360–1465), promulgator of the Kerala Dṛgganita School, and author of over a score of works on astronomy and astrology, holds a prominent place. He was a resident of Ālattūr, his house being situated on the northern bank of the river Bhāratappuzha, where he conducted his astronomical experiments and observations for over 55 years.

Parameśvara was a prolific writer. Some of his works still remain in manuscript form, while a few are yet to be found. His works on astronomy are:

1. *Dṛggaṇita*, (Computation True to Observation), CE 1431, his magnum opus, a practical manual, in two versions.
2. *Goladīpikā* I (Illumination on Spherics), in 302 verses, on spherical astronomy.
3. *Goladīpikā* II (Illumination on Spherics), in four chapters, on the same subject as above, but different from it.
4. *Grahaṇamaṇḍana* (Ornament on Eclipses), in two versions, of 89 and 100 verses.
5. *Grahaṇanyāyadīpikā* (Illuminator on Rationale of Eclipses) in 85 verses.
6. *Grahaṇāṣṭaka* (Octad on Eclipses), on the computation of eclipses.
7. *Vākyakaraṇa*, on methods for the derivation of several astronomical tables.

Parameśvara also commented on several standard works on astronomy that were popular in Kerala. These include: *Āryabhaṭīya* of Āryabhaṭa, *Laghubhāskarīya* and *Mahābhāskarīya* of Bhāskara I, *Mahābhāskarīya-Bhāṣya* of Govindasvāmin, *Laghumānasa* of Muñjāla, *Sūryasiddhānta, Līlāvatī* of Bhāskara II, and his own *Goladīpikā* II. All these commentaries except that on *Līlāvatī* have been published.

Although he was primarily an author of astronomical works, Parameśvara also wrote on astrology:

1. *Ācārasangraha*, a popular text.
2. *Jātakapaddhati*, in 44 verses.
3. Commentary on the *Muhūrtaratna* of Govinda.
4. Commentary on the *Jātakakarmapaddhati of Śrīpati.*

5. Commentary on the *Praśnaṣ-aṭapañcāśikā* of Pṛhuyaśas.

6. *Muhūrtāṣṭaka-dīpikā, Vākyadīpikā* and *Bhādīpikā,* mentioned by Parameśvara at the end of his commentary on the *Mahābhāskarīya.*

These works are not extant.

In his works Parameśvara evinces a refreshingly scientific outlook. He avers at the beginning of his *Grahanamandana* that he was setting out to compose the work after closely watching the movements and positions of the planets in the skies for a long time. At the beginning of his *Dṛggaṇita,* he says that, in real astronomy, computation should match observation. He enumerates a number of solar and lunar eclipses which he had observed between 1393 and 1432, and gives details about them in his *Grahaṇanyāyādīpikā.* He also points out the error between computed and observed readings and offers corrections and instructs that similar observations should be made at intervals and corrections enunciated for computation and observation to be identical.

K.V. Sarma

REFERENCES

Primary sources

Dṛggaṇita (Computation True to Observation). Ed. K.V. Sarma. Hoshiarpur: Vishveshvaranand Institute. 1963.

Goladīpikā I (Illumination on Spherics). Ed. T. Ganapati Sastri. Trivandrum: Trivandrum Sanskrit Series, No. 49. 1916.

Goladīpikā II (Illumination on Spherics). Ed. K.V. Sarma. Madras:Adyar Library and Research Centre. 1957.

Grahaṇāṣṭaka (Octad on Eclipses). Ed. K. V. Sarma. Madras: Kuppuswami Sastri Research Institute. 1959.

Grahaṇamaṇḍana (Ornament on Eclipses). Ed. K.V. Sarma. Hoshiarpur: Vishveshvaranand Institute. 1965.

Grahaṇanyāyadīpikā (Illuminator on Rationale of Eclipses). Ed. K.V. Sarma. Hoshiarpur: Vishveshvaranand Institute. 1966.

Jātakapaddhati. Ed. Kolatteri Sankara Menon. Trivandrum: Curator's Office. 1926.

Secondary Sources

Pingree, David. "Eclipse Observations of Parameśvara between 1393 and 1432." *Journal of the American Oriental Society* 87:337–39. 1967.

Pingree, David. *Census of the Exact Sciences in Sanskrit*. Philadelphia: American Philosophical Society. Series A, Vol. 4. 1981.

Sarma, K.V. *A Bibliography of Kerala and Kerala-based Astronomy and Astrology*. Hoshiarpur: Vishveshvaranand Institute. 1972.

See also: Eclipses – Astronomy

Pauliśa

Pauliśa, of Greek origin, is the originator of the *Pauliśa Siddhānta*, one of the five systems of astronomy of the early centuries of the Christian era. These were selectively redacted in the *Pañcasiddhāntikā* of Varāhamihira, the prodigious Indian astronomer–astrologer of the sixth century CE. According to a traditional verse attributed to the sage Kāśyapa, Pauliśa is one of the eighteen originators of Indian astronomical systems. Al-Bīrūnī, the Persian scholar, who sojourned in India from 1017 to 1030, stated that *Pauliśa Siddhānta* was written by Paulus-ul-Yunani, i.e. 'Paulus, the Greek'. Probably the Hindus prepared an Indianised Sanskrit *Pauliśa Siddhānta* on the basis of the Greek work. This original Sanskrit work is no longer available, and neither is the commentary on that work by Lāṭadeva which was referred to by Varāhamihira in *Pañcasiddhāntikā* I.3. However, the redaction of the *Siddhānta* in *Pañcasiddhāntikā* is fairly full and provides ample details about the nature and contents of the work. The Pauliśa system has certain things in common with the *Romaka Siddhānta* and with the *Vāsiṣṭha Siddhānta*, two of the other systems redacted in *Pañcasiddhāntikā*.

The epoch of the *Pauliśa Siddhānta*, which *Pañcasiddhāntikā* says is the same as that of *Romaka Siddhānta*, is the Hindu *caitra-śukla-pratipad* (first day of the bright fortnight of the month of Citrā) in the Śaka year 427 elapsed (*Pañcasiddhāntikā* I.8–10), which corresponds to mean sunset at Yavanapura (modern Alexandria in Egypt), or modern Sunday, March 20, 505. It is interesting that the same moment was adopted by Varāhamihira as the epoch of the *Saura Siddhānta* redacted in the *Pañcasiddhāntikā* (i.e. the old *Śurya Siddhānta*). The Indian time then was 37–20 *nāḍīs* from mean sunrise in Ujjain, on Sunday, March 20, 505. (*Pañcasiddhāntikā* III.13). In *Pañcasiddhāntikā* III.13, it is stated that the *deśāntara-nāḍīs* (longitudinal difference in terms of time, expressed in *nāḍīs*) from Yavanapura to Ujjain is 7–20, while that to Varanasi is 9. The actual intervals according to modern calculations are 7–38 and 8–50, the difference being 18 *vināḍīs* and ten *vināḍīs*, respectively. One minute being equal to four *vināḍīs*, the difference works out only to 4.5 minutes and 2.5 minutes, which is remarkable.

At the commencement of the *Pañcasiddhāntikā* (I.4), Varāhamihira gives a pat to the *Pauliśa Siddhānta* for the accuracy of the *tithi* (lunar day) calculated according to it, though he adds that the *tithi* derived through the *Saura Siddhānta* is much more accurate. In fact, it is these two schools, the Pauliśa and the Saura, one representing the Greek and the other the Indian, that Varāhamihira depicts rather fully in *Pañcasiddhāntikā*, allotting them entire sections. Thus the entire third chapter of *Pañcasiddhāntikā* is devoted to the depiction of planetary computation and allied matters according to the *Pauliśa Siddhānta*, Chapter Five to the Moon's cusps, Chapter Six to lunar eclipses, Chapter Seven to solar eclipses, and much of the long Chapter Eighteen to the motion of the planets. Although the *Pauliśa Siddhānta* was based on a Greek original, the text was painstakingly Indianised, both in the matter of content and presentation.

K.V. Sarma

REFERENCES

Dikshit, S.B. *Bhāratīya Jyotish Sastra (History of Indian Astronomy)*. Trans. R. V. Vaidya. Calcutta: Positional Astronomy Centre, India Meteorological Department. 1981.

Dvivedi, Sudhakara. *Gaṇaka Taraṅgiṇī or Lives of Hindu Astronomers*. Ed. Padmakara Dvivedi. Benares: Jyotish Prakash Press. 1933.

Pañcasiddhāntikā of Varāhamihira. Trans. T.S. Kuppanna Sastry. Ed. K.V. Sarma. Madras: P.P.S.T. Foundation. 1993.

The Pañcasiddhāntikā of Varāhamihira. Ed. and trans. G. Thibaut and Sudhakara Dvivedi. Varanasi: Chowkhamba Sanskrit Series Office. 1968.

The Pañcasiddhāntikā of Varāhamihira. Ed. and trans. O. Neugebauer and David Pingree. Copenhagen: Munksgaard. 1970.

Pingree, David. "History of Mathematical Astronomy in India." In *Dictionary of Scientific Biography*, vol. 15. New York: Scribners. 1978. pp. 545–54.

See also: Varāhamihira – Astronomy – al-Bīrūnī – *Deśāntra*

Physics

Early Indian thinkers developed a number of theoretical systems, which centred around two main themes: elements and atoms. Based upon a relatively broad review of the ancient philosophies, there is a rich scientific tradition in India.

ANCIENT INDIA (2000 BCE–CE 800)

Early Indian explanations about the physical makeup of the universe were religious and philosophical in nature. The oldest literary record of this period, the *Ṛgveda*, presents several fundamental concepts related to physical science such as *ap* (primeval water), which was considered the basic element of matter. Gradually, the doctrine of five fundamental elements emerged, as seen in the Upaniṣadic literature of around 700 BCE. The five elements were *pṛthvī* (earth), *ap* (water), *tejas* (fire), *vāyu* (air), and *ākāśa* (a non-material substance). The Sanskrit terms have a wider connotation than the English translation, so it is essential to present the original terms.

Much of this early literature was concerned with the attributes associated with each of these elements. They had both common as well as distinctive attributes, many coincident with the five senses, as follows: *pṛthvī* (earth) sound, touch, colour, taste, and odour; *ap* (water) sound, touch, colour, and taste; *tejas* (fire) sound, touch, and odour; *vāyu* (air) sound and touch; and *ākāśa* (non-material) sound. Different combinations of these five elements yielded certain products. The human body is a good example. The human embryo has energetic principles which are separated into form by *vāyu* (air). *Tejas* (fire) subsequently transforms the embryo. *Ap* (water) maintains moisture while *pṛthvī* (earth) gives it shape and size. Finally *ākāśa* expands the embryo and develops it.

Many of today's scholars question the notion that scientific inquiry only found fertile ground in the minds associated with the Greek tradition. In fact, a number of diverse systems of scientific inquiry were developing in India coincident with the Greek philosophies by 500 BCE. Those which presented five fundamental elements were the *Sāṃkhya, Nyāya,* and *Vaiśeṣika*. The *Jaina, Bauddha,* and *Cārvāka* schools, like many of the Greek thinkers, presented four fundamental elements. It is difficult to say with certainty which school or culture developed these ideas first, therefore who influenced whom, or if they developed independently.

The *Nyāya-Vaiśeṣika* school, prominent among early Indian systems, was popularised by an individual who came to be known as Kaṇāda. He and his followers extended the concept of five elements and built a comprehensive theory of atoms. The first four elements of this system (*pṛthvī, ap, tejas,* and *vāyu*) were material and considered either eternal or temporal. The fifth element, *ākāśa*, was considered eternal only. The eternal form consisted of imperceptible atoms, while the temporal form arose when these atoms joined to create perceptible products. This view of five elements was part of a much larger conceptual picture, that of *dravya* (substance). There were nine types of substances. In addition to the five elements or substances already mentioned, there was *dik* (space), *kāla* (time), *ātman* (self), and *manas* (mind). Each of these substances was considered inseparable from its respective set of attributes. In other words, they were one and the same.

The word used for atom was *aṇu* or *pramāṇu*. The *Nyāya-Vaiśeṣika* school held that the atom was indestructible, indivisible, without magnitude, spherical, and in constant motion. Two atoms of the same substance could join to form a dyad. Atoms of different substances could not join together, yet they could play a supportive role in the combination of materially compatible atoms. The dyad was regarded as too small to be perceived. The smallest visible structure was a triad (three dyads). This structure was referred to as *trasareṇu* or *tryaṇuka* which was about the size of a speck floating through a sunbeam. The principle of causality was crucial to the *Nyāya-Vaiśeṣika* school. Atoms were the material cause for the dyad, the effect. The dyads were the cause for the production of a triad which was another effect. The individual atoms lost their causative property once the triad was formed. The reason for atoms joining in the first place was attributed to *aḍṛṣṭa*, an unseen force.

Jaina philosophers held that atoms were both cause and effect. They theorised that atoms joined to form aggregates in response to attractive and repulsive forces which were inherent characteristics of the atoms themselves. Jaina atoms were all of one class. There was no distinction, qualitatively or quantitatively, between types of atoms such as earth-atoms, water-atoms, etc. According to the Buddhists, the atom was indivisible, unable to be analysed, invisible, inaudible, unable to be tested, and intangible. Neither were these atoms eternal. The Buddhists did not speak of atoms in terms of particles. They thought of them as a force of energy.

The *Nyāya-Vaiśeṣika* school was distinct from other Indian schools in that it placed much more emphasis on the attributes of matter. There were five general qualities possessed by all nine substances: number, dimension, distinctness, conjunction, and disjunction. Other qualities, more closely associated with modern physics included, *gurutva* (gravity), *dravatva* (fluidity), *snigdha* (viscosity), and *sthitisthāpaka* (elasticity). Gravity, or the cause of falling, was not considered a force but a quality which resided in a whole object. No apparent correlation between gravity and the mass of a particular object was presented. Fluidity was a quality of only three substances – earth, water, and fire – and was of two kinds. Natural fluidity was a specific quality of water. Incidental fluidity was associated with fire in the case of some melted substances. Viscosity was specific to water, causing cohesion and smoothness. Elasticity was only a quality of earthy substances as in the branch of a tree which was caused to return to its original condition if displaced. Another fundamental concept considered by ancient Indian thinkers was that of motion.

The *Nyāya-Vaiśeṣika* concept of *karma* (motion) was represented by five actions: *utkṣepaṇa* (throwing upwards), *avakṣepaṇa* (throwing downwards), *prasāraṇa* (expansion), *ākuñcana* (contraction), and *gamana* (going). Only one kind of motion was considered possible at a time and a substance experienced motion only for a moment. This motion was subsequently destroyed or rendered ineffective once completed. The motion of free atoms made possible by *aḍṛṣṭa* (unseen force) caused the material world to be formed.

Among Indian physical concepts, that of *ākāśa* should not be overlooked. Its special quality was sound. Though considered a non-material substance, it was believed to play a role in the formation of material objects. A modern physics concept which some have equated with *ākāśa* is that of ether, which is conceived to be a vast expanse or continuum. Sound moved through *ākāśa* like ripples across the surface of a pond. Unlike the ripples, sound moved through *ākāśa* in a succession of points. The first sound caused the second, and once the second sound was created, the first was destroyed.

Heat and light were understood in relation to one of the five basic elements, *tejas* (fire). When an object was heated it went through a series of distinct changes, each one lasting a prescribed number of moments. A clay pot for example, if heated, would in the first moment experience the production of atomic motion. The second moment would be characterised by the destruction of the clay's original colour. In the third and fourth moments one would find that the colour red would be produced followed by the destruction of the atomic motion created earlier. A new type of atomic motion, creative in nature, occurred in the fifth moment. A series of two disjunctions between the atoms and *ākāśa* was followed by a conjunction in the sixth through eighth moments. The final two moments consisted of the formation of a red coloured dyad followed by a triad. The ultimate source of all heat was thought to be the Sun. Light, though often associated with heat in many modern physical systems, possessed many qualities, distinctive from heat, in the early Indian systems.

Light was thought to consist of rays which emanated from the eyes just as a candle casts its light throughout a room. If an obstruction prevented the rays from touching an object, then that object simply could not be perceived. Mirrors were thought to possess particular attributes of colour which caused the light rays from one's eyes to return to one's face, upon striking the surface.

MEDIEVAL INDIA (CE 800–1800)

It is a common misconception that there was a lack of scientific inquiry during the Middle Ages. While India experienced great change during this time, such as the introduction of Islam, there was a continued advance and assimilation of scientific information with an ever-increasing number of collaborators. New cultural influences, changing technologies, and language barriers challenged as well as fostered scientific advances.

Physics was viewed as a distinct branch of study during the early part of the Middle Ages. Though experiments were still quite rare, use of this scientific technique began. A large number of texts were translated into different languages. A scholar named Ḥunayn ibn Isḥāq translated Aristotle's works into Syriac and his son Isḥāq ibn Ḥunayn followed suit, translating Euclid's *Elements*. Ideas were shared, expanded upon, and refined. A number of original inquiries were also initiated.

Among the more notable individuals describing physics research in India as well as Middle Asia at this time was Abū Rayḥān al-Bīrūnī. He authored many books devoted to the scientific achievements of the Indian people.

The latter part of this period experienced an influx of European influences. Educational institutions, scientific journals, and professional societies each played a role in the development of what was to become modern India. New political and economic structures, in the face of new language barriers between the educated few and the illiterate masses, threatened to disassemble the rich scientific heritage of the Indian people. Yet India assimilated the old and the new to emerge strong with an outstanding future of scientific inquiry before it. Much of that future rested in the hands of a few visionaries who understood the significance of science in India's future.

Modern India (CE 1800–Present)

The outstanding efforts of many have contributed to India's rich past and promising future. The following few individuals are some of the physicists who have exemplified the level of achievement found throughout modern India.

Jagdish Chandra Bose (1858–1937) was a biophysicist who explored the response of plants and animals to electrical stimulation. Like many pioneering physicists active at the turn of the century, Bose used instruments of his own design in his work. He was elected a Fellow of the Royal Society of London in 1920. C.V. Raman (1888–1970) investigated optics, including diffraction, molecular scattering of light, and magneto-optics. He was awarded the Nobel Prize in 1930 for the discovery which bears his name, the Raman Effect. This pioneering work described the molecular scattering of light and explained, among other things, why the ocean appears blue. Finally, Jawaharlal Nehru, the first Prime Minister of India, worked hard to foster what he described as a "scientific temper". He believed that for science to succeed in modern India, as it had in its past, there had to be strong support for the scientific enterprise from all segments of society.

The history of physics in India is a vast subject, offering extensive opportunities for further study. It is hoped that this brief presentation has encouraged some to explore the ancient philosophies, medieval refinements, and modern achievements.

William T. Johnson

References

Divatia, A.S. "History of Accelerators in India." *Indian Journal of Physics* 62A (7):748–774. 1988.

Mitra, A.P. *Fifty Years of Radio Science in India*. New Delhi: Indian National Science Academy. 1984.

Panda, N.C. *Maya in Physics*. Delhi: Motilal Banarsidass. 1991.

Pingree, David. *Census of the Exact Sciences in Sanskrit*. Philadelphia: American Philosophical Society. 1970.

Rahman, A., M.A. Alvi, S.A. Khan Ghori, and K.V. Samba Murthy. *Science and Technology in Medieval India–A Bibliography of Source Materials in Sanskrit, Arabic and Persian*. New Delhi: Indian National Science Academy. 1982.

Rao, C.N.R. and H.Y. Mohan Ram, eds. *Science in India*: 50 *Years of the Academy*. New Delhi: Indian National Science Academy. 1985.

Romanovskaya, T.B. "The Interrelations Between India and Middle Asia in the Field of Physics in the Middle Ages." In *Indo-Soviet Seminar on Scientific and Technological Exchanges between India and Soviet Central Asia in Medieval Period, Proceedings, Bombay: November* 7–12, 1981. New Delhi: Indian National Science Academy. 1981.

Sachau, Edward C., ed. *Alberuni's India*. London: Kegan Paul, Trench, Trubner & Co. Ltd. 1914.

Subbarayappa, B.V. "The Physical World: Views and Concepts." In *A Concise History of Science in India*. Ed. D.M. Bose. New Delhi: Indian National Science Academy. 1971.

Venkataraman, G. *Journey into Light: Life and Science of C.V. Raman*. Bangalore: Indian Academy of Sciences. 1988.

See also: al-Bīrūnī – Atomisim

Pi in Indian Mathematics

Pi or π is the most interesting number in mathematics, and its history will remain a never-ending story. It occurs in several formulas of mensuration and is variously involved in many branches of mathematics, including geometry, trigonometry, and analysis.

The earliest association of π is found in connection with the mensuration of a circle. The fact that the perimeter or circumference of any circle increases in proportion to its diameter was noted quite early. In other words, in every circle, perimeter/diameter = constant, or $p/d = \pi_1$, where π_1 is the same for all circles.

After knowing the perimeter, the area of the circle was often found by using the sophisticated relation

$$\text{area} = \left(\frac{p}{2}\right)\left(\frac{d}{2}\right) = \frac{pd}{4}.$$

But the earliest rules for determining area were of the form, area = $(kd)^2$, where k is a constant prescribed variously. Both these methods imply that the area of a circle is proportional to the square of its diameter (or radius r), or area = $\pi_2 r^2$.

We know that π_2 is the same as π_1, but this was not always known. Similarly, π_3 may be defined from the volume of a sphere. In this article, the symbol π is used to denote all the above three values, as well as for their common value, which is now known to be not only an irrational but a transcendental number.

Since the Indus Valley script has not been deciphered successfully, we cannot say any final thing about the scientific knowledge of India of that time (about the third millennium CE). Some conjectures about the value of π used in the *Ṛgveda* (about the second millennium CE) have been made. Definite literary evidence is available from texts related to Vedāṅgas, especially the *Śulbasūtras* which contain much older traditional material. In the *Baudhāyana Śulbasūtra*, the oldest of them, the perimeter of a pit is mentioned to be three times its diameter, thereby implying $\pi_1 = 3$. This simplest approximation is found in almost all ancient cultures of the world. In India, it is found also in classical religious works such as *Mahābhārata* (Bhīṣmaparva, XII: 44), and certain *Purāṇas*, as well as in some Buddhists and Jaina canonical works.

Different approximations to π_2 are implied in the various Vedic rules for converting a square into a circle of equal area and vice-versa. If r is the radius of the circle equivalent to a square of side s, the usual *Śulba* rule is to take $r = s(2 + \sqrt{2})/6$, which implies the approximation $\pi_2 = 18(3 - -2\sqrt{2}) = 3.088$. Recently, a new interpretation of the *Mānava*

Śulba Sūtra has yielded the relation $r = 4s/5\sqrt{2}$, thereby implying $\pi_2 = \frac{25}{8} = 3.125$, which is the best *Śulba* approximation found so far.

The ancient Jaina School preferred the approximation $\pi = \sqrt{10}$, which they considered accurate and from which they derived the value $\pi = \sqrt{3^2+1} \approx 3 + \frac{1}{6} = \frac{19}{6}$. This value ($\sqrt{10}$) continued to be used in India not only by the Jainas but by others, such as Varāhamihira, Brahmagupta, and Śrīdhara, even when better values were known.

With Āryabhaṭa I (born CE 476), a new era of science began in India. In the *Āryabhaṭīya*, he gave a fine approximation of π_1, surpassing all older values. It contains the rule that the perimeter of a circle of diameter 20,000 is close to 62,832, so that $\pi_1 = \frac{62832}{20000} = 3.1416$ *nearly*, which is correct to four decimal places, and he still calls it close and not exact.

This value had a respectful place in Indian mathematics and exerted great influence. How Āryabhaṭa obtained it is not known. It was known in China, but evidence of borrowing lacks documentary support. On the other hand, the two typically Indian values $\sqrt{10}$ and $\frac{62832}{20000}$ appear in many subsequent Arabic works.

In India, the Archimedian value $\frac{22}{7}$ for π first appeared in the lost part of Śrīdhara's *Pāṭī* (ca. CE 750). A Jaina writer, Vīrasena, quotes a peculiar rule in his commentary *Dhavalā* (CE 816). It is equivalent to $\rho = 3d + (16d + 16)/113$. If we leave out the redundant dimensionless number +16 in the brackets, this rule will imply a knowledge of the value $\pi_1 = \frac{355}{113} =$ which was known in China to Zu Chongzhi (CE 429–500). In explicit form, this value is found in India much later, e.g. in the works of Narāyaṇa II, Nīlakaṇṭha, and others. The simplified or reduced form $\frac{3927}{1250}$ of Āryabhaṭa's value of π occurs in the works of Later Pauliśa, Lalla, Bhaṭṭotpala and the great Bhāskara II.

Most significant contributions on the computations of π were made by the mathematics of the late Āryabhaṭa School of South India. Mādhava of Saṇgamagrāma (ca. CE 1340 to 1425), the first great scholar and founder of the School, gave the value π_1 = 2827, 4333, 8823, 3)/(9 $\times 10^{11}$) as known to the "learned men". This value yields an approximation correct to eleven decimals. Mādhava also knew the series $\frac{\pi}{4} = 1 - \frac{1}{3} + \frac{1}{5} - \frac{1}{7} + \ldots$ which was rediscovered in Europe in 1673 by Leibniz.

In discovering various series for π and in evolving techniques for improving their convergence, a great theoretical breakthrough was already attained in sixteenth-century India.

R.C. Gupta

References

Gupta, R.C. "Aryabhata I's Value of π." *Mathematics Education* 7(1): 17–20. 1973 (Sec. B).

Gupta, R.C. " Some Ancient Values of Pi and Their Use in India." *Mathematics Education* 9(1): 15. 1975 (Sec. B).

Gupta, R.C. " On the Values of π from the Bible." *Gaṇita Bhāratī* 10: 51–58. 1988.

Gupta, R.C. "New Indian Values of π from the Mānava Śulba Sūtra," *Centaurus* 31: 114–125. 1988.

Gupta, R.C. "The Value of π in the Mahabharata." *Gaṇita Bhāratī* 12: 45–47. 1990.

Gupta, R.C. "Sundararāja's Improvements of Vedic Circle–Square Conversions." *Indian Journal of History and Science* 28(2): 81–101. 1993.

Hayashi, Takao, T. Kusuba, and M. Yano. "Indian Values for π Derived from Āryabhaṭa's Value." *Historia Scientiarum* 37: 1–16. 1989.

Marar, K. Mukunda, and C.T. Rajagopal. "On the Hindu Quadrature of the Circle." *Journal of the Bombay Branch of the Royal Asiatic Society* (New Series) 20: 65–82. 1944.

Rajagopal, C.T., and M.S. Rangachari. "On Medieval Kerala Mathematics." *Archive for History of Exact Sciences* 35(2): 91–99, 1986.

Smeur, A.J.E.M. "On the Value Equivalent to π in Ancient Mathematical Texts: A New Interpretation." *Archive for History of Exact Science* 6 (4): 249–270. 1970.

See also: Śulbasūtras – Varāhamihira – Brahmagupta – Śrīdhara – Āryabhaṭa – Nīlakaṇṭha Pauliśa – Bhāskara – Mādhava

Precession of the Equinoxes

Precession of the equinoxes, which in Indian astronomy is called *ayanacalana* (shifting of the solstices), refers to the slow but continuous backward movement of the point of intersection of the ecliptic (which is a fixed circle) and the celestial equator (which keeps on moving backwards). This effectively means that the first point of Aries, which is the traditional point of the commencement of the Indian year, is really shifting backwards continuously. This shifting is, however, so slow, being a meagre 50.2 seconds per annum, that it comes to be noticed only when it accumulates over several years, and takes about 25,800 years to complete one circle.

Such a shift seems to have been noticed in India even during the Vedic times, as is evidenced by the Vedic priests' changing the beginning of their year backwards, from the constellation Mṛgaśiras (Delta Orionis) to the next previous constellation Rohiṇī (Alpha Tauri), and again backward, to Kṛittikā (Eta Tauri) in the course of time. But no measurement of it was made. Precession was not apparent during the time of astronomer

Āryabhaṭa (CE 499), since at that time the First point of Aries coincided with the equinox. However, later astronomers noted it and also measured its rate. The first Indian astronomer who gave a rule for finding the value of precession seems to be Devācārya, who, in his *Karaṇaratna*, I.36 (CE 689), gives the rate of precession which works out to 47 seconds per annum. The magnitude of the precession, called *ayana-aṃśa* (degrees of precession) of a celestial body would be the angular distance between its computed longitude and the First point of Aries on the zodiac (which Hindu astronomy takes as fixed).

Hindu astronomers did not conceive the shifting of the equinoxes as a continuously regressing phenomenon, which it really is. They understood it as an oscillatory motion, with the beginning of the Kali era (3102 BCE) as the 'zero-precession year', when the Sun and other planets were at the First point of Aries (Aśvinī), i.e. at the end of Zeta Piscium (Revatī). According to the *Sūryasiddhānta*, an oscillation of amplitude 27° to and fro would take 7200 years to complete. Thus, during the first 1800 years, the equinox moved forward by 27°, and during the next 1800 years it moved backward, coming back to the zero position at the end of the Kali year 3600 (which corresponds to CE 499, when Āryabhaṭa composed his work *Āryabhaṭīya*). It will continue to regress for 1800 years more, till CE 2299, when its forward motion for 1800 years would commence, again reaching the zero position at the end of 7200 years, covering, in all, 108°. Certain other texts, such as the *Karaṇaratna* and *Vākya-karaṇa* take the amplitude of the oscillation as 24°, the period for one complete oscillation being 7380 years. Among Indian astronomers, it was only Muñjāla (CE 932) who conceived precession as a continuous motion.

K.V. Sarma

References

Dikshit, Sankar Balakrishna. "Ayana Calana or Displacement of Solsticial Points." In *Bhāratīya Jyotish Śāstra (History of Indian Astronomy)*. trans. R.V. Vaidya. Part II. New Delhi: Director General of Meteorology. 1981. pp. 205–21.

The Karaṇaratna of Devācārya. Ed. Kripa Shankar Shukla. Lucknow: Lucknow University. 1979.

Sastry, T.S. Kuppanna. "The Concept of Precession of the Equinoxes." In *Collected Papers on Jyotisha*. Tirupati: Kendriya Sanskrit Vidyapeetha. 1989. pp. 126–28.

Putumana Somayājī

Putumana Somayājī (ca. 1660–1740), author of the work *Karaṇapaddhati* (Methodology for Astronomical Manuals) was a *nampūtiri* (the appellation of the Brahmin community of Kerala in South India), who belonged to a family which bore the name Putu-mana (New House). He was of Ṛgvedic denomination, and secured the surname Soma-yāji by having performed the Vedic *Soma* sacrifice. His real name remains unknown. The date of Soma-yājī is surmised on the basis of the chronogram which gives the date of completion of his *Karaṇapaddhati* (as 17,65,653) of the Kali era, stated in the *kaṭapayādi* notation of expressing numerals, which is in the year CE 1732. Again, in the concluding verse, Somayāji states that he hailed from the village named Śivapura, which has been identified as Covvaram in Central Kerala, where his descendants still reside.

Somayāji was a prolific writer, mainly on astronomy and astrology, his only work in a different discipline being *Bahvṛcaprāyaścitta*, a treatise which prescribes expiations (*prāyaścitta*) for lapses in the performance of rites and rituals by Bahvṛca (Ṛgvedic) Brahmins of Kerala. In addition to his major work, Somayāji is the author of several other works. In *Pañca-bodha* (Treatise on the Five), he briefly sets out computations at the times of *Vyatīpāta* (unsavoury occasion), *Grahaṇa* (eclipse), *Chāyā* (measurements based on the gnomonic shadow), *Śṛṅgonnati* (elongation of the Moon's horns), and *Mauḍhya* (retrograde motion of the planets), all of which are required for religious observances. His *Nyāyaratna* (Gems of Rationale), available in two slightly different versions, depicts the rationale of eight astronomical entities: true planet, declination, gnomic shadow, reverse shadow, eclipse, elongation of the Moon's horns, retrograde motion of the planets, and *Vyatīpāta*. Three short tracts on the computation of eclipses, including a *Grahaṇāṣṭaka* (Octad on Eclipses), are ascribed to Somayāji. He also composed a work called *Veṇvārohāṣṭaka* (Octad on the Ascent on the Bamboo), which prescribes methods for the computation of the accurate longitudes of the Moon at very short intervals. A commentary in the Malayalam language on the *Laghumānasa* of Muñjāla is also ascribed to him. On horoscopy, Somayāji wrote a *Jātakādeśa-mārga* (Methods of Making Predictions on the Basis of Birth Charts), which is very popular in Kerala.

The *Karaṇa-paddhati*, in ten chapters, is his most important work. The work is not a manual prescribing computations; rather it enunciates the rationale behind such manuals. Towards the beginning of the work, the author states that he composed the book to teach how the several multipliers, divisors, and R sines pertaining to the different computations and the like are to be derived. Thus, the work is addressed not to the almanac maker but to the manual maker. All the topics necessary to make the daily almanac are not treated in *Karaṇapaddhati*, whereas several other items not pertaining to manuals are dealt with. The work takes as its basis the parameters and postulates of the Parahita system of astronomy promulgated by Haridatta, except for the section on eclipses, where the more accurate system from the *Drgganita* of Parameśvara is taken as the basis.

In chapter I, the *Karaṇapaddhati* sets out the planetary parameters, the computation of the mean planets, and the corrections to be applied thereto. Chapter II is devoted to the derivation, by means of pulverisation (*kuṭṭaka*), of the multipliers and divisors necessary for planetary computations. Chapter III is concerned with various aspects of the computation of the Moon, and Chapter IV deals with miscellaneous matters relating to the determination of suitable epochs for commencing computations. Chapter V mentions methods for correlating observed results with computed results, while chapter VI depicts such matters as the circle and the circumference, and the R sines pertaining to various entities for different purposes. Chapter VII is devoted to the rationale of the derivation of the mnemonics relating to the epicycles, their R sines, and allied matters. Chapter VIII describes varied derivations using the gnomonic shadow. Chapter IX is devoted to stellar declination and latitudes. The last chapter is concerned with certain derivations from the shadow of the gnomon.

As a work on astronomical rationale and exposition of the logic of several practices, the *Karaṇpaddhati* is highly important. A very significant statement made by the Putumana Somayāji, towards the close of Chapter V, is that the conception of aeons and eras, the measures therefore, and the various derivations based on them are not really true; they are only a means to compute the positions of celestial bodies and matters related to them. What is really important is the correlation of computation with observation, and to affect the same a practical astronomer is authorised to make changes as necessary. This is a highly significant statement, maybe even revolutionary, which was made in a forthright manner by an orthodox Hindu astronomer.

K. V. Sarma

References

Karaṇa-paddhati of Putumana Somayājī. Ed. P.K. Koru. Cherp: Astro Printing and Publishing Co. 1953.

Karaṇa-paddhati of Putumana Somayājī. Ed. S.K. Nayar. Madras: Government Oriental Manuscripts Library. 1956.

Sarma, K.V. "Putumana Somayājī, an Astronomer of Kerala and his Hitherto Unknown Works." In *Proceedings of the 18th All India Oriental Conference*, Annamalainagar. 1955. pp. 562–64.

Walsh, C.M. "On the Quadrature of the Circle, and the Infinite Series of the Proportion of the Circumference to the Diameter Exhibited in the Four Sastras, the Tantra Sangraham, Yucti Bhasha, Carana Paddhati, and Sadratnamala." *Transactions of the Royal Asiatic Society of Great Britain and Ireland* 3(3): 509–23. 1835.

See also: Parameśvara – Haridatta – Eclipses

R

Rainwater Harvesting

Throughout history people have lived in areas where there are few rivers and where the direct collection of rainwater from roofs, paved courtyards, hillsides, or rock surfaces is one of the best available methods for securing a water supply. By extending this principle to provide water for crops, early civilisations practiced agriculture much further into the semi-desert areas of Arabia, Sinai, North Africa, India, and Mexico than has been possible in modern times – and this is not explicable by changes in climate.

Agriculture in the Old World originated in climatically dry regions in the Middle East and may have depended to some degree on rainwater running off nearby slopes almost from the start. Evidence is lacking until a later period, however, when some of the most striking applications of rainwater harvesting were related to crop production in the Negev Desert between 200 BCE and CE 700. One technique would be to dig a channel across a hillside to intercept water running downslope during storms. The water would be directed onto fields which, in the Negev, were carefully levelled and enclosed by bunds. Further west, steeper hillsides were used in Morocco, with cultivation on flat terraces formed behind stone retaining walls. In Tunisia, French travellers in the nineteenth century noted fruit trees being grown at the downslope end of small bunded rainwater catchment areas or microcatchments.

In India, one common technique is simply to build a bund across a gently sloping hillside, so that runoff flows originating from rainfall collect behind the bund, where water is left standing until the planting date for the crop approaches; then the land is drained, and the crop sown. This land behind the bund which is seasonally flooded and then later planted with a crop is known as an ahar in Bihar, or a *khadin* in Rajasthan. Although some *ahars* may be only one hectare in extent with a bund 100 metres long, others are very large and account for 800,000 hectares of cultivation in Bihar state. In desert areas of Rajasthan, there are many *khadins* of twenty hectares or more, some of them having been first constructed in the fifteenth century CE.

In North America, research on the modern potential of runoff farming methods has been stimulated by the realisation that people living in what is now Mexico and the south-western United States prior to European settlement had methods of directing rainwater from hillsides on to plots where crops were being raised, thereby making productive agriculture possible in an otherwise unpromising semi-arid environment. On lands occupied by the Hopi and Papago peoples in Arizona, fields were predominantly on alluvial valley soils below hillsides or gullies from which water could flow to the crops during rainstorms. Sites were chosen so that only minimal earthworks were needed to spread the water over the fields. These were short lengths of bund referred to as spreader dikes.

Arnold Pacey

REFERENCES

Bradfield, M. *The Changing Pattern of Hopi Agriculture*. London: Royal Anthropological Institute. 1971.

Evenari, M., L. Shanan, and N. Tadmor. *The Negev: The Challenge of a Desert.* 2nd ed. Cambridge, Massachusetts: Harvard University Press. 1982.

Kutsch, H. *Principal Features of a Form of Water-Concentrating Culture* (in Morocco). Trier: Geographisch Geselkschaft Trier. 1982.

Pacey, Arnold, and Adrian Cullis. *Rainwater Harvesting*. London: Intermediate Technology Publications. 1986.

See also: Irrigation

Ramanujan

Srinivasa Ramanujan was born on 22 December 1887 in the home of his maternal grandmother in Erode, India, a small town located about 250 miles south-west of Madras. Soon thereafter, his mother returned with her son to her home in Kumbakonam, approximately 160 miles south–southwest of Madras. Ramanujan's father was a clerk in a cloth merchant's shop, and his mother took in local college students to augment the family's meagre income.

Ramanujan's mathematical talent was recognised in grammar school, and he won prizes, usually books of English poetry, in recognition of his mathematical skills. At the age of fifteen, Ramanujan borrowed G. S. Carr's *Synopsis of Pure Mathematics* from the local Government College in Kumbakonam. This unusual book, written by a Cambridge tutor to teach students, contained approximately five thousand theorems, mostly without proofs, and was to serve as Ramanujan's primary source of mathematical knowledge.

With a scholarship, Ramanujan entered the Government College in Kumbakonam in 1904. However, by this time, he was completely absorbed with mathematics and would not study any other subject. Consequently, at the end of his first year, Ramanujan failed all of his exams, except mathematics. He lost his scholarship and therefore was unable to return to college.

For the next five years, working in isolation, Ramanujan devoted himself to mathematics. He worked on a slate, and because paper was expensive, recorded his mathematical discoveries without proofs in note-books. During this time, he attempted once more to obtain a college education, at Pachaiyappa College in Madras, but his singular devotion to mathematics, and illness, deterred him again.

Having married S. Janaki in 1909, Ramanujan sought employment in 1910. For over a year, he was privately supported by R. Ramachandra Rao, as he gradually became known in the Madras area for his mathematical gifts. In 1912, Ramanujan became a clerk in the Madras Port Trust Office, and this was to be a watershed in his career, for the manager, S. Narayana Aiyar, and the Chairman, Sir Francis Spring, took a kindly interest in Ramanujan and encouraged him to write English mathematicians about his work.

On 16 January 1913, Ramanujan wrote to the famous English number theorist and analyst, G.H. Hardy. He and his colleague J.E. Littlewood examined the approximately sixty mathematical results communicated by Ramanujan and were astounded by his many beautiful and original claims. Hardy strongly encouraged Ramanujan to come to Cambridge, so that his mathematical talents could be fully developed. At first, Ramanujan was reluctant to accept the invitation, because of orthodox Brahmin beliefs that crossing the seas makes one unclean, but on 17 March 1914 Ramanujan sailed for England.

During the next three years Ramanujan achieved worldwide fame for his mathematical discoveries, some made in collaboration with Hardy. However, in the spring of 1917, Ramanujan became ill and was confined to nursing homes for the next two years. Tuberculosis, lead poisoning, and a vitamin deficiency were among the many diagnoses made, but a more recent examination of Ramanujan's symptoms points to hepatic amoebiasis. In 1919, he returned to India with the hope that a more favourable climate and more palatable food would restore his health. However, his condition worsened, and on 26 April 1920, Ramanujan passed away.

After Ramanujan's death, Hardy strongly urged that Ramanujan's notebooks be edited and published with his *Collected Papers*. Two English mathematicians, G.N. Watson and B.M. Wilson, devoted over ten years to proving the approximately three to four thousand theorems claimed by Ramanujan in his notebooks, but they never completed the task. It was not until 1957 that an unedited photostat edition of Ramanujan's notebooks was published. In 1977, B.C. Berndt, with the help of Watson and Wilson's notes, began to devote all of his research efforts toward editing the notebooks, and by early 1994 four volumes were published. In 1976, G. E. Andrews discovered a sheaf of 140 pages of Ramanujan's work, now called the "lost notebook", in the library at Trinity College, Cambridge. Andrews has now proved many of the 600 claims made in this "notebook", arising from the last year of Ramanujan's life.

Ramanujan made many beautiful discoveries in several areas of number theory and analysis, in particular, the theory of partitions, probabilistic number theory, highly composite numbers, arithmetical functions, elliptic functions, modular equations, modular forms, q-series, hypergeometric functions, asymptotic analysis, infinite series, integrals, continued fractions, and combinatorial analysis. His influence can be traced to many areas of contemporary mathematics; this is evident in the proceedings of major conferences commemorating Ramanujan on the one-hundredth anniversary of his birth. Although much of Ramanujan's work is quite deep, many of his original discoveries can be understood with a background of only high school mathematics. In particular, his several results on solving systems of equations, representing integers as sums of powers, and approximating π are elementary. In the past few decades, as more of Ramanujan's results have been unearthed, his already great reputation has soared even more.

Most biographical sketches of Ramanujan's life rely chiefly on the obituaries written by P.V. Seshu Aiyar and R. Ramachandra Rao, and the writings of Hardy. However, Robert Kanigel's biography is by far the most complete and detailed description of Ramanujan's life. Much can also be learned from Ramanujan's letters to Hardy, his family, and friends.

Bruce C. Berndt

References

Andrews, George E., Richard A. Askey, Bruce C. Berndt, K. G. Ramanathan, and Robert A. Rankin. *Ramanujan Revisited*. Boston: Academic Press. 1988.

Berndt, Bruce C. *Ramanujan's Notebooks. Parts I–V.* New York: Springer Verlag. 1985, 1989, 1991, 1994, forthcoming.

Berndt, Bruce C., and Robert A. Rankin. *Ramanujan: Letters and Commentary.* Providence: American Mathematical Society. 1995.

Hardy, G. H. *Ramanujan*. Cambridge: Cambridge University Press. 1940 (reprinted New York: Chelsea, 1978).

Kanigel, Robert. *The Man Who Knew Infinity*. New York: Charles Scribner's Sons. 1991.

Ramanujan, Srinivasa. *Collected Papers*. Cambridge: Cambridge University Press, 1927 (reprinted New York: Chelsea, 1962).

Ramanujan, Srinivasa. *Notebooks*. Bombay: Tata Institute of Fundamental Research. 1957.

Ramanujan, Srinivasa. *The Lost Notebook and Other Unpublished Papers*. New Delhi: Narosa. 1988.

Rationale in Indian Mathematics

Rationale in Hindu mathematics and astronomy is expressed by the terms *Yukti* and *Upapatti*, both meaning 'the logical principles implied'. It is characteristic of the Western scientific tradition, from the times of Euclid and Aristotle up to modern times, to enunciate and deduce using step by step reasoning. Such a practice is almost absent in the Indian tradition, even though the same background tasks of collecting and correlating data, identifying and analysing methodologies, and arguing out possible answers, have to be gone through before arriving at results. However, in the final depiction, only the resultant formulae would be given, and that too in short, crisp aphoristic form, leaving out details of all the background work. Commentaries generally content themselves with explaining the text of the formulae and adding examples. This tendency towards selective depiction of results has resulted not only in blacking out the background but also in not understanding the mental working of the Indian scientist. It also throws into oblivion the methodologies that had evolved. For this reason, many Indian advances have been branded as unoriginal and borrowed.

The situation is, however, relieved to some extent by the presence of a few commentaries which took pains to explain elaborately the methodologies adopted by the original author and also set out the rationales of the formulae he enunciated. Among such commentaries mentioned:

Siddhāntadīpikā on Govindasvāmi's *Mahābhāskarīya-Bhāṣya* by Parameśvara (1360–1465),

Āryabhaṭīya-Bhāṣya by Nīlakaṇṭha Somayājī (b. 1444),

Yuktidīpikā on Nīlakaṇṭha Somayaji's *Tantrasangraha,* by Śaṅkara Vāriyar (1500–1560),

Vāsanābhāṣya on Bhāskarācārya's *Siddhāntaśiromaṇi* by Nṛsiṃha Daivajña (1621),

Marīcī, again on Bhāskarācārya's *Siddhāntaśiromaṇi* by Munīśvara (1627),

A few texts devoted solely to the depiction of rationale are also known, such as the *Yuktibhāṣā* (Mathematical 'rationale in language' Malayalam) by Jyeṣṭhadeva (1500–1610), *Jyotirmīmāṃsā* (Investigations on Astronomical Theories) by Nīlakaṇṭha Somayāji (b. 1444), *Rāśigolasphuṭānīti* (True Longitude Computation on the Sphere of the Zodiac) by Acyuta Piṣāraṭi (1600), and *Karaṇapaddhati* (Methods of Astronomical Calculations) by Putumana Somayāji (1660–1740).

However, what is more significant is the occurrence of a number of short tracts giving mathematical and astronomical rationale which are available, some in print and several others in the form of manuscripts. These tracts take up individual topics of importance, analyse the technical principles involved therein, compare the procedures adopted in different texts, and often suggest revisions. To cite an example, the work *Gaṇitayuktayaḥ* (Rationales of Hindu Astronomy) contains a set of 27 tracts providing rationalistic exegeses on several topics including parallaxes of latitude and longitude, elevation of the Moon's horns, constitution of epochs for new astronomical manuals, planetary deflections, and equation of the centre. It is also noteworthy that some of these exegeses establish the originality of the methodologies and formulae depicted by Indian scientists.

K.V. Sarma

REFERENCES

Āryabhaṭīya of Āryabhaṭa with the Bhāṣya of Nīlakaṇṭha Somayājī. Trivandurm: Oriental Research Institute and Manuscripts Library. 1930, 1931, 1957.

Gaṇitayuktayaḥ: Rationales of Hindu Astronomy. Ed. K.V. Sarma. Hoshiarpur: Vishveshvaranand Institute. 1979.

Jyotirmīmāṃsā of Nīlakaṇṭha Somayājī. Ed. K.V. Sarma. Hoshiarpur: Vishveshvaranand Institute. 1977.

Mahābhāskarīya of Bhāskarācārya with the Bhāṣya of Govindasvāmin and the Super-commentary Siddhāntadīpikā of Parameśvara. Ed. T.S. Kuppanna Sastri. Madras: Government Oriental Manuscripts Library. 1957.

Rāśigolasphuṭānīti: True Longitude Computation on the Sphere of Zodiac. Ed. K.V. Sarma. Hoshiarpur: Vishveshvaranand Institute. 1977.

Sarma, K.V. *A History of the Kerala School of Hindu Astronomy*. Hoshiarpur: Vishveshvaranand Institute. 1972.

Srinivas, M.D. "The Methodology of Indian Mathematics and its Contemporary Relevance." *PPST Bulletin* (Madras) 12: 1–35. 1987.

Rockets

Fire arrows (agni-baṅa) were used in India and other civilisations for thousands of years. Rockets are different in that they are self-propelled. It is widely agreed that the first recorded use of rockets comes from China in the 11th century CE. The invention travelled rapidly (presumably through the Mongols) to Europe, where it was first mentioned in 1258 CE and was experimented with and used up to the 15th century. The Moghuls in India also used it during the late 15th and early 16th centuries. However rockets fell into disuse with the increasing accuracy and power of artillery. The re-emergence of the rocket as a significant military weapon during the 18th century in the princely state of Mysore in South India is a fascinating little episode in the history of technology in India, with an interesting sequel in 19th century Britain and Europe. Haidar Ali, a bold officer in the army of the Raja of Mysore, and his son Tipu Sultan, used the rocket frequently in various battles in South India, including the four 'Anglo-Mysore Wars' fought in the second half of the 18th century (Haidar's father had already commanded 50 rocketmen for another South Indian prince). The interest in the events of the late 18th century arises from two facts: the balance of industrial power began to shift from Asia to Europe in that period, and interesting accounts have been left behind by European observers.

The rockets used by the Mysoreans consisted of a metal cylinder ('casing') containing the combustion powder ('propellant'), which was tied to a long bamboo pole or sword that provided the required stability to the missile. It bears a strong resemblance to the much smaller 'rocket' that is a part of the fireworks that are still commonly seen during the Indian festival of lights, Deepavali. Two specimens preserved in the Royal Artillery Museum, Woolwich Arsenal in England have these dimensions:

- Casing 58 mm outer diameter and 254 mm long, tied with strips of hide to a straight 1.02 m long sword blade.
- Casing 37 mm outer diameter and 198 mm long, tied with strips of hide to a bamboo pole 1.9 m long.

These rockets had a higher thrust and range than anything in use at that time in Europe, as confirmed by later tests in England. The range is often quoted as about 1000 yards. There are however other accounts that speak of rockets that generally weighed 3.5 kg, tied to 3 m bamboo poles, and with a range of up to 2.4 km; this was by European standards an outstanding performance for the time.

The superior performance of these rockets cannot be attributed to the propellant, which was a standard material like gunpowder. There was nothing unusual about the aerodynamics either; it turns out that their superiority lay in the material employed for the casing. The casing was a metal cylinder made of hammered soft iron. Although it was crude, it represented a considerable advance over earlier technology, as European rockets of the time had combustion chambers made of some kind of pasteboard. For example, Geissler in Germany used wood, covered with sailcloth soaked in hot glue. The use of iron (which at that time was of much better quality in India than in Europe), increased bursting pressures, which permitted the propellant to be packed to greater densities; this is what gave the rockets their outstanding performance.

There was at that time a regular Rocket Corps in Tipu's Mysore Army with a strength of about 5000, with several units of rather more than a hundred men each. There are accounts that mention the skill of the Mysorean operators in giving the rockets an elevation that depended on the varying dimensions of the cylinder and distance to the attack target. Furthermore, the rockets could be launched rapidly using a wheeled cart with three or more rocket ramps.

The first account we have of the use of these rockets is in the Battle of Pollilur, fought on 10 September 1780 during the Second Anglo-Mysore War near a small village about 180 km east of Bangalore. Haidar and Tipu achieved a famous victory in this battle, and it is widely held that a strong contributory cause was that one of the British ammunition tumbrels was set on fire by the Mysore rockets, a scene that is celebrated in a famous mural that can be seen at the summer palace in Tipu's capital Sri-ranga-pattana. Writing about this war, Sir Alfred Lyall, a British historian of the early 20th century, remarked that, as a consequence of this defeat, 'The fortunes of the English in India had fallen to their lowest water-mark.'

A celebrated victim of such a rocket attack was Colonel Arthur Wellesley (later Duke of Wellington and the hero of Waterloo), who suffered a traumatic encounter with rockets in a mango grove just outside Sri-ranga-pattana in 1799. From these and other accounts it is clear that the British were caught off guard by the Indians' use of rockets, which at the least caused great fear and confusion. They started developing their own rockets in the early years of the 19th century. The programme began when several Indian rocket cases were collected and returned to Britain for analysis. Further development was chiefly the work of Col. (later Sir) William Congreve, who tested the biggest skyrockets then available in London. Their range was found to be about 500–600 yards, less than half that of the Mysore rockets. He then developed his own, using the facilities of the Royal Laboratory at Woolwich Arsenal.

Congreve used these rockets during the Napoleonic wars and in several engagements during the Anglo-American War of 1812, sometimes with little and on other occasions with great effect. They were still rather unreliable and inaccurate, but had greater range than cannons and could even be fired from rowboats. It was a spectacular but unsuccessful attack on Fort McHenry that led to the reference to 'the rockets' red glare' in the US national anthem, which began as a patriotic song composed by Francis Scott Key, who was witness to the attack.

One major reason for interest in this episode is that it occurred during a time of global transition in geopolitics, economics and technology. Clearly, even in the late eighteenth century there were several Indian products technologically superior to Western equivalents, and both sides recognised this. But the British effort that followed had the sophistication of research and development today. Scientific principles were applied, designs made, products developed and tested, and all of this was carefully documented – a process alien to Indians of that time. The Indian rockets were well made but not standardised, being the creation of traditional artisans.

Roddam Narasimha

References

Narasimha, R. "Rocketing from the Galaxy Bazaar." *Nature* 400: 123. 1999.

Narasimha, R. *Rockets in Mysore and Britain, 1750–1850 A.D.* PD DU 8503. Bangalore National Aerospace Laboratories. 1985.

S

Salt

As India is one of the oldest civilisations, it is no wonder that salt was produced in ancient India and that one finds its mention in ancient scriptures. The salt industry flourished as a cottage industry for centuries. The word for salt is *Lāvaṇa* in Sanskrit. A passage from *Arthaśāstra*, a book dealing with the history of the Mauryan period (300 BCE), says that salt manufacture was even at that distant date supervised by a state official named Lavanadhyaksa and the business was carried out under a system of licences granted on the payment of fixed fees or part of the output. This tradition, handed down from Hindu kings of old, is followed even now, with variations, by the Government of India through its Salt Department. The history of many countries shows connections with salt; in India salt was used by Mahatma Gandhi as a tool to win independence, when he completed his famous 400-km march to the sea at Dandi on 6th April 1930. The development of the Indian salt industry will now be briefly examined in two periods, pre- and post-independence.

Salt was prepared in ancient times and until British rule from sources like seawater, subsoil and lake saline water, rock salt deposits, and water extracts of saline soils (particularly in Uttar Pradesh) mainly by solar evaporation. If required, artificial evaporation was resorted to. The salt was produced in the coastal regions of Bengal, Bombay, Madras and the Rann of Kutch and in the inland regions of Rajasthan, Uttar Pradesh, and Central India from saline brines of lakes and as rock salt in Punjab. The salt industry provided a source of revenue to the rulers in the respective regions, and it received protection and encouragement from them. However, the industry faced a sort of setback and discouragement from the British rulers who not only raised the taxes and levies as early as 1768 but also later started importing salt around 1835–36 from European countries, Aden, and other places. The salt industry in Bengal province first felt the impact of the British policy which then affected the salt industry not only in other parts of their empire but also in Goa, which was under Portuguese rule. The quantity of salt imported, mainly in Bengal, in the pre-independence period ranged between 0.4 and 0.6 million tons annually.

However the British rulers reconsidered their drastic policies around 1930–31, which helped to revive the Indian salt industry, and salt production picked up. An Indian chemical engineer, Kapilram Vakil, brought about changes and improvements in salt production methods, particularly from seawater, based on scientific principles. Before establishing India's first large-scale marine salt farm at Mithapur in 1927 with the support of the Maharaja of Baroda, he had earlier (1919–20) studied in detail the status of the salt industry in the eastern states of Bengal and Orissa. Thus production of salt in large salt works established on the basis of scientific knowledge began in India before independence.

After Independence, the majority of rock salt producing areas went to Pakistan. Salt is now mainly produced from seawater, subsoil brines, and inland lake brines. India exports salt; the imports are around 8000 to 15,000 tons per year in recent years and it is mostly as rock salt from Pakistan.

Efforts have to be made to improve the quality of salt produced from both seawater and subsoil brines in the field itself, and this is one of the challenges faced by the Indian salt industry. The industry is highly labour intensive and thus has a low output. It is necessary to upgrade the technologies, adopt mechanisation, and improve transportation, loading, and port facilities. This is another challenge to be tackled by the Indian salt industry, an ancient industry that must adapt to changing times.

S.D. Gomkale

REFERENCES

Aggarwal, S.C. *The Salt Industry in India*. Government of India. 1976.

Annual Reports of the Salt Department, Ministry of Industry, Government of India.

Śaṅkara Vāriyar

Śaṅkara Vāriyar (ca. 1500–1560) was a brilliant expositor of astronomy and mathematics who hailed from Kerala in the south of India. He belonged to the *vāriyar* community which was professionally assigned to certain peripheral duties in temples. Śaṅkara was a direct disciple of the well-known Kerala astronomer Nīlakaṇṭha Somyājī (b. 1444). Among other teachers whom he mentions in his works are Dāmodara (ca. 1450–1550), Jyeṣṭhadeva (ca. 1500–1610) and Citrabhānu (ca. 1475–1550). Śaṅkara mentions that his commentary on the work *Pañcabodha* was completed on the day 16,92,972 of the Kali era, which occurs in CE 1534, from which his date is definitely ascertained.

Śaṅkara Vāriyar wrote an advanced astronomical manual entitled *Karaṇasāra*, to which he has added a gloss of his own. But he is better known for his elaborate commentaries on such standard works as the *Līlāvatī* of Bhāskarācārya, the *Tantra-saṅgraha* of Nīlakaṇṭha Somayājī, and the anonymous *Pañcabodha*. The two commentaries, the one on the *Līlāvatī*, called *Kriyākramakarī* (Sequential Evolution of Mathematical Procedures) and that on *Tantrasaṅgraha* called *Yuktidīpikā* (Light on Astronomical Rationale), are highly significant writings in Indian mathematical literature. After giving the meaning of each textual verse or group of verses from these two texts, Śaṅkara sets out the background and evolution of the enunciation commented on, the different aspects thereof, and the step-by-step derivation of the relevant formula or procedure. This exposition of rationale, couched in simple verses and often running to several pages, serves to give an exposure to Indian mathematical thinking and open up the normally unexpressed mental working of the Indian mathematician, which has led most modern historians of mathematics to presume that much of Indian mathematics and astronomy was borrowed and not original. The rationale elaborated here relates among other things, to the summations of series, the circle and the irrationality of π and pulverisation in the field of mathematics; and the theory of epicycles, ascensional differences, rising of the signs, and problems based on the gnomic shadow, in astronomy. These commentaries are valuable also for the information on earlier authors like Mādhava of Saṅgamagrāma and their theories.

K.V. Sarma

REFERENCES

Līlāvatī of Bhāskarācārya with *Kriyākramakarī of* Śaṅkara and Nārāyaṇa. Ed. K.V. Sarma. Hoshiarpur: Vishveshvaranand Vedic Research Institute. 1975.

Tantrasaṅgraha of Nīlakaṇṭha Somayāji with *Yuktidīpikā* and *Laghuvivṛti* of Śaṅkara. Ed. K.V. Sarma. Hoshiarpur: Vishveshvaranand Vishva Bandhu Institute of Sanskrit and Indological Studies, Punjab University. 1977.

See also: Nīlakaṇṭha Somyạji – Rationale in Indian Mathematics

Śatānanda

Śatānanda, son of Śaṅkara and author of the popular astronomical manual *Bhāsvatī*, was a resident of Puruṣottamapurī, the modern city of Puri, in Orissa. The *Bhāsvatī* was written in the Śaka year 1021, CE 1099. The epochal constants are also given for Puri, instead of for Ujjain as is normal in astronomical manuals. In 82 pithy verses distributed in eight chapters, the *Bhāsvatī* sets out the several computations required for preparing the daily almanac. There is a traditional statement that eclipses computed according to the *Bhāsvatī* would be exact. Towards the beginning of the work, the author avers that he was composing the work on the basis of the (Old) *Sūryasiddhānta* condensed by Varāhamihira in his *Pañcasiddhāntikā*. There is also a pun in the word *Bhāsvatī*, based on the above, since *bhāsvān* means *Sūrya* (Sun). There is a pun on the word *Śatānanda* as well; the literal meaning of the word is "one who revels in hundreds", and in this work the author used the centesimal system for commencing the epochal position and specified several of the multipliers and divisors in computation in terms of hundreds. There are several recensions of the work and some manuscripts add an *Uttara-Bhāsvatī*.

Śatānanda introduced certain other innovations in his work. For the computation of Mean planets, he took not the *ahar-gaṇa* (number of elapsed days), but the *Varṣa-gaṇa* (number of elapsed years). As the commencement of the year, he did not take the Mean *Meṣādi* (Aries ingress), but the True *Meṣādi*, which is advantageous in certain aspects. Another speciality of Śatānanda's is that, as mentioned above, he adopted the centesimal system for commencing epochal positions and for specifying multipliers and divisors. The positions of the Sun and the Moon are stated in terms of *nakṣatras* (constellations) and not in *rāśis*, which is normally the case in texts of this type. Then again, Śatānanda took CE 528 as the zero precession year and the rate of precession as one minute per annum.

The popularity of Śatānanda's work in North and North-east India is attested to by the presence of a large number of manuscripts of *Bhāsvatī* in these regions and by the fact that most of his commentators hail from there. Some of the principal commentators are Aniruddha (*Śiśubodhinī*, 1495), Mādhava (*Mādhavī*, 1525), Acyuta (*Ratnamālā*, ca. 1530), Kuvera Miśra (*Ṭīkā*, 1685), Rāmakṛṣṇa (*Tattvaprakāśikā*, 1739), and Yogindra (1742).

K.V. Sarma

REFERENCES

Dikshit, S.B. *Bharatīya Jyotish Sastra (History of Indian Astronomy)* Trans. R.V. Vaidya. Pt. II: *History of Astronomy During the Siddhantic and Modern Periods*. Calcutta: Positional Astronomy Centre, Indian Meṭerological Department. 1981.

Dvivedi, Sudhakara. *Gaṇaka Taraṅgiṇī: Lives of Hindu Astronomers*. Ed. Padmakara Dvivedi. Benares: Jyotish Prakash Press. 1933. pp. 33–34.

Pingree, David. *Jyotiḥśāstra: Astral and Mathematical Literature*. vol. IV, Fasc. 4 of *A History of Indian Literature*. Ed. Jan Gonda. Wiesbaden: Otto Harrassowitz. 1981.

Śrīman-Śatānanda-viracita Bhāsvatī with Sanskrit and Hindi commentaries by Matr Prasad Pandeya. Ed. Ramajanma Mishra. Varanasi: Chaukhamba Sanskrit Samsthan. 1985.

See also: Varāhamihira – *Sūryasiddhānta* – Lunar Mansions – Precession of the Equinoxes – Astronomy

Science as a Western Phenomenon

Philosophers, historians, and sociologists of science all accept as a basic postulate that science is essentially Western. This postulate is still conditioning contemporary scientific ideologies. This article analyses the characteristics, history, and validity of this doctrine by means of a confrontation with one of the non-Western scientific contributions: science written in Arabic.

Classical science is essentially European, and its origins are directly traceable to Greek philosophy and science; this tenet has survived intact through numerous conflicts of interpretation over the last two centuries. Almost without exception, the philosophers accepted it. Kant, as well as Comte, the neo-Kantians as well as the neopositivists, Hegel as well as Husserl, the Hegelians and the phenomenologists as well as the Marxists, all acknowledge this postulate as the basis of their interpretations of Classical Modernity. Even until our time, the names of Bacon, Descartes, and Galileo (sometimes omitting the first, and sometimes adding a number of others), are cited as so many markers on the road to a revolutionary return to Greek science and philosophy. This return was understood by all to be both the search for a model and the rediscovery of an ideal. One might impute this unanimity to the philosophers' zeal to pass beyond the immediate data of history, to their wish for radical insight, or to their effort to seize what Husserl describes as "the original phenomenon (*Urphänomen*) which characterises Europe from the spiritual point of view". One would expect that the position taken by those who have stuck more closely with the facts of the history of science would be quite different, but such is not the case. This same postulate is adopted by the historians of science as a point of departure for their work, and especially for their interpretations. Whether they interpret the advent of classical science as the product of a break with the Middle Ages, whether they defend the thesis of continuity without breaking or cutting, or whether they adopt an eclectic position, the majority of historians agree in accepting this postulate more or less implicitly.

Today, in spite of the works of many scholars on the history of Arabic and Chinese science, in spite of the wide representation of non-Western scientists in *Dictionary of Scientific Biography*, the works of the historians rest on an identical fundamental concept: in its modernity as well as in its historical context, classical science is a work of European man alone. Furthermore, it is essentially the means by which this branch of humanity is defined. Occasionally the existence of a certain practical science in other cultures might be acknowledged; nevertheless, it rests outside history, or is integrated into it only to the extent of its contributions to the essentially European sciences. These are only technical supplements which do not modify the intellectual configuration or the spirit of the latter in any way. The image given of Arabic science constitutes an excellent illustration of this approach. Essentially it consists of a conservatory of the Greek patrimony, transmitted intact or enriched by technical innovation to the legitimate heirs of ancient science. In all cases, scientific activity outside Europe is badly integrated into the history of the sciences; rather, it is the object of an ethnography of science whose translation into university study is nothing more than Orientalism.

The effects of this doctrine are not limited to the domain of science, its history, and its philosophy. It is at the centre of the debate between modernism and tradition. As was the case in eighteenth-century Europe, we find in certain Mediterranean and Asian countries of today, that science (which is qualified as European) is identified with modernism. Our purpose here is not to redress wrongs, nor to oppose to that science qualified as European an alleged Eastern science. It is simply a matter of understanding the significance of the European determination of the concept of classical science, grasping the reasons for it and measuring its importance.

We shall begin by sketching the history of this view of European science and then estimate its effects. We shall limit ourselves to posing the problem and advancing several hypotheses, and we also add these two restrictions: the only non-European science considered is one which was produced by various cultures, by scholars of different beliefs and religions, all of whom wrote their science principally, if not exclusively, in Arabic. As for the tenets of the history of the sciences, we shall most often cite those of the French historians.

The concept of a European science is already present in the works of the historians and philosophers of the eighteenth century. In the debate of the Ancients and the Moderns, scholars and philosophers referred to science to define modernity where one combines reason and experience. Historical induction intended to give its concrete determinations to this dogmatic debate, so as to render the superiority of the Moderns indisputable. But the West was already being identified as Europe, and "Oriental wisdom" was already counterpoised against the natural philosophy of the post-Newtonian West, such as we find in Montesquieu's *Persian Letters* (1721).

Classical science is European and Western only to the degree that it represents a stage in the continuous and regular development of humanity. The *Discours Préliminaire* of Abbé Bossut in Diderot's *Encyclopédie Méthodique* offers an illustration of this concept. Dividing the history of the progress of the exact sciences into three periods, this tableau allows conjecture, alleged facts, and facts to intermingle. Its initial postulate is that "... all of the eminent peoples of the ancient world liked and cultivated mathematics. The most distinguished among them are the Chaldeans, the Egyptians, the Chinese, the Indians, the Greeks, the Romans, the Arabs, etc. ... in modern times, the western nations of Europe". Classical science is European and Western because, writes Abbé Bossut, "... the progress made by the western nations of Europe in the sciences from the sixteenth century to our times utterly effaces those of other peoples."

The concept of Western science changed in nature and extent at the turn of the nineteenth century. With what Edgar Quinet called the "Oriental Renaissance", the conceptualisation was completed in its anthropological dimension in the last century. This Oriental Renaissance ended by discrediting science in the East.

If it is true that the eighteenth-century concept still survived here and there, from the first years of the nineteenth century the materials and ideas of Oriental studies contributed the most to the makeup of the historical themes of the different philosophies. In Germany as well as in France, the philosophers adopted Oriental studies for diverse reasons in accordance with an identical representation: the East and the West do not oppose each other as geographical, but as historical positivities; this opposition is not limited to a period of history, but goes back to the essence of each term. In this regard, *Lessons on the History of Philosophy* and other works of Hegel can be invoked. Also at this time, as is shown by the French Restoration philosophers, the themes of the "call of the East" and the "return to the Orient" appear, which translate as a reaction against science, and more generally, against Rationalism. But it is with the advent and growth of the German philological school that the notion of science as a Western phenomenon was regarded as having been endowed with the scientific, and no longer purely philosophical, support which had been lacking until then.

This influence also extended into mythological and religious studies. For example, Friedrich von Schlegel distinguishes two classes of language: the flexional Indo-European languages, and others. The former are "noble", the latter less perfect. Sanskrit, and consequently, German, considered the closest to it, is "... a systematic language and perfect from its conception"; it is "... the language of a people not composed of brutes, but of clear intelligence". There is nothing surprising in this; with the advent of the German school we are already in the realm of classifying mentalities. From now on everything is in place for affecting the passage from the history of languages to history-through-languages.

A. Kuhn and Max Müller developed the comparative study of religions and myths around the middle of the century. The classification of mentalities is perfected. It is from the basis of these tenets and dating from this period that one of the most important efforts to establish the notion of science as Western and European in an allegedly scientific manner is elaborated. This project achieves its full extent in France in the work of Ernest Renan.

For Renan, civilisation is divided between Aryans and Semites; the historian only has to evaluate their contributions in a differential and comparative manner. The notion of race would constitute the foundation of historiography. By race, one meant the whole of the "... aptitudes and instincts which are recognisable solely through linguistics and the history of religions". In the last analysis, it is for reasons attributable to the Semitic languages rather than the Semites themselves that they did not and could not have either philosophy or science. "The Semitic race", writes Renan, "is distinguished almost exclusively by its negative traits: it has neither mythology, epic poetry, science, philosophy, fiction, plastic arts nor civil life". The Aryans, whatever their origin, define the West and Europe at one and the same time. Arabic science is, "... a reflection of Greece, combined with Persian and Indian influences".

The historians of science borrowed not only their representation of the Western essence of science from this tradition, but also some of their methods for describing and commenting on the evolution of science. Thus, they applied themselves to discovering the concepts and methods of science and to following their genesis and propagation through philological analyses of the terms and on the basis of the texts at their disposal. Like the historian of myths or of religion, the historian of the sciences must be a philologist as well. In France, the situation is such that philosophers borrow Renan's interpretation and, often, even his terminology. Even though historians have already abandoned this brand of anthropology, they nevertheless preserve and propagate a series of inferences engendered by it. These can be enumerated as follows:

(1) Just as science in the East did not leave any consequential traces in Greek science, Arabic science has not left any traces of consequence

in classical science. In both cases, the discontinuity was such that the present could no longer recognise itself in its abandoned past.

(2) Science subsequent to that of the Greeks is strictly dependent upon it. According to Duhem, "... Arabic science only reproduced the teachings that it received from Greek science". In a general fashion, Tannery reminds us that the more one examines the Hindu and Arabic scholars, "... the more they appear dependent upon the Greeks ... (and) ... quite inferior to their predecessors in all respects".

(3) Whereas Western science addresses itself to theoretical fundaments, Oriental science, even in its Arabic period, is defined essentially by its practical aims.

(4) The distinctive mark of Western science is its conformity to rigorous standards; in contrast, Oriental science in general, and Arabic science in particular, lets itself be carried away by empirical rules and methods of calculation, neglecting to verify the soundness of each step on its path. The case of Diophantus illustrates this idea perfectly: as a mathematician, said Tannery, "... Diophantus is hardly Greek". But when he compares the *Arithmetics* of Diophantus to Arabic algebra, Tannery writes that the latter "... in no way rises above the level achieved by Diophantus."

(5) The introduction of experimental norms which, according to historians, totally distinguishes Hellenistic science from classical science, is solely the achievement of Western science.

Thus it is to Western science alone that we owe both the concept and experimentation. We are not going to oppose this ideology to another. We propose simply to confront some of these elements with the facts of the history of science, beginning with algebra and concluding with the crucial problem of the relationships between mathematics and experimentation.

ALGEBRA

As with the other Arabic sciences, algebra had practical aims, a flair for calculation, and an absence of rigorous standards. It is precisely this that allowed Tannery to make his statement. Bourbaki took this as his authorisation to exclude the Arabic period when he retraced the evolution of algebra. The historical writings of the modern mathematician Dieudonné are significant; between the Greek prehistory of algebraic geometry and Descartes, he finds only a void, which, far from being frightening, is ideologically reassuring. Some historians cite al- Khwārizmī, his definition of algebra and his solution of the quadratic equation, but it is generally to reduce Arabic algebra to its initiator. This restriction misconstrues the history of algebra, which in actuality does not show a simple extension of the work of al- Khwārizmī in the West, but an attempt at theoretical and technical overtaking of his achievements. Moreover, this overtaking is not the result of a number of individual works, but the outcome of genuine traditions. The first of these traditions had conceived the particular project of arithmetising the algebra inherited from al-Khwārizmī and his immediate successors. The second one, in order to surmount the obstacle of the solution by radicals of third and fourth degree equations, formulated in its initial stage a geometric theory of equations, subsequently to change viewpoint and study known curves by means of their equations. In other words, this tradition engaged itself explicitly in the first research in algebraic geometry.

As we have said, the first tradition had proposed arithmetising the inherited algebra. This theoretical program was inaugurated at the end of the tenth century by al-Karajī, and is thus summarised by one of his successors, al-Samaw'al (d. 1176): "to operate on unknowns as the arithmeticians work on known quantities".

The execution is organised into two complementary stages. The first is to apply the operations of elementary arithmetic to algebraic expressions systematically; the second is to consider algebraic expressions independently from that which they can represent so as to be able to apply them to operations which, up to that point, had been restricted to numbers. Nevertheless, a program is defined not only by its theoretical aims, but also by the technical difficulties which it must confront and resolve. One of the most important of these was the extension of abstract algebraic calculation. At this stage, the mathematicians of the eleventh and twelfth centuries obtained some results which unjustly are attributed to the mathematicians of the fifteenth and sixteenth centuries. Among these are the extension of the idea of an algebraic power to its inverse after defining the power of zero in a clear fashion, the rule of signs in all its general aspects, the formula of binomials and the tables of coefficients, the algebra of polynomials, and above all, the algorithm of division, and the approximation of whole fractions by elements of the algebra of polynomials.

In a second period, the algebraists intended to apply this same extension of algebraic calculation to irrational algebraic expressions. Al-Karajī questioned how to operate by means of multiplication, division, addition, subtraction, and extraction of roots on irrational quantities. To answer this question the mathematicians gave, for the first time, an algebraic interpretation of the theory contained in Book X of the *Elements*. This book was regarded by Pappus, as well as by Ibn al-Haytham much later, as a geometry book, because of the traditional fundamental separation between continuous and discontinuous magnitudes. With the school of al-Karajī, a better understanding of the structure of real algebraic numbers is achieved.

In addition, the works of this algebraic tradition opened the route to new research on the theory of numbers and numerical analysis. An examination of numerical analysis, for example, reveals that after renewing algebra through arithmetic, the mathematicians of the eleventh and twelfth centuries also effected a return movement to arithmetic to search for an applied extension of the new algebra. It is true that the arithmeticians who preceded the algebraists of the eleventh and twelfth centuries extracted square and cube roots, and had formulas of approximation for the same powers. But, lacking an abstract algebraic calculation, they could generalise neither their results, their methods, nor their algorithms. With the new algebra, the generalisation of algebraic calculation became a constituent of numerical analysis which, until then, had only been a sum of procedures, if not prescriptions. It is in the course of this double movement which is established between algebra and arithmetic that the mathematicians of the eleventh and twelfth centuries achieved results which are still wrongly attributed to the mathematicians of the fifteenth and sixteenth centuries. This is the case with the method attributed to Viète for the resolution of numerical equations, the method ascribed to Ruffini-Horner, the general methods of approximation, in particular that which D. T. Whiteside designates by the name of al-Kāshī-Newton, and finally, the theory of decimal fractions. In addition to methods, which were to be reiterative and capable of leading in a recursive manner to approximations, the mathematicians of the eleventh and twelfth centuries also formulated new procedures of demonstration such as complete induction.

We have just seen that the algebraist arithmeticians from the end of the tenth century among others elaborated the concept of polynomials. This tradition of algebra as the "arithmetic of unknowns", to use the expression of the time, opened the road toward another algebraic tradition which was initiated by Umar al-Khayyām (eleventh century), and renewed at the end of the twelfth century by Sharaf al- Dīn al-Ṭūsī. While the former formulated a geometric theory of equations for the first time, the latter left his mark on the beginnings of algebraic geometry.

The immediate predecessors to al-Khayyām, such as al-Bīrūnī, al-Māhānī, and Abū al-Jūd, had already been able, in contrast to the Alexandrian mathematicians and precisely because of the concept of the polynomial, to treat the problems of solids in terms of third degree equations. But al-Khayyām was the first to address these unpondered questions: can one reduce the problems of straight lines, planes, and solids to equations of a corresponding degree, on the one hand, and on the other, re-order the group of third degree equations to seek, in the absence of a solution by factoring, solutions which can be reached through means of the intersection of auxiliary curves? To answer these questions, al-Khayyām is led to formulate the geometric theory of equations of a third or lesser degree. His successor, al-Ṭūsī, did not delay in changing perspective; far from adhering to geometric figures, he thought in terms of functional relations and studied curves by means of equations. Even if al-Ṭūsī still solved equations by means of auxiliary curves, in each case the intersection of the curves is demonstrated algebraically by means of their equations. This is important, since the systematic use of these proofs introduces into the practice instruments which were already available to the mathematical analysts of the tenth century: affine transformations, the study of the maxima of algebraic expressions, and with the aid of what will later be regarded as derivatives, the study of the upper bounds and lower bounds of roots. It is in the course of these studies and in applying these methods that al-Ṭūsī grasps the importance of the discriminant of the cubic equation and gives the so-called Cardan formula just as it is found in the *Ars Magna*. Finally, without enlarging any further on the results which were obtained, we can say that both on the level of results as well as that of style, we find al-Khayyām and al-Ṭūsī fully in the field allegedly pioneered by Descartes.

If we exclude these traditions and justify this exclusion by invoking the practical and computational aims of the Arab mathematicians and an absence of rigorous standards of proof in their work, we can say that the history of classical algebra is the work of the Renaissance.

Among the mathematical disciplines, algebra is not a unique case. To varying degrees, trigonometry, geometry, and infinitesimal determinations are likewise illustrative of the preceding analysis. In a more general sense, optics, statistics, mathematical geography, and astronomy are also no exception. Recent works in the history of astronomy render Tannery's understanding of the Arab astronomers and the interpretations which he gives of them manifestly outmoded, if not erroneous. But since we assigned ourselves the task of examining the doctrine of the Western nature of classical science, we shall restrict our discussion to an essential component of this doctrine, experimentation.

EXPERIMENTATION

In fact, is not the cleavage between the two periods of Western science, the Greek period and the Renaissance, often marked by the introduction of experimental norms? Undoubtedly the general agreement of the philosophers, historians, and sociologists of science stops here; the divergences become apparent as soon as they attempt to define the meaning, the implications, and the origins of these experimental norms. The origins are linked in one case to the current of Augustinian–Platonism, in another to the Christian tradition, and particularly to the dogma of Incarnation, in a third case to the engineers of the Renaissance, in a fourth to the *Novum Organum* of Francis Bacon, and finally, in a fifth, to Gilbert, Harvey, Kepler, and Galileo. Some of these attitudes superimpose upon one another, become entangled or contradictory, but they all converge on one point: the occidental nature of the new norms.

Nevertheless, as early as the nineteenth century, historians and philosophers such as Alexander von Humboldt in Germany and Cournot in France diverge from this predominating position to attribute to the Arab period the origins of experimentation. It is difficult to analyse the origins or the beginnings of experimentation correctly, since no study has been made of the interrelations of the different traditions and the different themes to which the concept of experimentation has been applied. Perhaps it would be in writing such a history; especially a history of the term itself, that one could give an accounting of the multiplicity of uses and ambiguities of the concept. For this analysis two histories are needed: the history of the relationship between art and science and that of the links between mathematics and physics.

With the history of the relationship between science and art, we are in a position to understand when, why, and how it became accepted that knowledge can emanate from demonstrations and from the rules of practice at the same time, and that a body of knowledge possesses the stature of a science while, at the same time, it is conceived in its possibilities of practical realisation with an external purpose. The traditional opposition between science and art seems likely to be the work of the intellectual currents of the Arabic period. Certainly one fact is striking: whether we are dealing with Muslim traditionalists, rationalist theologians, scholars of different fields, and even philosophers of the Hellenistic tradition such as al-Kindī or al-Fārābī, all contribute to the weakening of the traditional differentiation between science and art. In other respects, this general trait is at the origin of the opinion of some historians regarding the practical spirit and realistic imagination of the Arab scholars. Knowledge is accepted as scientific without its conforming either to the Aristotelian or the Euclidean scheme. This new concept of the stature of science promoted the dignity of scientific understanding of disciplines which traditionally were confined to the domain of art, such as alchemy, medicine, pharmacology, music, or lexicography. Whatever might be the importance of this concept, it could only lead to an extension of empirical research and to a diffuse notion of experimentation. One does witness the multiplication and systematic use of empirical procedures: the classifications of the botanists and the linguists, the control experiments of the doctors and alchemists, and the clinical observations and comparative diagnostics of the physicians. But it was not until new links were established between mathematics and physics that such a diffuse notion of experimentation acquired the dimension that determines it, a regular and systematic component of the proof. Primarily it is in Ibn al-Haytham's work in the field of optics where the emergence of this new dimension can be perceived.

With Ibn al-Haytham the break is established with optics as the geometry of vision or light. Experimentation had indeed become a category of the proof. The successors of Ibn al-Haytham, such as al-Fārisī, adopted experimental norms in their optical research, such as that performed on the rainbow. What did Ibn al-Haytham understand by experimentation? We will find in his work as many meanings of this word and as many functions served by experimentation as there are links between mathematics and physics. A thorough look at his texts indicates that the term and its derivatives belong to several superimposed systems, and are not likely to be discerned through simple philological analysis. But if attention is fixed on the content rather than the lexical form, one can distinguish several types of relationships between mathematics and physics which allow one to spot the corresponding functions of the idea of experimentation. In fact, the links between mathematics and physics are established in several ways; even if they are not specifically treated by Ibn al- Haytham, they underlie his work and are amenable to analysis.

As for the field of geometric optics, which was reformed by Ibn al-Haytham himself, the only link between mathematics and physics is a similarity of structures. Owing to his definition of a light ray, Ibn al-Haytham was able to formulate his theory on the phenomena of propagation, including the important phenomenon of diffusion, so that they relate perfectly to geometry. Then several experiments were devised to assure technical verification of the propositions. Experiments were designed to prove the laws and rules of geometrical optics. The work of Ibn al-Haytham attests to two important facts which are often insufficiently stressed: first of all, some of his experiments were not simply designed to verify qualitative assertions, but also to obtain quantitative results; in the second place, the apparatus devised by Ibn al-Haytham, which was quite varied and complex, is not limited to that of the astronomers.

In physical optics one encounters another type of relationship between mathematics and physics and therefore a second meaning of experimentation. Without opting for an atomistic theory, Ibn al-Haytham states that light, or as he writes, "the smallest of the lights", is a material thing, external to vision, which moves in time, changes its velocity according to its medium, follows the easiest path, and diminishes in its intensity depending on its distance from its source. Mathematics is introduced into physical optics at this stage by means of analogies established between the systems of movement of a heavy body and those of the reflection and refraction of light. This previous mathematical treatment of the concepts of physics permitted them to be transferred to an experimental plane. Although this situation on the experimental level might be somewhat approximate in nature, it nevertheless furnishes a level of existence to ideas which are syntactically structured, but semantically indeterminate, such as Ibn al-Haytham's scheme of the movements of a projectile.

A third type of experimentation, which was not practiced by Ibn al-Haytham himself but was made possible by his own reform and his discoveries in optics, appears at the beginning of the fourteenth century in the work of his successor al-Fārisī. The links established between mathematics and physics aim to construct a model and to reduce by geometric means the propagation of light in a natural object to its propagation in an artificial object. The problem is to define for propagation, between the natural and the artificial object, some analogical correspondences which were genuinely certain of mathematical status. For example, they built a model of a massive glass sphere filled with water to explain the rainbow. In this case, experimentation serves the function of simulating the physical conditions of a phenomenon that can be studied neither directly nor completely. The three types of experimentation studied all reveal themselves both as a means of verification and as furnishing a plane of material existence to ideas which are syntactically structured. In the three cases, the scientist must realise an object physically in order to handle it conceptually. Thus, in the most elementary example of rectilinear propagation, Ibn al-Haytham does not consider any arbitrarily chosen opening of a black box, but rather specific ones, in accordance with specific geometric relationships, in order to realise as precisely as possible his concept of a ray.

To recapitulate several points:

(1) The tenet of the Western nature of classical science which was launched in the eighteenth century owes to the Orientalism of the nineteenth century the image that we now recognise.

(2) On the one hand, the opposition between East and the West underlies the critique of science and rationalism in general; on the other, it excludes the scientific production of the Orient from the history of science both *de facto* and *de jure*. An absence of rigour is invoked, as well as the computational attributes and the practical aims of science written in Arabic, to justify this effective debarment from the history of science.

(3) This tenet reveals a disdain for the data of history as well as a creative capacity for ideological interpretation, which are admitted as evidence for ideas that raise many more problems than they solve. Thus we have the notion of a Scientific Renaissance, when in several disciplines everything indicates that there was merely a reactivation. These pieces of pseudo-evidence quickly become conceptual bases for a philosophy or sociology of science, as well as the departure point for theoretical elaborations in the history of science.

We must ask ourselves if the moment has not arrived to abandon this characterisation of classical science and its still lively traces in the writing of history, to restore to the profession of the historian of science the objectivity required of it, to ban the clandestine importation and diffusion of uncontrolled ideologies, to refrain from all reductionist tendencies which favour similarities at the expense of differences, and to be wary of miraculous events in history. The neutrality of the historian is not an a priori ethical value; it can only be the product of patient work which will not be duped by the myths which the East and West have engendered. Above all, it is necessary to cast out the periodisation everywhere admitted in the history of science. The term used for classical algebra or classical optics, for example, will integrate the works which extended from the tenth to the seventeenth centuries. Consequently, it will re-align not only the idea of the classical sciences but also that of medieval science. The classical sciences will then reveal themselves as the product of the Mediterranean which was the hub of exchanges among all civilisations of the ancient world. Only then will the historian of science be able to enlighten the debate over modernism and tradition.

Science as a Western Phenomenon: Postscript

The sixteen years since the publication of the original French edition of the text above has been a very fertile period for the study of the history of Islamic science. Indeed, we have witnessed an unprecedented rebirth of this discipline. Texts have been written and translated, new collections have appeared, and specialised reviews and journals have been published. These have offered historians the possibility of developing and comparing their research findings with facts. The task remains huge, and we are only at the beginning, but at least this new growth of information puts to rest the argument of ignorance.

With all this new information, one would have expected historians and philosophers to rectify the impressions and ideas they had inherited from the nineteenth century. One would have thought that the doctrine of Western science which we have described and analysed here would have disappeared along with the props on which that doctrine was based. Indeed we were beginning to see a growing tendency to break with this doctrine and its implications. Then, for reasons which are extraneous to science and its history, images of Islamic society – if not of Islam itself – arose, according to which it was seen as irrational and intolerant and thus a society foreign to science. The aging doctrine was naturally given new life because of these images. How is it possible, under these conditions, to reconcile such an image of Islamic society with scientific results obtained from the heart of that same society? It was enough to back up the preceding doctrine with another, the doctrine of double marginality: with regard to the society which saw the development of science, and with regard to the history of the sciences. Thus, one could still write in 1992, "We must remember that at an advanced level the foreign sciences had never found a stable institutional home in Islam", or "Greek learning never found a secure institutional home in Islam, as it was eventually to do in the universities of medieval Christendom" (Lindberg, 1992). As for the second marginality, we have already described how it works. Thus, we are back where we started and the doctrine of the Westernness of science is saved. Undoubtedly, these ideological views are beginning to give way, weakened by new research findings. And even if they are still capable of slowing down the acceptance of facts, it is not widespread, and it certainly will not last much longer.

Roshdi Rashed

References

al-Khayyām. *L'algèbre d'Omar Alkhayyāmī*. Ed. F. Woepcke. Paris: B. Duprat. 1851.

al-Samaw'al. *al-Bāhir en algèbre*. Ed. R. Rashed. Damascus: University of Damascus. 1972.

Cournot, A. A. *Considérations sur la marche des idées et des èvenements dans les temps modernes*. Paris: Vrin. 1973.

Dieudonné, J. *Cours de géométrie algébrique*. Paris: Presses universitaires de France. 1974.

Duhem, P. *Le système du monde*. 10 vols. Paris: A. Hermann. 1913–1959.

Encyclopédie méthodique. Paris: Pancoucke. 1784.

Hegel, G.W.F. *Lectures on the History of Philosophy*. New York: Humanities Press. 1968.

Kojève, A. "The Christian Origin of Modern Science." In *Mélanges Alexandre Koyré*. Paris: Hermann. 1964, vol. II. pp. 295–306.

Lindberg, David. *The Beginning of Western Science*. Chicago: University of Chicago Press. 1992. p. 182.

Montesquieu. *Oeuvres complètes*. Paris: Gallimard. 1964.

Rashed, R. "L'extraction de la racine n^{iene} et l'invention des fractions décimales." *Archive for History of Exact Sciences* 18(3): 191. 1978.

Rashed, R. "Résolution des équations numériques et algèbre: Sharaf al-Dī al- Tasi, Viète." *Archive for History of Exact Sciences* 12(3): 244. 1974.

Renan, E. *Histoire général et système comparé des langues sémitiques*. Paris: Imprimerie impériale. 1863.

Renan, E. *Nouvelles considérations sur le caractère générale des peuples sémitiques*. Paris: Imprimerie impériale. 1859.

Sarton, G. *The Incubation of Western Culture in the Middle East*. Washington: Library of Congress. 1951.

Schlegel, F. "Essai sur la langue et la philosophie des Indiens." In *The Philosophy of History*. London: Bell & Daldy. 1871.

Tannery, P. *La Géométrie grecque*. Paris: Gauthier-Villars. 1887.

See also: Western Dominance – East and West

Sexagesimal System

The sexagesimal system was an ancient system of counting, calculation, and numerical notation that used powers of sixty much as the decimal system uses powers of ten. Rudiments of the ancient system survive in vestigial form in our division of the hour into sixty minutes and the minute into sixty seconds. Origins of the system of counting are essentially irretrievable because they lie in the period before the invention of writing. They are associated with the ancient Sumerians, whose language incorporated the only known system of sexagesimal counting. The Sumerians were probably also the inventors of the system of calculation, which seems to have its origin in a system of counters (small clay objects also referred to as tokens or calculi) representing the units $1, 10, 60, 60 \times 10, 60^2, 60^2 \times 10$, and 60^3. This counter-calculation system may go back as far as ca. 3500 BCE. The method of numerical notation apparently originated from the system of counters by adapting it to a form that could be represented in writing when the first writing system was invented ca. 3000 BCE.

By 2000 BCE this had developed into a system of sexagesimal place notation that functioned like our decimal system but with several important distinctions. Whereas the decimal system uses ten unique symbols, plus a period or comma to separate fractions from integers, the fully developed sexagesimal system of notation used only two unique symbols (those for 1 and 10) and essentially ignored our distinction between fractions and integers. These features, as well as lack of a symbol for zero, reflect its origins in the system of calculating with counters: zero would not have been represented in a system of counters (obviously, one does not count something that does not exist), and unique symbols for the integers 1–60 would have been excluded for practical reasons. Lack of a distinction between fractions and integers goes back to the fact that fractions are essentially alien to prehistoric systems of counting. Names for small quantities that we would call fractions are based on subdivisions of the weight system (*mina* and *shekel*) and were clearly not thought of as fractions in our sense of the word. In keeping with this picture is the symbol for medial zero that appears in Babylonian mathematical texts of the late first millennium BCE. This so-called zero is really nothing but a "spacer" symbol and reflects the practice (occasionally observable in surviving texts) of leaving a blank space where we would place a zero.

During the third millennium BCE, this Sumerian sexagesimal system was adopted, along with other features of Sumerian culture, by semitic-speaking Akkadians, and the Akkadian language, which – like Sumerian – was written in cuneiform script on clay tablets, served as the vehicle for its diffusion and preservation. As an essential part of the education of scribes, it was still very much alive when Greeks conquered the Near East in the time of Alexander the Great. The two centuries of Greek presence in Mesopotamia (330–129 BCE) facilitated the incorporation of Babylonian mathematical astronomy into the Western tradition, and with it also came knowledge of the Babylonian sexagesimal system. This, however, seems never to have been used systematically by Greeks outside of Babylonia. Division of the circle into 360 degrees is a vestige of the Babylonian system that has been transmitted to us by the Greeks.

Modern knowledge of the existence of an ancient sexagesimal system goes back as far as the recovery of Greek mathematics, beginning in the Renaissance. Little, however, was known about its true character until the middle of the nineteenth century when decipherment of cuneiform writing revealed the system in its developed form of the first millennium BCE. Speculations about its origins have tended to postulate conscious creation, as opposed to gradual evolution. The 360-day year and choice of 60 as base (because 1, 2, 3, and 5 are all prime factors) are among these theories, all equally without evidence. Only in the twentieth century has the Sumerian system of counting become relatively well understood, and the role of clay counters as prototypes of number symbols has only become clear in the last decade.

Marvin A. Powell

References

Powell, Marvin A. *Sumerian Numeration and Metrology*. Ph. D. Dissertation, University of Minnesota. 1971.

Powell, Marvin A. "The Origin of the Sexagesimal System: The Interaction of Language and Writing." *Visible Language* 6: 5–18. 1972.

Powell, Marvin A. "The Antecedents of Old Babylonian Place Notation and the Early History of Babylonian Mathematics." *Historia Mathematica* 3: 417–439. 1976.

Powell, Marvin A. "Metrology and Exact Sciences in Ancient Mesopotamia." In *Civilizations of the Ancient Near East*. Ed. J. Sasson et al. New York: Scribners. 1995.

Sphujidhvaja

Sphujidhvaja was apparently of Greek descent, and flourished in Western India under the patronage of the Kṣatrapa ruler Rudradāman II. He wrote the extensive genethlialogical manual entitled *Yavanajātaka* (Horoscopy of the Greeks) in CE 270, which shows the position and influence of the stars at one's birth. Towards the close of his work, Sphujidhvaja says that, before him, in CE 150, the great Greek genethlialogist Yavaneśvara redacted into Sanskrit prose a Greek astrological work, so that it could be studied by those who did not know Greek, and that he, Sphujidhvaja is composing a versified redaction of the work of Yavaneśvara. The work reveals Sphujidhvaja as a competent scholar, a master of Sanskrit versification, and an expert genethlialogist.

Sphujidhvaja states that he composed the work in 4000 verses. But the only manuscript of the work available today contains only 2300 verses. In this imperfect manuscript, the first few sections are numbered, but not so the subsequent ones, which number, in all, 79. Possibly several sections are lost. In these 79 sections, the work covers a large number of aspects of horoscopy and natural astrology, including:

- Zodiacal signs and planets, their icons, nature, and relationships;
- Iconography of *horās* and decons;
- Astrology of conception, birth, and nature;
- Horoscopes;
- Planetary placements, and combinations affecting human beings;
- Prediction on the basis of questions;
- Reconstruction of lost horoscopes;
- Omens and dreams; and
- Military astrology.

The several aspects of each of these topics are looked at from different points of view, and intimations, indications and predictions based thereon are stated categorically.

It is interesting that, although based ultimately on a Greek text, there was substantial Indianisation effected in the *Yavanajātaka*. This was accomplished by using Indian equivalents to Greek terms, adopting Hindu deities in place of Greek ones, incorporating the names of Hindu castes and professional orders, mentioning local manners and customs, and the like. *Yavanajātaka* was looked upon as authority by all later Indian genethlialogists and used as such even in the foremost of Indian texts like *Bṛhajjātaka* and *Bṛhatsaṃhitā* of Varāhamihira, *Sārāvalī* of Kalyāṇavarman, *Praśnavidyā* of Bādarāyaṇa, and many others.

The significance of *Yavanajātaka* stems from another point as well. The work refers to contemporary social orders, professions, religious classes and groups, items of ordinary use, manners and customs, dress, and a host of other things related to the life and society of the times. The information provided by the work on these subjects makes it a good source for the study of the culture and civilisation of India during the early centuries of the Christian era.

K.V. Sarma

References

Bṛhajjātaka of Varāhamihira with the commentary of Bhaoṭṭotpala. Bombay: Jnanadarpana Press. 1874.

Bṛhatsaṃhitā by Varāhamihirācārya, with the commentary of Bhaoṭṭotpala. 2 pts. Ed. Avadha Vihari Tripathi. Varanasi: Varanaseya Sanskrit University. 1968.

Pingree, David. "The Yavanajātaka of Sphujidhvaja." *Journal of Oriental Research* (Madras) 31(1–4): 26–32. 1961–62.

Sārāvalī of Kalyāṇavarman. Ed. V. Subrahmanya Sastri. Bombay: Nirṇayasāgar Press. 1928.

The *Yavanajātaka of Sphujidhvaja*. Ed. and trans. David Pingree. Cambridge, Massachusetts: Harvard University Press. 1978.

See also: Astrology – Varāhamihira – Yavaneśvara

Śrīdhara

The mathematical works of Śrīdhara were very popular and made him quite famous. In spite of his great popularity, some controversies have been raised about his life, work, and time, such as whether he was a Hindu or a Jaina, and whether he wrote before or after Mahāvīra (ninth century CE). Some uncertainties exist because Śrīdhara's works are not fully extant. Often he is confused with other authors of the same name. Here we shall give views which are now generally accepted.

Like so many ancient Indian authors, Śrīdhara did not provide any information about himself in his works. Other sources have not been helpful in finding any glimpse of his personal life. So we do not know his parents or teachers, or even where he was born, educated, or worked. But some evidence shows that he was a Saivite Hindu. An example in his *Pāṭīgaṇita* is about the payment for the worship of the five-faced Hindu god Śiva. He starts his *Triśatikā* with a homage to the same god. However, in a manuscript which has a possible commentary and additional examples written by an apparently Jaina writer, the word "Jinam" is found in place of "Śivam". This led to the claim that Śrīdhara was a Jaina, but other evidence does not support this view.

More serious is the controversy about Śrīdhara's time. That he lived after Brahmagupta is evident from the fact that he literally quoted (and criticised) a rule from Brahmagupta's *Brāhmasphuṭasiddhānta* (CE 628). The most significant question in this connection is whether Śrīdhara wrote before or after Mahāvīra, whose date of ca. CE 850 is certain. There are similarities in the works of these two mathematicians. Most scholars believe that indications generally place Śrīdhara before Mahāvīra. A new fact has come to light recently; David Pingree found that a rule of Śrīdhara's had been quoted by Govindasvāmin (about CE 800–850) in his commentary on the *Horāśāstra* of Parāśara (under 14.97). This latest and direct evidence once again supports the generally accepted view of placing Śrīdhara in the eighth century or ca. CE 750.

Śrīdhara is known to be the author of the following three works on mathematics:

(1) *Pāṭīgaṇita* (on arithmetic and mensuration);

(2) *Pāṭīgaṇita-sāra* (an epitome of the above); and

(3) *Bījagaṇita* (on algebra).

The *Pāṭīgaṇita* is a standard Indian treatise on practical mathematics meant "for the use of the people". It is also called *Bṛhat-Pāṭī* (Bigger Pāṭī) and *Navaśatī* (Having Nine Hundred) because it is believed to have nine hundred stanzas. Unfortunately, it is not extant in full; the available text contains only two hundred fifty-one stanzas. In terms of topics, treatment of definitions, logistics, mixtures of things, and series is available in full, but that of the plane figures is incomplete. Many other treated topics are mentioned in the list of contents, but they are totally missing from the manuscript. These were excavations, piles of bricks, sawn pieces of timber, heaps of grain, shadows, and the mathematics of zero (whose loss is quite sad).

The *Pāṭīgaṇita-sāra* is also called *Triśatī* or *Triśatikā* because it was a "Collection of 300 verses". It was the author's own summary of the larger work on the subject.

Śrīdhara's *Bījagaṇita* (Algebra) seems to be lost completely. We know about it from a statement of Bhāskara II (CE 1150) who also quoted a rule from it. The rule gives a method for solving any quadratic equation and has also been quoted by others. Many other rules from different works of Śrīdhara have also been quoted by various writers. This shows the popularity of his works.

A recently discovered work called *Gaṇita-Pañcaviṁśī* is also stated to be from the pen of Śrīdhara, but it may not be his genuine work. Similarly, a number of astronomical, astrological, and other works authored by different persons of the same name have been ascribed wrongly. Confusions were created both by the similarity of names and also by the proximity of their dates.

In India, ten has been the base of counting since very early times. But the number and names of decuple terms used in the decimal numeration were at variation throughout the ancient period. Later on, the decuple terms were used to denote the notational places when the positional system was developed, and their number was standardised to eighteen (which was a traditionally sacred number). Śrīdhara gave a definite list of eighteen terms which became standard in Indian mathematics. It runs thus: *eka, daśa, śata, saharsa, ayuta, lakṣa, prayuta, Koṭi* ($= 10^7$), *arbuda, abja* (or *abda*), *kharva, nikharva, mahāsaroja, śaṅku* (or *śankha*), *saritāṁpati, antya, madhya,* and *parārdha* ($= 10^{17}$).

Among the arithmetical rules discussed in the *Pāṭīgaṇita, Sūtra* 49–50 gives a formula for finding the time in which a sum lent out at simple interest will be paid back by equal monthly instalments. Under the *Sūtra* 63–64 he presents the famous problem of a Hundred Fowls, in which we have to solve the indeterminate equations in integers.

For mensurations related to a circle, Śrīdhara used the approximation $\pi = \sqrt{10}$, a very ancient Indian value. But there is some evidence to show that he used $\pi = \frac{22}{7}$ in the lost part of *Pāṭīgaṇita*.

Another of Śrīdhara's great achievements was his mensuration of the volume of a sphere.

R.C. Gupta

References

Asthana, Usha. *Ācārya Śrīdhara and his Triśatikā*. Ph.D. thesis, Lucknow University. 1960.

Datta, B. "On the Relation of Mahāvīra to Śrīdhara." *Isis* 17:25–33. 1932.

Dvivedi, Sudhakara, ed. *Triśatikā of Śrīdhara*. Benares: Chowkhamba Sanskrit Book Depot. 1899.

Ganitanand. "On the Date of Śrīdhara." *Gaṇita Bhāratī* 9: 54–56. 1987.

Pingree, David. "The Gaṇitapañcaviṁśī of Śrīdhara" (with text). In *Ludwik Sternbach Felicitation Volume*. Ed. J.P. Sinha. Lucknow: Akhila Bharatiya Sanskrit Parishad. 1979. pp. 887–909.

Ramanujacharia, N. and G.R. Kaye. "The Triśatikā of Śrīdharācārya" (with translation). *Bibliotheca Mathematica Series* 3, 13: 203–217. 1912–1913.

Shukla, K.S., ed. and trans. *The Pāṭīgaṇita of Śrīdharācārya*. Lucknow: Lucknow University. 1959.

Singh, Sabal, "Time of Śrīdharācārya." *Annals of the Bhandarkar Oriental Research Institute* 21: 267–272. 1950.

See also: Mahāvīra – Algebra – Mathematics – Bhāskara

Śrīpati

Śrīpati was the most prominent Indian mathematician of the eleventh century. He was the son of Nāgadeva and lived in Mahārāṣṭra. He flourished during the period from CE 1039–1056.

General appreciation for Śrīpati's fame as a mathematician is based on his arithmetic, *Gaṇitatilaka*, and the two mathematical chapters of his astronomical work entitled *Siddhāntaśekhara*. The thirteenth chapter, the *Vyktagaṇitādhyāya*, contains arithmetical rules, series, mensuration, and shadow reckoning. The fourteenth chapter, the *Avyaktagaṇitādhyāya*, is one of the few extant Hindu works on algebra. The only edition (by Kapadia) of the *Gaṇitatilaka* contains a valuable commentary by Siṁhatilaka (ca. CE 1275).

Besides including in his works selected rules of his predecessors, Śrīpati enriched Indian mathematics by giving some improved and some original rules. The *Gaṇitatilaka* contains the earliest known treatment of simple addition and subtraction in any Indian work. Also included is the earliest version of our angular method of addition used to check the accuracy of a sum and simplify addition. The *Vyktagaṇitādhyāya* gives an improved rule for the volume of an excavation.

The *Avyaktagaṇitādhyāya* contains most of Śrīpati's original ideas and rules. In it, he presents the idea of an extensive system of symbolism in algebra, the only Hindu treatment of the cubing of signed numbers, and explicit recognition of the nature of imaginary quantities that is second only to Mahāvīra's. He also provides the earliest versions of our ordinary method of solving simple linear equations by using inverse operations and the earliest formal treatment of factorisation. The rules not only give the ordinary method of factoring based on successive division, but also an additional method for factoring a non-square number by expressing it as the difference of two squares.

Śrīpati also displayed his mastery of indeterminate equations by giving an improved rule for solution of the factum and original rules for the solution of the pulveriser. He also described an original method to obtain rational solutions of the square-nature.

All Śrīpati's works are in verses. The *Siddhāntaśekhara* contains only rules, but the *Gaṇitatilaka* is written in an autocratic style of teaching strategy, in which a rule is followed by plentiful exercises. The book is quite secular in nature. According to the author's testimony, it was written for public use. It contains arithmetical and commercial rules, and some problems solvable by simple linear, quadratic, and radical equations. A garland problem and numerous other fanciful problems included in the book make it pleasurable and interesting. The practical and aesthetic values of the *Gaṇitatilaka* cannot be underestimated.

Śrīpati, however, also had some weaknesses. His mathematics of division by zero is all wrong. In the *Vyaktagaṇitādhyāya*, he uses the term *caturbūja* to mean a square, a quadrilateral in general, a cyclic quadrilateral, a quadrilateral with equal altitudes, and an isosceles trapezoid. Because of this inconsistency, his rules on mensuration of quadrilaterals are hard to interpret.

Kripanath Sinha

References

Kapadia, H.R. *Gaṇitatilaka by Śrīpati*. Baroda: Gackwad's Oriental Series No. 78. 1937.

Misra, Babuaji. *Siddhāntaśekhara by Śrīpati, Pt. II.* Calcutta: Calcutta University Press. 1947.

Sinha, Kripanath. "Śrīpati: An Eleventh-Century Indian Mathematician." *Historia Mathematica* 12: 25–44. 1985.

See also: Algebra – Mathematics – Mahāvīra

Śulbasūtras

The *Śulbasūtras* are manuals which prescribe the construction of different types of fire altars. Every householder was instructed, by the Vedic religion, to maintain a sacred fire, primarily for daily worship and offerings. These fires, called *gārhapatya* (domestic), *dakṣiṇa* (southern), and *āhavanīya* (oblatory), were to be maintained, without their ever going out, in altars of different designs, such as circular, semi-circular, square, rectangular, and triangular. Seasonal and special worships required altars with elaborate designs like that of an eagle with outstretched wings. The size, shape, direction, position, and the number and measure of the bricks used, and also the increase in the measure, for extraneous reasons, were all specified in the *Śulbasūtras*. These traditional practices are referred to in the *Ṛgveda*, the earliest of the Vedas, and elaborated in the *Yajurveda*. Later they came to be written down as manuals, supplementing the regular texts depicting sacrifices, and also in independent texts called *Śulbasūtras*. Adherents of different Vedas and Vedic schools had different *śulba* texts, named after the authors of these texts and pertaining to the Veda which they advocated. This is how the *śulba* texts named after Āpastamba, Bodhāyana, Kātyāyana, Mānava, Maitrāyaṇa, Varāha, Hiraṇyakeśin, Satyāṣaḍha, Vādhūla, and Laugākṣi, pertaining to one or the other of the schools of the *Yajurveda*, and *Maśaka Śulbasūtra* pertaining to the *Sāmaveda* came into being.

The *Śulbasūtras*, meaning pithy aphoristic statements (*sūtra*) for work with string (*śulba*), prescribes, by means of addition, subtraction, multiplication, division, and squaring, simple rules not only for the construction of circles, squares, rectangles, triangles, wheel-shapes, trapezia, and rhombi, but also for extending these figures by specific proportions as required in different sacrifices. It also specifies methods to reduce a circle to a square of equal area, and vice versa. A fine approximation of the value of the root of two is contained in the rule given in the *Baudhāyana-Śulbasūtra*: "Increase the measure of the side (of a square) by its third part, and the third part by its fourth part. The fourth part is decreased by its own thirty-fourth part. (The approximate diagonal will result)." The rule gives the approximation:

$$\text{Root } 2 = 1 + \tfrac{1}{3} + \tfrac{1}{3\times4} - \tfrac{1}{3\times34},$$

i.e. $\frac{577}{408}$, i.e. 1.4142157, which is very approximate to the correct value of the root of 2.

The *Śulbasūtras* specify or give geometrical constructions for the following: (1) to draw a straight line at a right angle to a given line; (2) to construct a square on a given side; (3) to construct a rectangle of given sides; (4) to construct an isosceles trapezium of given base, top and altitude; (5) to construct a square equal to the sum of two squares;

(6) to construct a square equal to the difference of two squares; (7) to construct a square equal to a given rectangle; (8) to construct a rectangle with a given side and equal in area to a given square: (9) to construct a square equal in area to a given isosceles triangle; (10) to construct a square equal in area to a rhombus; (11) to construct a square equal in area to a given circle; and (12) to construct a circle equal in area to a given square.

It would seem that the sacrificial hall of the Vedic Indians formed, as it were, the workshop and laboratory to formulate and develop their geometry.

K.V. Sarma

References

Amma, T.A. Sarasvati. *Geometry in Ancient and Medieval India*. Delhi: Motilal Banarsidass. 1979.

Datta, B. and A.N. Singh. "Hindu Geometry." *Indian Journal of History of Science* 18: 121–88. 1980.

Gupta, R.C. "Baudhāyana's Value of Root 2." *Mathematics Education* 6B: 77–79. 1972.

Gupta, R.C. "Vedic Mathematics from the *Śulbasūtras*." *Indian Journal of Mathematics Education* 9 (2): 1–10. 1989.

Sen, S.N. and A.K. Bag, Ed. and Trans. *The Śulbasūtras*. New Delhi: Indian National Science Academy. 1983.

van der Waerden, B.L. *Geometry and Algebra in Ancient Civilizations*. Berlin: Springer-Verlag. 1983.

See also: Geometry – Baudhāyana

Sūryasiddhānta

The *Sūryasiddhānta* is a complete work on Hindu astronomy and is more popular and widely studied in North India than in the South. In order to enhance its prestige and antiquity, it is stated in the text itself (1.29) that it had been communicated by a representative of the God Sun to Maya, several thousand years ago. However, both internal and external evidence show that the work was composed between CE 600 and 1000. In about five hundred verses, distributed through fourteen chapters, the work deals with all aspects of Hindu astronomy, and also cosmology, geography, astronomical instruments, and time reckoning. It follows the midnight day reckoning.

The contents of the work are comprehensive. Chapter I speaks about the circumstances that led to the composition of the work, the aeons and aeonary revolutions of the planets, the principles underlying the computation of the planets and the nodes, the position of the planets at the beginning of the current aeon, the Prime meridian and local time, and the inclination of the orbits of the planets. The time units given in the work are more in conformity with the *Purāṇas* than with other texts on Hindu astronomy. Chapter II deals with the computation of the true motion and the true longitudes of the planets. Chapter III is devoted to the determination of the directions, place and time, and also the calculation of the precession of the equinoxes. It is noteworthy that the *Sūryasiddhānta* is the earliest available Indian text which contains a discussion of the calculation of the precession of the equinoxes. Chapters IV and V deal in detail with lunar and solar eclipses and their computation. Chapters VII and VIII are concerned with the conjunction of one planet with another, and a planet with a star, including their observation. Chapter IX deals with the determination of the heliacal rising of the planets and Chapter X with the phases of the moon. Chapter XI is concerned with the phenomenon of *pāta*. Cosmology and geography of the worlds and of the Earth occur in Chapter XII. The construction of the armillary sphere and its working, and a mention of the main astronomical instruments form the subject matter of Chapter XIII. Chapter XIV enumerates and defines nine types of reckoning time, such as lunar, solar, sidereal, tropical, etc.

The *Sūryasiddhānta* is indebted to earlier astronomers like Āryabhaṭa (b. 476) and Brahmagupta (b. 598) in certain matters like the inclination of the planetary orbits, the tabular Sines, etc. It is also to be noted that the *Sūryasiddhānta* does not include chapters on arithmetic, algebra, or astronomical problems, which are generally included in texts of this type.

The popularity of the *Sūryasiddhānta* is clear from the very large number of manuscripts of the work found throughout the land. About 35 commentaries on the work, written by scholars from different regions of India, and about 20 works, including planetary tables and manuals, based in the *Sūryasiddhānta*, have been identified. The popularity of the work is also reflected by the number of almanacs prepared on the basis of the *Sūryasiddhānta*.

K.V. Sarma

References

Dikshit, S.B. *Bharatiya Jyotish Sastra. (History of Indian Astronomy)*. Pt. II. Trans. R. V. Vaidya. Calcutta: Positional Astronomy Centre, India Meteorological Department. 1981.

Pingree, David. *Jyotiḥśāstra: Astral and Mathematical Literature*. Wiesbaden: Otto Harrassowitz. 1981.

The Sūryasiddhānta: A Text-book of Hindu Astronomy. Trans. with notes by Ebenezer Burgess. Delhi: Motilal Banarsidass. 1989.

The Sūryasiddhānta with the Commentary of Parameśvara. Ed. Kripa Shankar Shukla. Lucknow: Department of Mathematics and Astronomy, Lucknow University. 1957.

See also: Astronomy – Precession of the Equinoxes – Armillary Spheres – Astronomical Instruments – Āryabhaṭa – Brahmagupta

Suśruta

Suśruta is the author of one of the three major Ayurvedic works, the *Suśrutasaṃhitā*. He was the son of Viśvāmitra and a student of Divodāsa Dhanvantarī, the King of Kāśi (now Varanasi) who flourished prior to 1000 BCE.

It was the practice in ancient times to engage military physicians who accompanied the kings and army commanders to the battlefield to treat wounded soldiers. Therefore, there was a need to train surgeons. One of the eight specialised branches of Āyurveda is surgery (*śalya-tantra*), and Suśruta's work stands foremost in this field.

The *Suśrutasaṃhitā* consists of 186 chapters which are divided into six sections as follows:

1. forty-six chapters in the *Sūtra* section dealing with the principles and practices of surgery, and a description of food and drinks;
2. sixteen chapters in the *Nidāna* section dealing with the diagnosis of important surgical ailments;
3. ten chapters in the *Śārīra* section dealing with the creation of the Universe, embryology, obstetrics and anatomy;
4. forty chapters in the *Cikistā* section dealing with the treatment of surgical ailments;
5. eight chapters in the *Kalpa* section dealing with toxicology;
6. sixty-six chapters in the *Uttara* section dealing with the diagnosis and treatment of the diseases of the eye, ear, nose, and throat, paediatrics, seizures by evil spirits, and internal diseases in general.

Though the primary object of this work is to deal with surgery, it can be seen from the above that all eight specialised branches of Āyurveda are covered in it. According to Suśruta, any one who wants to be highly skilled in surgery should be thoroughly versed in anatomy by dissecting a dead body. He describes the selection of dead bodies, their preservation, and different methods of examining various parts of the body. He describes different types of blunt and sharp instruments, sixty different steps to be followed for healing wounds, plastic surgery, and operative procedures for the removal of stones in the urinary tract. According to Suśruta, there can be no more virtuous act for a physician or surgeon than to alleviate human suffering.

Bhagwan Dash

REFERENCES

Bhishagratna, Kunjālal. *An English Translation of the Suśrutasaṃhitā*. Varanasi: Chowkhamba Sanskrit Series Office. 1991.

Kutumbíah, P. *Ancient Indian Medicine*. Bombay: Orient Longmans. 1962.

Mukhopadhyaya, Girindranath. *History of Indian Medicine*. New Delhi: Oriental Books Reprint Corporation. 1974.

Sankaran, P.S., and P.J. Deshpande. "Suśruta." In *Cultural Leaders of India: Scientists*. Ed. V. Raghavan. New Delhi: Publication Division, Ministry of Information and Broadcasting, Government of India. 1976. pp. 44–72.

Śarmā, Priyavrata. *Indian Medicine in Classical Age*. Varanasi: Chowkhamba Sanskrit Series Office. 1972.

Suśruta, *Suśrutasaṃhitā*. Varanasi: Chaukamba Orientalia. 1980.

See also: Medical Ethics – Medicine: *Āyurveda*

T

Technology and Culture

General accounts of the history of science and technology (or, more narrowly, of inventions) are scarce. The few that are available are also of fairly recent origin: obviously, the idea of a history of science (where science has been identified with Galilean science) and technology (identified with industrial technology) could not have appeared much earlier than this century. Not many people even know that the word "scientist" was first used by William Whewell in 1833.

Also, most available histories have remained the work of western scholars. This has not been an entirely happy circumstance. On the contrary, it has afflicted these histories with certain methodological and other infirmities which have had the effect of reducing them to mythological works. This is especially so when they are studied with regard to aspects of the history of science, technology, and medicine in the non-Western world.

One of the first is a history of technology and engineering written by Dutch historian R.J. Forbes. Forbes's work appeared in 1950 under the title *Man the Maker*. In it, he conceded that technology was the work of humankind as a whole, and that "no part of the world can claim to be more innately gifted than any other part". A few years thereafter, Forbes produced his rich and prodigiously detailed *Studies in Ancient Technology* which set out a remarkable description of the different technologies of Asia, Africa, pre-Colombian America, and Europe. However, it is in *The Conquest of Nature* that his Eurocentric assumptions came to the fore: in that work (as the title itself indicates), Forbes went on to subsume the technological experience of people from diverse cultures under a philosophical anthropology that was unmistakably Western, if not Biblical – the domination of nature myth originating in Genesis. And after a discussion about the grievous consequences of a seriously flawed modern technology, he ended his book promising redemption from the technological genie through the Christian event of Easter. How does one prescribe a text like this to Hindu, Chinese, or Arab readers?

Another influential work of about the same period is *A History of Western Technology* by a German scholar, Friedrich Klemm. In it, Klemm provided a picture of technological development in the West in which non-Western ideas and inventions had no hand at all. The English translation which appeared in 1959 barely mentions Joseph Needham's work on China in the bibliography. Klemm could not have substantiated his interpretation of Western technological development unless he consciously played down non-western technology. In fact, the only quote on Chinese technology in Klemm's book is from the *Guan yin zi*, the work of a Daoist mystic of the eighth century CE: Klemm used it to prove why the alleged religiously collared oriental rejection of the world in China could not have provided a stimulus for the emergence of science and technology in that country.

This distorting Eurocentric perspective continued to hold sway even over the more standard (five volume) *A History of Technology* edited by Charles Singer, E.J. Holmyard and A.R. Hall. The first volume of this work appeared in the same year as Joseph Needham's *Science and Civilisation in China*, and the editors themselves acknowledged that up to the period of the Middle Ages in Europe, China had the most sophisticated fund of technological expertise. Three of the Singer volumes dealt with pre-industrial technology, where logically China (and India) should have been given major space and Western technological development would have appeared in proper perspective in the nature of an appendix. However, Chinese, Indian, and other technologies were ignored and Western technology made the focus of the exercise.

In addition to manifesting such ignorance of non-Western science and technology, these histories suffered from another methodological limitation: they restricted themselves to a record of artefacts and machines disembodied from the latter's social and cultural contexts. The problem was eventually recognised by some Western scholars themselves.

These histories are evidence that the western scholars associated with them proved incapable of stepping out of their cultural cages, either knowingly or involuntarily. Either way, this eroded the credibility of their work as it exhibited both their lack of objectivity and their general incompetence when called upon to deal with societies other than theirs.

They show that our dominant descriptive and evaluatory ideas of technology and culture both in the Western and non-Western world have been formulated over the past couple of centuries with reference to the West's experience of these phenomena. Concepts and categories reflected from a limited area of human experience have been indiscriminately used to explain and assess the rest of the world.

New frameworks are therefore inevitable. We are in the post-traditional, post-colonial, post-modern age. But unless the outmoded intellectual environment that engendered this subjective and tunnel-visioned output is rigorously dissected, analysed, and then jettisoned, the new frameworks needed for the alternative histories and encyclopaedias intending to take their place are in danger of turning into copies of the old.

There are two preliminary aspects of this intellectual mal-development that need elucidation. First, there is the perception of humankind as *homo faber*, a tool-making animal, which is basically a reflection of fairly recent Western experience with the machine. Fascinated by the bewildering profusion of tools and machines, Western historians began to look at the ability to produce these as a special field with its own history and set out to create a distinct species of man in the image of *homo faber*. This scholarly creation had its repercussions in encouraging the overestimation of the singularity or uniqueness of Western culture in comparison with others (although all cultures are unique and incommensurate). The elaborate, embarrassing exercise in culture-narcissism soon became routine since it was not to be challenged for nearly a century. (It is important to point out that the *homo faber* idea is quite recent to humankind: it is consistently absent in not just other cultures, but even within a large part of the West itself).

For instance, it was taken for granted that the system of production that got generated in the last century and a half in the West was the only one with any significance simply because in the light of the present — and to all appearances—it had apparently emerged as the dominant one. Therefore, its past was the only one worth considering. This notion was in turn bolstered by another: a self-generating model of technological development in which the historian attempted to trace the evolution of modern science and technology by working backwards to the experiences and ways of thinking characteristic of Mediterranean antiquity. Thus the roots of modern technology were shown to be exclusively founded on the work of Greek and Roman thinkers, mathematicians, engineers and observers of nature with no input from any other culture areas or people.

This brings us to the second aspect we have alluded to above, and this concerns the relationship between knowledge and power and the impact of this on interpretations of technology and culture. Throughout history, knowledge has generally remained closely linked with interests. Even when encyclopaedias, for instance, have traditionally sold themselves on the Francis Bacon principle that "knowledge is power", they too have continued to reflect an undeclared, equally influential, political principle — that "power is knowledge".

The intrusion of Europeans into non-European societies and the gradual establishing of political dominance and inequality between societies stimulated the inauguration of a new discourse about such societies. Political dominance came to be as routinely and unabashedly expressed in the form of knowledge as it was through the barrel of guns. Edward Said has already written on the invention of the discourse on "orientalism" and its direct political uses. But there are less controversial discourses that have had even larger repercussions only now being acknowledged. As a result, much academic knowledge in the Western world about the non-Western world, particularly the latter's technology traditions, remains not only distorted or contaminated by the ethnic concerns, goals, theories, obsessions, and peculiar assumptions of Western scholars and universities; it is still largely defined, legitimised, and decided by them irrespective of whether there is any concurrence from the non-Western world.

The combination of these two aspects proved deadly: the emerging conception of Western man alone as *homo faber*, once it took firm roots within the situation of political dominance, rendered any appreciation of technique elsewhere—technique not necessarily reflected only in tools or machines — difficult and often impossible. In fact, the combination helped inaugurate its very own dark age. For it generated among Western (and not a few non-Western) scholars several major assumptions concerning technology and culture. We shall discuss three of these.

The first emerged in relation to Western man's attitude towards the past, particularly with regard to pre-industrial technology. *Homo faber* exercised his new found power over the past by deriding it: this is reflected in the rewriting of history from today's perspective in which the past is seen as mere prelude to the present. Earlier technological innovations are considered primitive precursors of later developments. Here we have a good example of the parochialism of the modern/Western mind as it proceeds to take experiences of technology and culture exclusive not just to the late twentieth century but to extremely small segments of the world population and makes these the basis for investigating, analysing, assessing, and judging the general activities of human societies over hundreds of years. This was the case even when such societies were not so technologically enamoured, dependent, or controlled as some of them seem to be now.

The second assumption relates to humankind's so-called unique propensities for technology when compared with that of the animal world, an uncritical theory best summed up in a single word: speciesism. After deciding on the issue of the comparative technological competence of all living species in its own favour, the West came to the conclusion that the rest of creation, because inferior, was expendable if so required to further its own scheme of things.

But it is the third assumption that concerns us most seriously here: it is the idea that Western man can be equally distinguished from non-Western societies as well on the ground that the latter, like the animal and other "lower" species, also lacked technological development as it emerged in the West.

This idea was appropriately reflected in academia in the emergence of two new sciences: the discipline of sociology, which focused on so-called advanced societies and their flair for technology; and the subject of anthropology which occupied itself with non-Western cultures, limited to primitive or pre-industrial tools. Anthropology's political origins have been blandly asserted by Claude Levi-Strauss in his controversial Smithsonian lecture:

> Anthropology is not a dispassionate science like astronomy, which springs from the contemplation of things at a distance. It is the outcome of a historical process which has made the larger part of mankind subservient to the other, and during which millions of innocent human beings have had their resources plundered and their institutions and beliefs destroyed, whilst they themselves were ruthlessly killed, thrown into bondage, and contaminated by diseases they were unable to resist. Anthropology is daughter to this era of violence: its capacity to assess more objectively the facts pertaining to the human condition reflects, on the epistemological level, a state of affairs in which one part of mankind treated the other as an object.

It is within such an imperialist context that the histories and technological experience of non-Western societies could be written off or ignored: the latter, after all, were conquered peoples. When technology is seen through an anthropological prism, the emerging picture is bound to be far removed in character from a scenario that emerges from a sociological perspective. What is more, it is bound to be even more far removed from reality itself.

Some impression of that reality is discernible in the period before political dominance began to corrupt the objectivity of knowledge. Before the so-called "voyages of discovery", though non-Europeans were conceived as fantastic, wild, opulent, even monstrous, they were rarely considered inferior or backward; and even the actual European encounter with the scientific, technologic and medical traditions of non-Western societies was different from what eventually became the stuff of politically directed myths. In fact, from the day that the Portuguese mariner Vasco da Gama landed in India until almost three centuries later, Asia had a larger and more powerful impact on Europe than is normally recognised. Donald Lach has appropriately titled the first volume in his *Asia in the Making of Europe* "The Century of Wonder". It was not without reason that an Englishman of the time addressed the Indian Emperor by describing himself as "the smallest particle of sand, John Russel, President of the East India Company with his forehead at command rubbed on the ground". Nor can we forget that the first presents offered by da Gama to the King of Calicut included some striped cloth, hats, strings of coral beads, wash basins and jars of oil and honey. The king's officers naturally found them laughable.

It would take a few more decades before the Europeans landing in the Indian subcontinent would notice anything beyond gold and spices. But by 1720 and for a period of up to a hundred years, a new category of observers came visiting, some from newly formed learned societies in England. Their detailed reports were a result of the European quest for useful knowledge in different fields.

In his pioneering volume, *Indian Science and Technology in the Eighteenth Century*, the Indian historian Dharampal includes several accounts from these observers which describe among others the Indian techniques of inoculation against smallpox and plastic surgery. (While the first was eventually banned by the English, the latter was learnt, adopted, and developed). The accounts also document Indian processes like the making of ice, mortar, and waterproofing for the bottoms of ships; water mills, agricultural implements like the drill plough, water harvesting and irrigation works, and the manufacture of iron and of a special steel called *wootz*.

More techniques (like those involved, for instance, in the manufacture of Indian textiles) are described in *DeColonizing History* (Alvares, 1991) and *Science and Technology in Indian Culture* (Rahman, 1984). But even this documentation, impressive as it is, is now recognised to be but the tip of the proverbial iceberg.

The Chinese, like the Indians at Calicut, had a similar experience with an embassy and its gifts from London. The edict of Qian Long to the embassy is worth quoting: "There is nothing we lack, as your principal envoy and others have themselves observed. We have never set much store on strange or ingenious objects, nor do we need any of your country's manufactures" (Fairbank, 1971).

Immediately after the encounter, the graph of European reaction rises with esteem and wonder; and then, as political conquest and overlordship increase, the graph alters course and begins to record increasing denigration. A remarkable transformation of image thus takes place as the political relationship between Europe and non-European societies changes to the advantage of the former, rendering the Europeanisation of the world picture almost an act of divine will.

By 1850, political dominance over the non-Western world was clearly installing distorted ideas not only about that part of the world but rebounding to distort Western man's image of himself as well. Already by 1835, for instance, the British had acquired a flattering notion of their own civilisation (Victorian England was seen to be at the top of the pyramid of civilisation) and a thorough-going contempt for Asia.

This contempt finds expression in the famous Minute of Lord Babington Macaulay:

> I have never found one amongst them (the orientalists) who could deny that a single shelf of a good European library was worth the whole native literature of India and Arabia ... It is, I believe, no exaggeration to say that all the historical information which has been collected from all the books written in the Sanskrit language is less valuable than what may be found in the most paltry abridgment used at preparatory schools in England. In every branch of physical or moral philosophy the relative position of the two nations is nearly the same.

Dharampal has produced an interesting record of these assessments of science and technology in India among Western observers as the relationship between India and Britain changed to Britain's advantage.

Regarding the question of Indian astronomy, he discusses the case of Prof. John Playfair, Professor of Mathematics at the University of Edinburgh and an academician of distinction. Playfair studied the accumulated European information then available on Indian astronomy and arrived at the conclusion that the Indian astronomical observations pertaining to the period 3102 years BCE appeared to be correct in every text. This accuracy could only have been achieved either through complex astronomical calculations by the Indians or by direct observation in the year 3102 BCE. Playfair chose the latter. Opting for the former would have meant admitting that "there had arisen a Newton among Brahmins to discover that universal principle which connects, not only the most distant regions of space, but the most remote periods of duration, and a De La Grange, to trace, through the immensity of both its most subtle and complicated operations."

Similar attitudes prevailed concerning the knowledge of how Indians produced *wootz*. J.M. Heath, founder of the Indian Iron and Steel Company and later prominently connected with the development of the steel industry in Sheffield, wrote "... iron is converted into case steel by the natives of India, in two hours and a half, with an application of heat that in this country, would be considered quite inadequate to produce such an effect; while at Sheffield it requires at least four hours to melt blistered steel in wind-furnaces of the best construction, although the crucibles in which the steel is melted, are at a white heat when the metal is put into them, and in the Indian process, the crucibles are put into the furnace quite cold".

However, Health would not admit that the Indian practice was based on knowledge "of the theory of operations", simply because "the theory of it can only be explained by the lights of modern chemistry".

By the beginning of this century, the Western mind had already convinced itself that Western science and philosophy were the only approach to metaphysical truth ever attained by the human species and that the Christian religion provided wisdom and insight incumbent on all people everywhere to believe.

The result is reflected in the output of academia: a "history of art" turned out to be nothing but a history of European art and a "history of ethics" a history of Western ethics. While European music was music, everything else remained mere anthropology. The contemporary evaluation of human activity in the West as compared with the non-Western world was unabashedly provided by the late Jacob Bronowski in the Ascent of Man in words almost echoing Macaulay in 1837:

> We have to understand that the world can only be grasped by action, not by contemplation. The hand is more important than the eye. We are not one of those resigned, contemplative civilisations of the Far East or the Middle Ages, that believed that the world has only to be seen and thought about and who practiced no science in the form that is characteristic for us. We are active; and indeed we know, as something more than a symbolic accident in the evolution of man, that it is the hand that drives the subsequent evolution of the brain. We find tools today made by man before he became man. Benjamin Franklin in 1778 called man a 'tool-making animal' and that is right.

Now, there were obviously perverse consequences of such a view: scholars in several non-western societies, schooled in an educational system imposed on their societies through the colonial establishment, readily incorporated similar ideas about their own histories. In an article in *Nature* 35 years ago, Joseph Needham had to chide a native scholar of Thailand for claiming that his own people had not made any contribution to science despite compelling evidence to the contrary. Nevertheless, the colonisation project succeeded in convincing many non-European intellectuals and scholars that only the West was active. They facilely accepted the idea that activity *per se* was desirable compared to judicious or necessary activity; that only the West was capable of thinking in the abstract sense. If this opinion were carried to its logical conclusion, it would appear that if the rest of humankind had survived for hundreds of years, this must be due to some form of manna falling providentially from the heavens.

The damage done by these years of extremely ideological scholarship and a ruinous ethnocentrism to the history of technology was bad enough. Predictably, the impression of an empty technological wilderness invented by Western scholarship about non-Western societies had a parallel, simultaneous, destructive impact on the assessment of their cultures as well. So insidious was the nature of this outrageous assumption regarding Western and non-Western abilities, that even Joseph Needham, Mark Elwin, Abdur Rahman and a host of other scholars participated in pointless debates which often took it for granted. One major debate, for example, focused on why China (and India) did not produce either modern science or an industrial revolution on the European pattern, especially since Chinese technology had already reached a level of sophistication not yet attained in any other part of the world as late as the fifteenth century.

Attempted answers compared and contrasted the internal conditions within Chinese society with those within Europe; the argument eventually succeeded in establishing the conclusion that no scientific or industrial revolution occurred in China because the social conditions in China were not the same as those within Europe. Thereafter, a host of cultural and social factors were dragged out of context and labelled probable "obstacles" either to the development of technology or modern science.

A critique of the three assumptions we have surveyed above therefore becomes compelling and inevitable, if we are to eschew their myriad fallacies in future. We shall take each in turn.

The idea that the past was merely a prelude, and a primitive one at that, may come naturally to anyone who has begun to feel that the present era of technical change is inevitable. Yet future societies may assess their past (our today) basing themselves on values other than those celebrating mere technical change. Already mindless technical change and built-in technological obsolescence have been assaulted by several global thinkers on the ground of ecological unsustainability and resource scarcity. It would also be wrong to think that because man did not have technology as he now does, he was necessarily impoverished. If there is anything the past gives us it is this positive impression of survival in all kinds of environmental scenarios. There is also evidence of more widely dispersed creativity when man was not submerged by technology than there is today. In many areas of human experience, we are yet to match even the technological achievements of the past which were driven by values other than mere complexity for its own sake or profit.

A similar argument may be used against the assumption that humankind is the only tool-making species there is. Several naturalists and ethologists have documented the diversity of nature's schemes at fabrication; most notably, Felix Paturi in his *Nature, Mother of Invention* and Karl von Frisch in *Animal Architecture*. Scholars like Lewis Mumford have gone further in stating quite bluntly that in their expression of certain technical abilities other species have for long been more knowledgeable than man.

> Insects, birds and animals, for example, have made far more radical innovations in the fabrication of containers, with their intricate nests and bowers, their geometric beehives, their urbanoid anthills, and termitaries, their beaver lodges, than man's ancestors had achieved in the making of tools until the emergence of homo sapiens. In short, if technical proficiency alone were sufficient to identify and foster intelligence, man was for long a dullard, compared with many other species.

Niko Tinbergen, another ethologist, after years of close observation of other species, has come to the following conclusion: "It was said that 1. animals cannot learn; 2. animals cannot conceptualise; 3. cannot plan ahead; 4. cannot use, much less make tools; 5. it was said they have no language; 6. they cannot count; 7. they lack artistic sense; 8. they lack all ethical sense." All of these statements, says Tinbergen, are untrue.

It cannot be said therefore that, in contrast with other species, humankind alone is a tool-maker. Thus the attempt to distinguish man from other living species because of his tool-making capacity is now seen to be a result of limited knowledge and unwarranted assumption of qualitative discontinuities between human beings and other species. It will also be useful to recall here that the ability to fabricate and organise is not a singular human trait – it is an intrinsic feature of nature since nature can exist only in a given form, whether at its most primary constituents at the sub-atomic level or even at the level of crystalline structures or the multiple tiers of a primary forest.

However, it is the third assumption – concerning the West's genius for technology and the rest of the world's incompetence in the same department – that contains the greatest mythological component of them all.

As we shall presently see, such an assumption has not only no historical basis, it is in fact contrary to historical and even to contemporary evidence. As for the gift of Greek rationality, suffice it to say that for two thousand years it gave no technological advantage to those who had it over those who did not. On the contrary, major scientific concepts, technological artefacts, tools, and instruments emerged in cultures that had nothing to do with either Greece or Rome.

The other problem with this assumption is it cannot even cope with the long established view that the science and technology traditions of most societies, particularly so of the West, are in significant ways mixed traditions. Even the little that we know about it indicates that the cross-cultural borrowing of techniques and technology is impressive. Thus a very large number of critical inventions from both India and China helped fill significant gaps in the technological development of the West. A simple example from Francis Bacon's work will suffice to illustrate this point. He wrote:

> It is well to observe the force and virtue and consequences of discoveries. These are to be seen nowhere more conspicuously than in those three which were unknown to the ancients, and of which the origin, though recent, is obscure and inglorious; namely, printing, gunpowder, and the magnet. For these three have changed the whole face and state of things throughout the world, the first in literature, the second in warfare, and the third in navigation; whence have followed innumerable changes; inasmuch that no empire, no sect, no star, seems to have exerted greater power and influence in human affairs than these mechanical discoveries.

Now all these three mechanical discoveries were Chinese. Yet here again Western scholars have found it hard to acknowledge their origin. Borrowing of techniques from India is easily documented as well.

The documentation of technology in other cultures is only beginning. For example, it was only in 1974 that Sang Woon Jeon's *Science and Technology in Korea* appeared. There is as yet no major record of technology in Africa or South America though there is now available a large volume of documented evidence that both areas were rich in tools and techniques, from metallurgy to textiles.

In India, the other large storehouse of useful and appropriate tools (some still in productive use), the most extensive documentation of technology has only recently commenced, sparked in part by Dharampal's *Indian Science and Technology in the Eighteenth Century* and the work of scholars like Abdur Rahman.

The immediate impact of these re-invigorated investigations, stimulated largely by political independence, is a fresh debate over the issue of technology and culture: the old assumption of one technology and one culture in which others are seen to make a few, presumably inconsequential, contributions, is in tatters. Whatever its own pretensions to be the only viable culture, the West is finally being seen by non-Western societies as only one among several: a certain balance between cultures gets restored even though economic inequality persists. In fact, in some cases the pendulum has swung to the other side with cultures unabashedly resuming their traditions. There has naturally been a reverberation in the climate of ideas.

Changes in perception of this kind have already come about in other academic disciplines. To cite just one example, world histories were once written as if Europe were the centre of the planet, if not the cosmos. There has been progress since: Geoffrey Barraclough and Leon Stavrianos, for instance, have both succeeded in producing comprehensive histories which avoid the older Eurocentric perspectives.

But even assuming we are able to produce, culture by culture, a fairly objective and comprehensive record of science, technology, and medicine, we would still be uncomfortably close to the pet obsession and perception of the present epoch. If there is anything the recent past has shown us, it is that we can be all too zealous judges in our own cause. We continue to celebrate uncritically our technological feats even when we know that the principal criterion of success for any species (and the human species is no exception) is primarily its ability to survive.

Therefore, it may be best not to get trapped in the debates on what is basically a sub-history: the history of slave-machines or automation or the recent machine-propensity of some cultures. After all, the *homo faber* concept is itself a distorted reflection of the natural endowments of the human species, an example of reductionist thinking. We know now that reductionism readily distorts knowledge, often pauperises it, but rarely enhances it.

What is required in the circumstances then is a paradigm shift. I would like to suggest this can be achieved by replacing the heavily loaded term "technology" (too close identified with externalised objects) with the more neutral word, "technique". Technique has a larger ambit than technology and does not necessarily express itself only in the form of tools or artefacts. For the moment, we may define it briefly as every culture's distinct means of achieving its purposes. The natural propensity of human beings is to rely on technique, not technology, for while it has been proven that we can survive without technology, we cannot survive without technique.

Thus there can be no technique without culture, no culture without technique. An investigation into a culture's techniques is bound to be considerably more difficult than the recording of a culture's artefacts. The important gain here would be that we would begin with a more democratic assumption: that there is no culture without a system of techniques. Such a postulate would inoculate us effectively against methodological, ethnocentric, and other fatal flaws the *homo faber* concept was both parent and heir to. It would nip in the bud any undesirable future forays into cultural narcissism or ethnocentric discourse.

If this is indeed so, the more logical assumption would be that every culture has relied on a corresponding system of techniques that has guaranteed survival. Understood in this way, it makes far better sense to talk of Western technique (even though today largely expressed in the form of technology), or African, Indian, Chinese, Maya, or Arabic techniques of survival (in which technology may not be given that importance for fairly valid reasons). But even a relatively low importance given to technology could never mean a poverty of technique. The idea that the human species is technique-natured could be empirically falsified if a human society could be found that lacked technique—and not just machines or artefacts.

Technique, then, is nothing but the permanent but dynamic expression of an individual culture. Cultures can only express themselves or survive through technique: the alternative is chaos. Non-human species may be guided in the exercise of technique by inflexible inborn patterns of behaviour. But the human world is as rigorously bound by the controls imposed by the symbolic universe that emerged as a substitute for weakened instinctual patterns. Myth, for instance, is technique. Interaction with (or manipulation of) nature may take place either through myth-making, scientific construction, and myriad other ways. All are expressions of the symbolic universe human beings inherit because they are human beings.

Thus we share the necessity of functioning through technique — not just through tools — with other living species — from the mammoth geobiological processes of Gaia to the cross-pollination of the rice plant to species of bird and animal, some of which, like the bower bird, are more prone to technology than others. Thus the so-called potter-wasp is known for its technique in constructing what we human beings culturally recognise as pots: however, the small vessels are the conclusion of technique: without it, there would be no "pot", no propagation and, therefore, no survival.

This will also explain why so-called primitive societies are often more complex in their social-cultural arrangements — their rich fund of botanical knowledge, slash-and-burn techniques, elaborate myths are as much an expression of technique – than modern societies.

Our new paradigm – based on a thorough-going analysis of technique—will enable us to concentrate more effectively on those aspects of human experience in non-western societies where there may be appropriate development of technology (as in India and China) but a superabundance yet of technique. A large number of these, particularly in India or in the Islamic cultures, may be located squarely within the domain of the sacred. They would be unintelligible outside such a framework of understanding.

We shall also observe in such societies that even where there is sophisticated technology, it retains an unobtrusive (not invasive) character. This can be seen from the merged outlines of Arab architecture to the irrigation works of South India or Sri Lanka.

An encyclopaedia of non-Western science, technology, and medicine may restrict its scheme to a bare description of the evolution of machines or artefacts incompetently covered by earlier conventional Western works, but it must do so guided by the background of the larger canvas of technique. Here the scholar will eventually examine theories of language in the same detailed manner as he would the culture and preparation of food or the control of breath – all extremely detailed sciences in India and China; he would examine irrigation and animal husbandry techniques, the domestication of cultivars of crop plants, record the elaborate knowledge of plants and of the human body, and seek to understand theories of cosmic phenomena and of the behaviour of annual events like the monsoons.

The aim of the historian is to describe the nature of this individual system and not place it within a hierarchical ordering of societies. His task is to document this immense richness, not endeavour to swamp and drive it into oblivion on the questionable assumption that Western technology is the only direction that human technique will take. The growing anxiety over Western technology is closely associated with the threat it is perceived to pose to the fate of the planet and to survival. We may have to examine its history clinically to diagnose why it has generated the kind of problems it poses for humankind. Here, only a proper study of technique and culture within non-Western societies (and not as Forbes hoped, the event of Easter) will bring some balance and provide urgent clues to the origins of what Jamal-ud-din described as the illness of occidentosis, the plague of the West.

Claude Alvares

REFERENCES

Alvares, Claude. *De-Colonizing History: Technology and Culture in India, China and the West: 1492 to the Present Day*. Goa: The Other India Press. 1991.

Barraclough, G. *An Introduction to Contemporary History*. London: Penguin. 1967–1974.

Dharampal. *Indian Science and Technology in the Eighteenth Century*. Delhi: Impex India. 1971.

Goody, J. *Technology, Tradition and the State in Africa*. Cambridge: Cambridge University Press. 1980.

Jeon Sang-woon. *Science and Technology in Korea*. Cambridge, Massachusetts: MIT Press. 1974.

Lach, Donald F and E. Van Kley. *Asia in the Making of Europe*. Chicago: University of Chicago Press. 1965.

Levi-Strauss, Claude. "The Scope of Anthropology." *Current Anthropology* 7(2): 112–123. 1966.

Mumford, Lewis. *The Myth of the Machine*. 2 vols. New York: Harcourt, Brace and World. 1967–1970.

Nasr, S.H. *Science and Civilization in Islam*. Cambridge, Massachusetts: Harvard University Press. 1968.

Nasr, S.H. *Islamic Science: an Illustrated Study*. London World of Islam Festival Publishing Co. 1976.

Needham, J. et al. *Science and Civilization in China*. Cambridge: Cambridge University Press. 1954.

Rahman, A., ed. *Science and Technology in Indian Culture – a Historical Perspective*. Delhi: NISTADS. 1984.

Said, Edward. *Orientalism*. London: Routledge & Kegan Paul. 1978.

Sardar, Z., ed. *The Touch of Midas: Science, Values and Environment in Islam and the West*. Manchester: Manchester University Press. 1983.

Stavrianos, L.S. *Global Rift*. New York: Morow. 1981.

Teng, Ssu-Yu. *China's Response to the West: A Documentary Survey, 1939–1923*. Cambridge. Massachusetts: Harvard University Press. 1954.

Tinbergen, N. "The Cultural Ape." TLS: *Times Literary Supplement*, 28 February 1975. p. 217.

See also: East and West – Colonialism and Science – Environment and Nature – Textiles – Metallurgy – Ethnobotany – Irrigation – Agriculture

Textiles

India may be described as one of the ancient centres of the cotton textile industry, since early evidence of cloth has been found in prehistoric archaeological sites. The spinning and weaving of cloth was very much a part of everyday life in ancient India. The loom is used as poetic imagery in several ancient texts. The *Atharvaveda* says that day and night spread light and darkness over the Earth as the weavers throw a shuttle over the loom. The Hindu God Vishnu is called *Tantuvardhan* or "weaver" because he is said to have woven the rays of the Sun into a garment for himself.

It is interesting to note that in the third or second century CE, when the cotton industry in India was in a flourishing state, in Europe cotton was still virtually unknown. The Greek scholar Herodotus thought that cotton was a kind of animal hair like sheep's hair. At the beginning of the Christian era, Indian textiles figure prominently in the trade with Rome. Arrian, the Roman historian, testifies to the export of dyed cloth from Masulia (Masulipatnam on the Coromandel coast), Poduca (Pondicherry), Argaru (Uraiyūr in Tanjavūr district, Tamil Nāḍu) and other places in south India. Legend has it that Indian cloth was purchased in Rome for its weight in gold. The quality of Indian dyeing too was proverbial in the ancient world, and in St. Jerome's Bible, Job says that wisdom is more enduring than the dyed colours of India. Indian textiles even passed into Roman vocabulary as is seen by the use as early as 200 CE of a Latin word for cotton, *Carbasina*, derived from the Sanskrit *kārpasa*.

The history of Indian textiles constitutes one of the most fascinating and at the same time tragic chapters in Indian history. In the sixteenth century the Portuguese first set foot on Indian shores and were followed in quick succession by the Dutch, English, and French. For the next hundred years "Indian cotton was king" and Europe was in the grip of what economic historians describe as "the calico craze". Indian textiles were used in the Middle East, Africa, and Europe not merely as dress material but also as coverlets, bedspreads, and wall hangings. The joint English sovereigns, William and Mary are described as having landed in England in 1689, resplendent in Indian calico. Daniel Defoe, the author of *Robinson Crusoe*, commented that Indian calico, which at one time was thought fit to be used only as doormats, was now being used to adorn royalty.

However, there was a dramatic reversal of fortune in the eighteenth century. The cotton revolution in England rendered redundant the products of Indian handlooms. The British crown imposed the first ban on Indian textiles in 1700 and repeatedly after that. By the end of the century, instead of Indian cloth being exported abroad, the machine-made cloth of Manchester and Lancashire flooded the Indian market. Around the same period, India was hit by one of the worst famines beginning in the late seventeenth century and continuing through the eighteenth century. The words attributed to Lord Bentinck, the Governor of India in the 1830s, that "the bones of the weavers are bleaching the plains of India" are a dramatic but apt description of the fate of the Indian weavers.

The eclipse suffered by the Indian textile industry lasted until the early twentieth century, until its grand revival under Gandhi, who initiated the *khādi* movement. The *charkhā*, or Indian spinning wheel, and *khādi*, or homespun cloth, became symbolic of the Indian struggle for independence. Foreign cloth was burnt in public squares and the Indian spinning wheel became a part of the home of every Indian patriot.

Since Independence, a sea change has occurred in the traditional Indian textile industry. The changeover to power looms and jet looms and the introduction of computer designs is setting new traditions in Indian textiles. In the course of its historical vicissitudes, the Indian textile industry has gone through a process of change as well as cultural assimilation.

Indian Textile Technology

The first process in the weaving of a cloth is warping and sizing, and in India this is done in the open. Bamboo sticks, about one hundred and twenty in number, are fixed upright in the street or what is called the warping grove, at a distance of a cubit from one another. Rows of women walk up and down the line, each carrying a wooden spindle in the left hand and a bamboo wand in the right. As they walk, they intertwine the threads between the split bamboos. These threads are then stretched horizontally from tree to tree, evenly washed with rice starch and carefully brushed. The right amount of tension in the warp is required to prevent the yarn from breaking while on the loom.

In India spinning was and still is almost exclusively the occupation of women. More specifically, this was the sole occupation of destitute women and widows. It is interesting that this corresponds to the English notion of the 'spinster' as one who has to spin for her livelihood since she has no one to support her.

The earliest looms in use in India were either the pit loom or the vertical loom. The *Atharvaveda*, probably compiled in the early pre-Christian era, says, "A man weaves it, ties it up; a man hath borne it upon the firmament. These pegs propped up the sky; the chants, they made shuttles for weaving ... (sic)." However, the most common type of loom in use was the horizontal pit loom in which the loom is placed inside an earthen pit and is operated with foot treadles. By depressing the pedal with one foot and raising the other, one set of threads get depressed and the weaving shed is formed through which the throw shuttle is shot across by hand. References to such looms are scattered throughout ancient and medieval inscriptions. Around the fifteenth century one begins to get reference to the draw loom. This consisted of several levers and so enabled the weaving of complex patterns. The introduction of the fly shuttle in the 1930s towards the end of British rule in India resulted in the partial mechanisation of loom technology and in another three decades this was followed by the introduction of the jacquard. Nowadays, partially mechanised looms, power looms, and jet looms are displacing the traditional Indian handloom.

Traditional Indian Costumes

Different types of cloth are worn and woven in the different parts of India, since this is a vast land with varying climates. Generally, men tend to wear a longish lower cloth of about one and a half yards in length, called *dhōti* or *lungi* in the north and *veshti* in south India, while the traditional upper cloth consists of a single piece of cloth called *aṅgavastra*. However, in hot weather, men generally go without the upper cloth. In many parts of northern India, men also wear a head gear against the dust and heat. This is especially true of desert regions like Rajasthān. The Indian women wear large skirts or loose trousers called *salwār* and longish or short jackets. Alternately, they wear a six-yard piece called a *sāri* and a blouse for the upper part. In the colder parts of India, such as the Himalayan mountain ranges and Kashmir, the garments are thicker and more elaborate, including warm woollen shawls and heavy jackets. It is noteworthy that in antiquity, stitched garments such as shirts, trousers and blouses were hardly ever worn in India. In the ancient sculptures and paintings such as the ones at Amarāvati or Brahadīśvaram, it is only the menials, palace attendants, common soldiers and dancing girls, all of them belonging to the lower echelons of society, who are depicted wearing stitched garments. Such garments are never depicted on the upper classes or royalty nor on the images of gods and goddesses. A plausible reason may be the association of impurity and pollution with stitched cloth.

Colours and Designs in Textiles

Traditional Indian textiles reflect the Indian ethos. There is an aura of religion and romance around Indian weaving. Everything is significant – the colours chosen, the motifs, and the wearing occasion. Crimson or shades of red are very auspicious and worn by women on the occasion of their marriage as well as by ceremonial priests in certain parts of India, such as the Mādhvā Brāhmins of Karnātaka. White represents purity and ochre, renunciation, and these are the colours worn by Hindu widows as well as ascetics. Yellow and green denote fertility and prosperity and are worn in the spring. Black is considered inauspicious, although pregnant women in south India wear black, perhaps to ward off the evil eye.

As late as the eighteenth century, colouring was done entirely through vegetable dyes such as madder and indigo, although now dyers have almost entirely switched over to chemical dyes except in the case of highly specialised textiles like the *kalamkāris*.

The earliest designs on Indian textiles seem to have been geometrical. A twelfth century Sanskrit text called the *Mānasōllāsa* described textiles designed with dots, circles, squares and triangles. The depiction of flora and fauna was related to religion and popular beliefs. The lotus, which has great spiritual significance in Hinduism, and the mango design are among the most popular Indian motifs. Swan, peacock, parrot and elephant are also commonly depicted. The tree of life, which symbolises fertility and prosperity, is another auspicious motif. All these designs are patterned on the loom itself and it may take a handloom weaver working on an ordinary frame loom as long as thirty days to weave an elaborate six-yard *sāri* with designs and gold lace. As the weaver weaves, he also sings the special loom songs, a tradition which has now almost entirely died out except perhaps in some interior weaving villages in Uttar Pradesh or the remote south. These loom songs tell of the glory of particular weaving castes or are full of esoteric religious metaphors describing god as the eternal weaver, weaving the web of life, and the human body as the cloth he has woven.

TEXTILE VARIETIES

Traditional Indian textiles are unique and unparalleled for their beauty. The *jāmdāni* is an elaborate textile which is woven with multiple shuttles and resembles tapestry work. Floral motifs called *bootis* in gold or silver lace are scattered over the body with heavy gold lace on the borders. The most striking of these designs is the *pannā hazāra*, literally a thousand emeralds, in which the flowers shimmer and gleam all over the sari. The Benarsi sāris called *Kiṁkhābs* woven in Uttar Pradesh are legendary for their loveliness, although Benaras in the north and Kāṅchipuram in the south were traditionally associated with pure cotton rather than silk. It was the British who introduced sericulture in Kāṅchipuram in the nineteenth century. Gadhwāl and Venkaṭagiri saris of Andhra and the Īrkal saris of Karnataka specialise in rich gold borders and heavy panel-like *pallūs* (that portion of the sāri which is draped over the shoulder). Another variety is the tie and dye (called variously *bandini, ikāt*, or *chungdi*) produced in Rajasthān, Orissa, Andhra Pradesh, and Madurai where the fibre is tie-dyed before weaving. A unique Andhra textile is the *telia*, which was soaked in oil before weaving and catered exclusively to the West Asian market because it was woven to suit desert conditions. This textile appeared in the sixteenth century with Muslim rule and died out with the collapse of the Islamic empires. Another textile which became popular in the Mughal period was the *mashroo* (also the *himroo*) in which cotton was used in the warp and silk in the weft. Initially these were used as Islamic prayer mats by the Mughal nobility who were forbidden by Islamic tenets to use any animal product. They therefore contrived the mashroo which enabled them to have their comfort without violating the religious tenet against the use of pure silk.

Textiles also form an important part of temple ceremonials such as the flag cloth hoisted in temples, the garments put on the deity, the cloth covering the chariots in which the deities are taken out on a procession and the ritual dance costumes. The *kalamkāri* cloth of Andhra, in which mythological stories are sketched minutely on cloth with a fine pen as well as the *Nādhadwāra pichwāis* of Rajasthan, are of this genre. In India it is also the practice among wandering groups of minstrels to render dramatic narrations of mythological stories, and the elaborately painted screens used on these occasions form an important aspect of traditional Indian textiles.

Vijaya Ramaswamy

References

Desai, Clena. *Ikat Textiles of India*. Bombay: Marg Publications. 1965.

Irwin, John. "A Bibliography of the Indian Textile Industry." *Journal of Indian Textile History*, part II. 1956.

Irwin, John and B. Catherine Brett. *Origins of Chintz*. London: **Her** Majesty's Stationery Office. 1970.

Ramaswamy, Vijaya. *Textiles and Weavers in Medieval South India*. New Delhi: Oxford University Press. 1985.

See also: Colonialism and Science

Time

The ancient Indians acknowledged their gratitude to Nature with elaborate rituals centred around Fire (*Agni*). The positions of fiery objects in the sky which were the prime agents of Nature decided when the Gods had to be propitiated. Even today, the temporal coordinates of important rituals like the sacred thread ceremony, marriages, and anniversaries are scheduled and announced to the gods in astronomical terms and not by the Gregorian calendar. The householder (*gṛihastha*) had to establish the sacrificial fire (*Agnyādhāna*) on the first day of the waning Moon, or perhaps on the full moon. But dissenting opinions always existed. The

Śatapatha Brāhmaṇa (Commentary of the Hundred Ways), for example, urged the householder to install the sacrificial fire when he so desired.

The *Śatapatha Brāhmaṇa* is the oldest available prose record of these ancient religious practices. It is next only to the *Ṛgveda* and its accompanying commentary (Brāhmaṇa), the *Aitareya*. These texts record the development of five centuries of fire rituals. The *Ṛgveda* has been variously dated from 4500 to 2500 BCE. According to the tradition of the age, the teacher (*guru*) conveyed knowledge orally to the student (*śiṣya*) who learned by rote. All astronomical works evolved over the centuries without the limitations of a script and adapting to contemporary conditions. Terminology also changed so much that the original Vedic texts and commentaries became obscure.

The *Ṛgveda* refers to a *Yuga* of four years (*samvatsara*) as did the Aztec and Egyptian calendars. Intercalary months were inserted in the middle and at the end of the *Yuga* to catch up with the seasons. Dirghatma, the son of Mamata, the first known astronomer of the Vedic era, recorded the movements of the Sun, Moon and planets over forty years and found that the lunar year (*Cāndravarṣa*) of twelve months (*Cāndramāsa*) with thirty days in each leaves a gap of 21 days in the four year *Yuga*. He therefore fixed the *Yuga* to be a unit of four times 365 days with the insertion of intercalary days.

Linking seasonal periodicity with the lunar cycle caused trouble for all ancient astronomers. The Vedic sages laid down four kinds of years: the lunar year of 354 days, the tropical year of 365.25 days, the civil (*sāvana*) year of 360 days, and the sidereal year of 366 days. The lunar year was allowed to retrograde through the seasons to come into coincidence at the end of thirty sidereal years; however some schools preferred to insert intercalary days for this purpose. The Egyptians also divided their year into 36 decans each of ten days, and it is now believed that these systems derived from an early Aryan civilisation.

Around 1200 BCE, the *Yuga* cycle of five sidereal years of 366 days seems to have become generally accepted. Lagadha, who lived near Srinagar in Kashmir, documented the new calendar around 900 BCE. His *Vedāṅga Jyotiṣa* is a compendium of astronomical rules that have been extant since perhaps 1200 BCE. As was the practice then, the work is written in Sanskrit verse to facilitate commitment to memory. The *Vedāṅga Jyotiṣa* names the five years of the *Yuga* as *Samvatsara, Parivatsara, Idavatsara. Anuvatsara*, and *Vatsara*. The *Yuga* began at the white half of the month of January (*Māgha*) and ended with the dark half of December (*Pauṣa*). The year was bisected according to the northward (*uttarāyaṇa*) and southward progressions (*dakshiṇāyaṇa*) of the Sun. Each *āyana* then comprised 183 civil days (*sāvana divasa*) measured from sunrise to sunrise. Each day was found to increase by one *prastha* during the *uttarāyana* while the night shortened by the same amount. This trend was reversed during the *dakshiṇāyaṇa*. Since it is also stated that the increase or decrease of daylight during one *āyana* is equal to six muhurtas, the *prastha* is 1.5738 of our present day minutes. This peculiar unit of time appears in the *Vishṇu Purāṇa* which predated Hellenic science, as well as in the later Jain work on astronomy, the *Sūryaprajnāpathi*. The *Sūryasiddhānta* also divides the 24-hour day into 21600 *prāna*, perhaps because a heavenly body transits one minute of arc in this period. Being related to the seasonal variation of the length of the day the *prastha* is 23.61 *prāṇa*.

By the end of the fifth century CE, at least five great texts (*Siddhāntas*) and thirteen minor texts (*Siddhāntikas*) on astronomy had been inscribed on dried palm leaves or on birch bark. The texts themselves were frustratingly obscure but commentaries and explanations were provided by the astrologers of the period. The great astronomer, Varāhamihira (ACE 500), who summarised the systems of astronomy in his Five Texts (*Pañca Siddhāntika*), admitted that their common source was the Text of the Forefathers (*Pitāmaha Siddhānta*). Distinct similarities observed between the verses of the *Vedāṅga Jyotiṡa* and of the *Pitāmaha Siddhānta* indicate that the *Pitāmaha Siddhānta* was a development from much earlier times. It is therefore incorrect to state that Indian astronomy began only in CE 500 with imported knowledge from Mesopotamia, Greece, or Persia.

After the missionary Padmasambhava took Buddhism over the Himalayan Mountains to Tibet and China, the pilgrim Yuan-Zhuang travelled in India from CE 629 to 645 and collected 657 Buddhist texts written in various Indian languages. In his *Abhidharma Shunzheng Lilun*, he describes units of time then prevalent in India. The day was divided into the forenoon; noon and afternoon while the night had three watches (*trimāya*). The shortest interval of time was the *kshaṇa*, the time taken by a woman to spin one *hsun* of thread; it also referred to an 'instant' or the blink of an eye. This term is still in colloquial use.

Instruments to measure time documented by Varāhamihira and others around CE 500 included the sundial, water clocks, and sand clocks. The sundial (*chāyā yantra*) was elementary in construction as compared to later European models where the time could be read directly off the dial. A vertical staff (*yaṡṭi* or *ṡhanku*) 12 digits (*aṅgulas*) long was fixed at the centre of a horizontal circle of radius equal to the length of the staff. This celestial circle represents the intersection of the celestial sphere with the horizontal plane. The shadow of the staff would cross the celestial circle at two points on an east–west line. A square (*caturaśra*) was then drawn circumscribing the circle, touching it at the four cardinal points. The eastern and western sides of the square were then graduated in digits. The projection of the shadow of the staff when the Sun was upon the equator (*viṣhuvadbha*) determined the latitude of the place. Further measurements on the shadow of the staff enabled the chronographers to compute the time of day. The main instrument (*nara yantra*) was a larger and more precise version of the sundial.

The *Siddāntha Śiromani* also describes simpler variants of the sundial such as the *yaṣṭi yantra*. Here the celestial circle is drawn as for the *chāyā yantra*. From tables existing at that time or from observations conducted over the previous years, the radius of the diurnal circle of the day (*dyujya*) was calculated from the declination of the Sun on that day. A second and smaller circle of this radius was then drawn on the floor concentric with the earlier celestial circle. The circumference of this diurnal circle was then divided into sixty equal parts representing the division of the daily solar time of revolution into sixty *nādīs*. The staff was set but not rigidly fixed at the centre, and to ascertain the time of the day, it was pointed towards the Sun in such a manner that it cast no shadow. The extremity of the staff then represented the position of the Sun on the celestial sphere. The distance from the tip of the staff to the point of sunrise or sunset on the celestial circle was measured. The arc subtended on the diurnal circle by this chord length gave, in *nādīs*, the time since sunrise or until sunset.

The wheel (*cakra*) was a hand-held instrument to obtain the altitude of the Sun and its zenith distance and so compute the time since sunrise. Even smaller instruments such as the arc (*dhanus*) were in use at that time.

Brahminaic rituals require not only the absolute time as determined by stellar and planetary position but also a stopwatch to regulate the course of various sacrifices. The water clock (*jala yantra*) served this purpose. A leaking water vessel, the equivalent of the Greek clepsydra, was the first version. This, however, proved to be very inaccurate as the vessel takes infinite time to empty completely. As an improvement, water leaking at a constant rate from a reservoir was collected in a pot of volume exactly one *prastha*. In due course the *prastha* became a unit of time.

The sinking water clock (*ghaṭi*) proved to be more accurate and was frequently in use in the early eleventh century and up to the arrival of the mechanical clocks from Europe. A hemispherical copper vessel (*kapāla*) six digits high and with a capacity of about 3.13 litres (60 *palas*) with a hole at the bottom to admit a gold pin four digits long and weighing approximately 0.183 gram (3.33 *māṣa*) was placed on water. The vessel sank in one *ghaṭi* or *ghaṭika* sixty times in a day. This clock is used even now in some temples and for orthodox rituals.

As the measurement of time in ancient India developed, astronomers and astrologers became powerful and influential. Varāhamihira observed that a king who did not honour an astrologer was destined to destruction; that neither a thousand elephants nor four thousand horses could accomplish as much as a single astrologer. Time keeping, so essential to astronomy, then evolved into an esoteric science.

Division of the day according to various sources.

The Puraṅas (? BCE)

15 nimeṣa = 1 kāṣṭa

30 kāṣṭa = 1 kalā

30 kalā = 1 muhūrta

30 muhurta = 1 day or nycthemeron

The Sūrya Siddāntha (< 900 BCE)

10 gurvakṣara = 1 prāṇa

1 prāṇa = 1 vināḍi

60 vināḍi = 1 nāḍi, nāḍikā or ghaṭikā

60 nāḍī = 1 day

The Vedāṅga Jyotiṣa (CE < 400)

5 akṣaras = 1 kāṣṭā

124 kāstās = 1 kalā = 305/201 prasthas

10.05 kalās = 1 nāḍikā

2 nāḍikās = 1 muhurta = 30.5 prasthas

30 muhurtas = 1 day

The traveller Yuan Zhuang (CE 629)

120 kṣaṇ as = 1 tatkṣaṇa

60 tatkṣaṇas = 1 lava

30 lavas = 1 muhūrta

5 muhūrtas = 1 'time'

6 'times' = 1 nycthemeron

The Jain saint Mahāvīrācārya (ca. CE 850)

7 Uchchhavas = 1 stoka

7 stokas = 1 lav

38.5 lav = 1 ghaṭi

2 ghaṭis = 1 muhūrta

30 muhūrtas = 1 day

15 days = 1 pakṣa

2 pakṣas = 1 māsa

2 māsa = 1 season

3 seasons = 1 āyana

2 āyanas = 1 year

The traveller al-Bīrūnī (CE 1030)

6 prāṇa = 1 caṣaka

15 caṣaka = 1 kṣaṇa

4 kṣaṇa = 1 ghaṭi

60 ghaṭi = 1 nycthemeron

Nataraja Sarma

REFERENCES

Sarma, Nataraja. "Measures of Time in Ancient India." *Endeavour* 16(4): 185–188. 1991.

Satya Prakash. *Founders of Sciences in Ancient India*. Delhi: Research Institute of Ancient Scientific Studies. 1965.

Shastri, A.M. *India as Seen in the Bṛihat Saṁhita of Varāhamihira*. Delhi: Motilal Banarsidass. 1969.

Sūrya Siddāntha with a commentary by Ranganātha. Ed. P. Gangooly; Trans. E. Burgess. Calcutta: University of Calcutta. 1935.

Thibaut, G. "Contributions to the Explanation of the Vedānga Jyothiśa." *Journal of the Asiatic Society of Bengal* 46: 411. 1877.

Watters T. *On Yuang-Chwang's Travels in India*. Delhi: M. Manoharlal. reprinted 1973.

See also: Astronomy – *Sūryasiddhānta* – Varāhamihira – Gnomon

Trigonometry

Trigonometry offers one of the most remarkable examples of transmission of the exact sciences in antiquity and the Middle Ages. Originating in Greece, it was transmitted to India and, with several modifications, passed into the Islamic world. After further development it found its way to medieval Europe.

The very term "sine" illustrates the process of transmission. The Greek word for "chord" ($\epsilon\ 'u\theta\epsilon\hat{\iota}a$, literally "a straight line [subtending an arc]") was translated into Sanskrit as *jīva* or *jyā* ("string of a bow") from the similarity of its appearance. The former word was phonetically translated into Arabic as *jyb*, which was vocalised as *jayb* (meaning "fold" in Arabic), and this was again translated into Latin as *sinus*, an equivalent to the English sine.

It was by tracking back along this stream of transmission that the first chord table ascribed to Hipparchus (fl. 150 BCE) was successfully recovered by G. J. Toomer in 1973 from an Indian sine table (compare Tables 1 and 2). Toomer showed that some numerical values ascribed to Hipparchus in the *Almagest* of Ptolemy (fl. CE 150) could only be explained by hypothesising the use of this reconstructed table.

According to this reconstruction, Hipparchus used 6875 "minutes" as the length of the diameter (1) of the base circle, in other words, as the greatest chord subtending the half circle (= R crd180° = 6875). This number is the result of rounding after dividing 21,600 minutes (360°) by the value of π = 3; 8, 30 (in this article we follow the convention: integer and fraction are separated by a semicolon, the former is in decimal form and the latter is in sexagesimal form with commas to separate the places.) In India 3438 "minutes", namely, the rounded half of D, was used as the length of the radius (R), which is the largest "half chord" (*jyārdha* or *ardhajyā*).

Thus the relation between the Greek chord and the Indian sine can be expressed as:

$$AB = 2AHR;\ \mathrm{crd}2\alpha = 2R\sin\alpha. \quad (1)$$

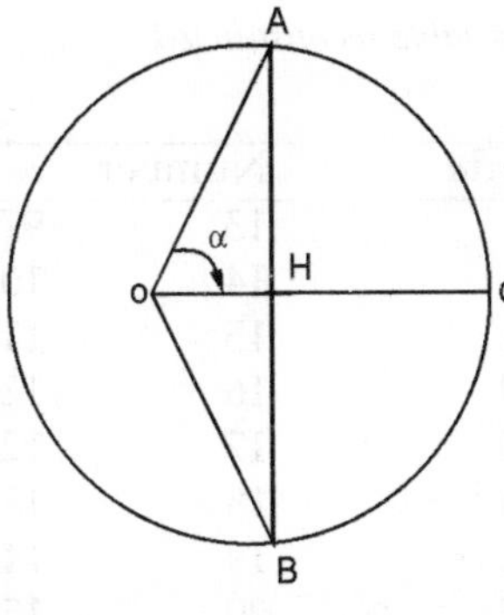

Plane trigonometry was the essential tool for mathematical astronomy in India. All the astronomical texts in Sanskrit either give a kind of sine table or presuppose one. On the other hand, trigonometry was studied only as a part of astronomy and it was never an independent subject of mathematics. Furthermore since they were not aware of spherical trigonometry, Indian astronomers developed the method called *chedyaka* in which the sphere was projected on to a plane.

Sine table with $R = 3438$

The earliest Indian sine table with $R = 3438$ is found in Āryabhaṭa's book on astronomy, *Āryabhaṭīya* (CE 499). It should be remembered that "table" here does not mean that the numbers are actually arranged in a tabular form, i.e. in lines and columns. As is usually the case with Sanskrit scientific texts, all the numbers are expressed verbally in verse. For brevity's sake, Āryabhaṭa gives only the tabular differences (Δ, the fourth column of Table 2). This is the standard sine table in ancient India. Exactly the same table is found in the *Jiuzhili*, a Chinese text on the Indian calendar written in CE 718 by an astronomer of Indian descent, but it did not have any influence on Chinese mathematics.

The values of sines were geometrically derived from $R\sin 90°(= R)$ and $R\sin 30°(= R)$ by two formulas:

$$R\sin(90° - \alpha) = \sqrt{R^2 - (R\sin\alpha)^2}, \tag{2}$$

$$R\sin\frac{\alpha}{2} = \sqrt{\left\{\frac{R - R\sin(90° - \alpha)}{2}\right\}^2 + \left(\frac{R\sin\alpha}{2}\right)^2}. \tag{3}$$

It is worth noting here that the first tabular sine was equated with the arc ($3;45° = 225'$) which it subtends. This means that, when a' is small enough, the approximation $\sin a' \approx a'$ can be applied.

Table 1 *Hipparchus chord table reconstructed*

Number	α	R crd$a$	Number	α	R crd$a$
1	7;30	450	13	97;30	5169
2	15	897	14	105	5455
3	22;30	1341	15	112;30	5717
4	30	1779	16	120	5954
5	37;30	2210	17	127;30	6166
6	45	2631	18	135	6352
7	52;30	3041	19	142;30	6510
8	60	3438	20	150	6641
9	67;30	3820	21	157;30	6743
10	75	4185	22	165	6817
11	82;30	4533	23	172;30	6861
12	90	4862	24	180	6875

What is more remarkable in Āryabhaṭa is that he gives an alternative method for computing tabular differences by means of the formula:

$$\Delta_{n+1} = \Delta_n - (\Delta_1 - \Delta_2)\frac{J_n}{J_1},$$

where $J_n = R \sin n\ (3;\ 45^o)$. The formula, after several centuries of misunderstanding, was correctly interpreted by a south Indian astronomer Nīlakaṇṭha (born in 1444), one of the most distinguished scholars belonging to the Mādhava school (see below). When suitable values of $(\Delta_1 - \Delta_2)$ and J_1 are used, this formula produces very good values for the rest of R sins. It seems that Āryabhaṭa's sine values were commuted by this second method rather than by the geometrical method.

Sine table with $R = 120$

There is another kind of Indian sine table which uses $R = 120$. The table is found in the *Pañcasiddhāntikā* of Varāhamihira, a younger contemporary of Āryabhaṭa. This table is closely related to the Greek chord table with $R = 60$ which is offered in Ptolemy's *Almagest*. Because of the relation (1) given above, all the numerical values in the chord table with $R = 60$ can be transferred directly to the sine table with $R = 120$. Table 3 compares the first four and the last four values in the *Pañcasiddhāntikā* with Ptolemy's corresponding ones. The fractional parts after the semicolon in both tables are expressed sexagesimally. It seems that Varāhamihira's values were the results of rounding the numbers in the second fractional place of a chord table similar to that of Ptolemy.

Table 2 *Indian sine table with* $R = 3438$

Number	α	$R \sin a$	Δ	Number	α	$R \sin a$	Δ
1	3;45	225	225	13	48;45	2585	154
2	7;30	449	224	14	52;30	2728	143
3	11;15	671	222	15	56;15	2859	131
4	15	890	219	16	60	2978	119
5	18;45	1105	215	17	63;45	3084	106
6	22;30	1315	210	18	67;30	3177	93
7	26;15	1520	205	19	71;15	3256	79
8	30	1719	199	20	75	3321	65
9	33;45	1910	191	21	78;45	3372	51
10	37;30	2093	183	22	82;30	3409	37
11	41;15	2267	174	23	86;15	3431	22
12	45	2431	164	24	90	3438	7

Table 3 *Comparison of the first four and last four values in the Pañcasddhāntikā with those of Ptolemy*

Number	$\alpha(°)$	Varāhamihira with R = 120 $R\sin a$	Ptolemy with R = 60 R crd2a
1	3;45	7;51	7;50,54
2	7;30	15;40	15;39,47
3	11;15	23;25	23;24,39
4	15	31;4	31;3,30
...	...	...	...
21	78;45	117;42	117;41,40
22	82;30	118;59	118;58,25
23	86;15	119;44	119;44,36
24	90	120	120

Table 4 *Mādhavaś sine table*

Number	$a(°)$	R sin a	Number	$a(°)$	R sin a
1	3;45	0224;50,22	13	48;45	2548;38,06
2	7;30	0448;42,58	14	52;30	2727;20,52
3	11;15	0670;40,16	15	56;15	2858;22,55
4	15	0889;45,15	16	60	2977;10,34
5	18;45	1105;01,39	17	63;45	3038;13,17
6	22;30	1315;34,07	18	67;30	3176;03,50
7	26;15	1520;28,35	19	71;15	3255;18,22
8	30	1718;52,24	20	75	3320;36,30
9	33;45	1909;54,35	21	78;45	3321;41,29
10	37;30	2092;46,03	22	82;30	3408;20,11
11	41;15	2266;39,50	23	86;15	3430;23,11
12	45	2430;51,15	24	90	3437;44,48

In these earlier Indian sine tables, only the twenty-four values in the first quadrant are given, with the interval of 3°45′. Although Indians knew and used cosines (*koṭijyā*), they had no need of tabulating them because they knew that cosines could be derived from sines by the relation (2) above. On the other hand they were interested in the versed sine (*śara* in Sanskrit, meaning "arrow" or *utkramajyā*, "sine of the reversed order", CH in Fig. 1) which is defined as:

$$R \text{ vers } \alpha = R - R\sin(90° - \alpha).$$

Using this relation, Brahmagupta (seventh century) simplified some formulas. For instance, formula (3) was rewritten as:

$$R\sin\frac{\alpha}{2} = \sqrt{\frac{D \times R\text{vers}\alpha}{4}}. \quad (3')$$

A table of versed sines can be easily obtained by adding Δs in Table 2 successively from the bottom upward (namely, in the "reversed order").

Brahmagupta computed anew 24 sines with $R = 3270$ in the *Brāhmasphu-ṭasiddhānta*. Elsewhere in this book and in the *Khaṇḍakhādyaka*, he offers a sine table with $R = 150$ and with the interval of $15°$. This small table generates remarkably correct sine values when his ingenious method of second order interpolation is applied.

BHĀSKARA II

An improved version of the traditional sine table was prepared by Bhāskara II (b. 1114). In the chapter "Derivation of Sines" (*Jyotpatti*) of his *Siddhāntaśiromaṇi* he introduces two new values:

$$R \sin 36^\circ = \sqrt{\frac{5-\sqrt{5}}{8}} R,$$

$$R \sin 18^\circ = \sqrt{\frac{\sqrt{5}-1}{4}} R,$$

With these two values and formulas (2) and (3) above, he obtains $R \sin 3n^\circ$ (where $n = 1, 2, 3, ..., 30$). Further he combines them with the approximate value

$$R \sin 1^\circ \approx 60'$$

using the new formula:

$$R \sin(\alpha \pm \beta) = \frac{R \sin \alpha R \cos \beta \pm R \cos \alpha R \sin \beta}{R}, \tag{4}$$

which is equivalent to the modern formula:

$$\sin(\alpha \pm \beta) = \sin \alpha \cos \beta \pm \cos \alpha \sin \beta.$$

Thus he could obtain sines for all the integer degrees of a quadrant. Formula (4) was unknown to Indians before Bhāskara II, while a chord version of the same formula was known to Ptolemy.

Trigonometry underwent a remarkable development in the early fifteenth century on the western coast of south India (the modern state of Kerala). The person who initiated this development was Mādhava (fl. ca. 1380/1420) of Saṅgamagrāma (near modern Cochin). His important works on astronomy and mathematics are now lost, but we know his achievements from the books of his successors. A sine table ascribed to him is quoted in Nīlakaṇṭha's commentary on the *Āryabhaṭīya* (Table 4).

A couple of verses, which are often quoted by the students of the Mādhava school and which are ascribed to Mādhava himself by Nīlakaṇṭha, give the method of computing sines. The method can be expressed as:

$$R\sin\theta = \theta - \frac{\theta^3}{3!R^2} + \frac{\theta^5}{5!R^4} - \frac{\theta^7}{7!R^6} + \frac{\theta^9}{9!R^8} - \ldots$$

With $R = 1$ this is equivalent to Newton's

$$\sin\theta = \theta - \frac{\theta^3}{3!} + \frac{\theta^5}{5!} - \frac{\theta^7}{7!} + \frac{\theta^9}{9!} - \cdots.$$

Similar power series for cosine and versed sine are ascribed to Mādhava.

Michio Yano

References

Gupta, R.C. "South Indian Achievements in Medieval Mathematics." *Gaṇita-Bhāratī* 9(1–4): 15–40. 1987.

Pingree, David and David Gold. "A Hitherto Unknown Sanskrit Work Concerning Mādhava's Derivation of the Power Series for Sine and Cosine." *Historia Scientiarum* 42: 49–65. 1991.

Toomer, Gerald J. "The Chord Table of Hipparchus and the Early History of Greek Trigonometry." *Centaurus* 18: 6–28. 1973.

Toomer, Gerald J. *Ptolemy's Almagest*. New York: Springer Verlag. 1984.

See also: Sexagesimal System – Āryabhaṭa – Mādhava – Varāhamihira – Brahmagupta – Śrīpati – Nīlakaṇṭha

V

Vākyakaraṇa

Vākyakaraṇa, apocryphally ascribed to Vararuci, is an astronomical manual produced in about CE 1300 which was very popular in South India, especially in Tamil Nadu and the adjoining regions. The work is so called because it was a *karaṇa* (astronomical manual) which used computational tables where the numbers are expressed in mnemonic sentences (*vākyas*), phrases, and words. It is also called *Vākyapañcādhyāyī,* because it contains *pañca adhyāyas* (five chapters). Until recently, when the *Nautical Almanac* came to be used in South India for the computation of the Hindu almanac with its 'five limbs' (*pañca-aṅga*): the lunar day, weekday, asterism, *yoga* and *karaṇa*, the *Vākyakaraṇa* was widely used for that purpose.

The author of the *Vākyakaraṇa* hailed from Kariśaila or Kānchī in Tamil Nadu, as he himself states in his work. He also states that he based his work on the writings of Bhāskara I (CE 629), the exponent of the school of astronomy promulgated by Āryabhaṭa (b. CE 476), and the works of Haridatta (CE 683), author of *Grahacāranibandhana* and *Mahāmārganibandhana.* The *Vākyakaraṇa* was elaborately expounded by Śundararāja (CE 1500) who had contacts with the Kerala astronomer Nīlakaṇṭha Śomayāji. The date of *Vākyakaraṇa* is determined as ca. CE 1300, on the basis of epochs which the author gives for the computation of the planets.

In five chapters, the *Vākyakaraṇa* deals with all aspects of astronomy required for the preparation of the Hindu almanac. Chapter I is concerned with the computation of the Sun, the Moon, and the Moon's nodes, and Chapter II with that of the planets. Chapter III is devoted to problems involving time, position, and direction, and other preliminaries like the precession of the equinoxes. The computation of the lunar and solar eclipses is the concern of Chapter IV. Chapter V is devoted to the computation of the conjunction of the planets, and of planets and stars.

For the computation of the Moon, the *Vākyakaraṇa* employs the 248 moon sentences of the ancient astronomer Vararuci. But for the five planets, Mars, Mercury, Jupiter, Venus, and Saturn, the author himself computed 82 tables devoted to the different planetary cycles, containing, in all, 2075 mnemonic sentences (*Kujādi-pañca-graha-mahāvākyas*).

The results obtained through the computations enunciated in the *Vākyakaraṇa* are not very accurate, going by modern standards, but they were accurate enough for the determination of auspicious times and other matters required in the routine life of orthodox Hindus. Besides, there was the saving of much time and labour in working with the simple methods advocated in the work.

K.V. Sarma

References

Candravākyāni (Moon Sentences of Vararuci). Ed. as Appendix II in *Vākyakaraṇa*. Ed. T.S. Kuppanna Sastri and K.V. Sarma. Madras: Kuppuswami Sastri Research Institute. 1962. pp. 125–34.

Kujādi-pañcagraha-mahāvākyāni (Long Sentences of the Planets Kuja etc.). Ed. as Appendix III in *Vākyakaraṇa* Ed. T.S. Kuppanna Sastri and K.V. Sarma. Madras: Kuppuswami Sastri Research Institute. 1962. pp. 135–249.

Vākyakaraṇa with the commentary *Laghuprakāśikā* by Sundararāja. Ed. T.S. Kuppanna Sastri and K.V. Sarma. Madras: Kuppuswami Sastri Research Institute. 1962.

See also: Bhāskara – Haridatta – Āryabhaṭa – Nīlakaṇṭha Śomayāji – Precession of the Equinoxes – Lunar Mansions

Values and Science

Is science value free? If "science", as Lord Rutherford is reported to have said, "is what scientists do" then scientists would have to be superhuman to keep their values out of what they do in their laboratories. We cannot clinically isolate our cultural and ethical (and through them our historical) baggage from our human activities. Values play an important part in what we select to do or not do and how we actually do it. In as far as science is a human activity, it is subject to the strengths and weaknesses of all human activities.

However, in Western tradition, up to quite recently, scientists were seen as quasi-religious supermen, heroically battling against all odds to discover the truth. Also, the truths they wrestled out of nature were said to be absolute, objective, value-free, and universal. The idea of scientists as dedicated hermit-like lone researchers is now obsolete. Nowadays, science is an organised, institutionalised, and industrialised venture. The days when individual scientists, working on their own, and often in their garden sheds, made original discoveries are really history. Virtually all science today is big science requiring huge funding, large, sophisticated, and expensive equipment, and hundreds of scientists working on minute problems. As such, science has become a unified system of research and application, with funding at one end and the end product of science, often technology, at the other.

Values enter this system in a number of ways. The first point of entry is the selection of the problem to be investigated. The choice of the problem, who makes the choice and on what grounds, is the principle point of influence of society, political realities of power, prejudice, and value systems on even the "purest" science. Often, it is the source of funding that defines what problem is to be investigated. If the funding is coming from government sources then it will reflect the priorities of the government – whether space exploration is more important than health problems of the inner city poor, or nuclear power or solar energy should be developed further. Private sector funding, mainly from multinationals, is naturally geared towards research that would eventually bring dividends in terms of hard cash. Some eighty per cent of research in the United States is funded by what is called the "military–industrial complex " and is geared towards producing both military and industrial applications.

Subjectivity thus enters science in terms of what is selected for research which itself depends on where the funding is coming from. But values also play an important part in what is actually seen as a problem, what questions are asked and how they are answered. For example, cancer rather than diabetes may be seen as a problem even though they may both claim the same number of victims. Here both political and ideological concerns, as well as public pressure, can make one problem invisible while focusing attention on another. Moreover, if, for example, the problem of cancer is defined as finding a cure then the benefits of the scientific research accrue to certain groups, particularly the pharmaceutical companies. But if the function of scientific research is seen as eliminating the problems of cancer from society, then another group benefits from the efforts of research: the emphasis here shifts to investigating diet, smoking, polluting industries, and the like. Similarly, if the problems of the developing countries are seen in terms of population, then research is focused on reproductive systems of Third World women, methods of sterilisation, and new methods of contraceptives. However, if poverty is identified as the main cause of the population explosion then research would take a totally different direction: the emphasis would have to shift to investigating ways and means of eliminating poverty, developing low cost housing, basic and cheap health delivery systems, and producing employment generating (rather than profit producing) technologies. The benefits of scientific research would go to the Third World poor rather than Western institutions working on developing new methods of contraceptives and companies selling these contraceptives to developing countries. Thus both the selection of problems and also their framing in a particular way are based on value criteria.

It can be legitimately argued that these factors are external to science, that within science, the scientific method ensures neutrality and objectivity by following a strict logic – observation, experimentation, deduction, and value-free conclusion. But scientists do not make observations in isolation. All observations take place within a well-defined theory. The observations, and the data collection that goes with them, are designed either to refute a theory or provide support for it, and theories themselves are not plucked out of the air. Theories exist within paradigms – that is a set of beliefs and dogmas. The paradigms provide a grand framework within which theories are developed and make sense, and observations themselves have validity only within specific theories. Thus, all observations are theory laden, and theories themselves are based on paradigms which in turn are burdened with cultural baggage. All of which raises the question: can there ever be such things as value-neutral "objective facts"? Studies of scientists working in laboratories have shown that the scientific method, by and large, is a myth. Researchers seldom follow it in the linear fashion that exists in the textbooks. Neither do they ascertain new "facts" suddenly out of the blue. And the same holds true of the laws of nature they are supposed to be discovering. It appears that scientists do not actually "discover" laws of nature; they manufacture them. Scientific knowledge advances by a process of manufacturing which involves thousands and thousands of workers assembling "facts" which through peer review and other procedures end as description or laws of nature.

Value judgments are also at the very heart of a common element of scientific technique: statistical inference. When it comes to measuring risks, scientists can never give a firm answer. Statistical inferences cannot be stated in terms of "true" or "false" statements. When statisticians test a scientific hypothesis they have to go for a level of "confidence". Different problems are conventionally investigated to different confidence-limits. Whether the limit is 95 or 99.9 per cent depends on the values defining the investigations, the costs, and weight placed on social, environmental, or cultural consequences. In most cases, the importance given to social and environmental factors determines the limits of confidence and the risks involved in a hazardous scientific endeavour. For example, when a chemical plant is placed in an area with an aware and politically active citizenry the risks are worked out to a high level of confidence. However, when one is located in an area where the citizens themselves are ignorant of the dangers and do not command political power, the confidence levels are much more relaxed. The people of Bhopal and Chernobyl know this to their cost.

But it is not just in its institutions and method that science is value laden. The very assumptions of science about nature, universe, time, and logic are ethnocentric. In modern science, nature is seen as hostile, something to be dominated. The Western "disenchantment of nature" was a crucial element in the shift from the medieval to the modern mentality, from feudalism to capitalism, from Ptolemaic to Galilean astronomy, and from Aristotelian to Newtonian physics. In this picture, "Men" stand apart from nature, on a higher level, ready to subjugate and "torture" her, as Francis Bacon declared, in order to wrestle out her secrets. This view of nature contrasts sharply with how nature is seen in other cultures and civilisations. In Chinese culture, for example, nature is seen as an autonomous self-organising entity which includes humanity as an integral part. In Islam, nature is a trust, something to be respected and cultivated and people and environment are a continuum – an integrated whole. The conception of laws of nature in modern science drew on both Judeo-Christian religious beliefs and the increasing familiarity in early modern Europe with centralised royal authority, with royal absolutism. The idea that the universe is a great empire, ruled by a divine logos, is, for example, quite incomprehensible both to the Chinese and the Hindus. In these traditions the universe is a cosmos to which humans relate directly and which echoes their concerns. Similarly, while modern science sees time as linear, other cultures view it as cyclic as in Hinduism or as a tapestry weaving the present with eternal time in the Hereafter as in Islam. While modern science operates on the basis of either/or Aristotelian logic (X is either A or non-A), in Hinduism logic can be fourfold or even sevenfold. The fourfold Hindu logic (X is neither A, nor non-A, nor both A and non-A, nor neither A nor non-A) is both symbolic as well as a logic of cognition and can achieve a precise and unambiguous formulation of universal statements without quantification. Thus the metaphysical assumptions of modern science make it specifically Western in its main characteristics.

The metaphysical assumptions of modern science are also reflected in its contents. For example, certain laws of science, as Indian physicists have began to demonstrate, are formulated in an ethnocentric and racist way. The Second Law of Thermodynamics, so central to classical physics, is a case in point: due to its industrial origins the Second Law presents a definition of efficiency that favours high temperatures and the allocation of resources to big industry. Work done at ordinary temperature is by definition inefficient. Both nature and the non-Western world become losers in this new definition. For example, the monsoon, transporting millions of tons of water across a subcontinent is "inefficient" since it does its work at ordinary temperatures. Similarly, traditional crafts and technologies are designated as inefficient and marginalised. In biology, social Darwinism is a direct product of the laws of evolutionary theories. Genetic research appears to be obsessed with how variations in genes account for differences among people. Although we share between 99.7 (unrelated people) and 100 per cent (monozygotic twins) of our genes, genetic research has been targeted towards the minute percentage of genes that are different in order to discover correlations between genes and skin colour, sex or "troublesome " behaviour. Enlightened societal pressures often push the racist elements of science to the sidelines. But the inherent metaphysics of science ensures that they reappear in new disguise. Witness how eugenics keeps reappearing with persistent regularity. The rise of IQ tests, behavioural conditioning, foetal research, and socio-biology are all indications of the racial bias inherent in modern science.

Given the Eurocentric assumptions of modern science, it is not surprising that the way in which its benefits are distributed and its consequences are accounted for are themselves ethnocentric. The benefits are distributed disproportionally to already over-advantaged groups in the West and their allies elsewhere, and the costs disproportional to everyone else. When scientific research improves the military, agriculture, manufacturing, health, or even the environment, the benefits and expanded opportunities science makes possible are distributed predominantly to already privileged people of European descent, while the costs are dumped on the poor, racial and ethnic minorities, women, and people located at the periphery of global economic and political networks. Science in developing countries has persistently reflected the priorities of the West, emphasising the needs and requirements of middle class western society, rather than the wants and conditions of their own society. In over five decades of science development, most of the Third World countries have nothing to show for it. The benefits of science just refuse to trickle down to the poor.

But modern science is not only culturally biased towards the West: it represents the values of a particular class and gender in Western societies. As feminist scholars have shown, science in the West has systematically marginalised women. Women, on the whole, are not interested in research geared towards military ends, or torturing animals in the name of progress, or working on machines that put one's sisters out of work. But more than that, even the least likely fields and aspects of science bear the fingerprints of androcentric projects. Physics and logic, the prioritising of mathematics and abstract thought, the so-called standards of objectivity, good method and rationality – feminist critique has revealed androcentric fingerprints in all. This is the case, for example, in the mechanistic model of early modern astronomy and physics, in modern particle physics, and in the coding of reason as part of ideal masculinity. The focus on quantitative measurements, variable analysis, impersonal and excessively abstract conceptual schemes is both a distinctively masculine tendency and one that serves to hide its own gendered character. Science has tried to hide its own masculine nature in other ways, by, for example, making women themselves objects in scientific investigation. It was not entirely accidental that sexology became a major science at the same time as women in the West were fighting for the vote and equal rights in education and employment. A number of studies have shown that scientific work done by women is invisible to men even when it is objectively indistinguishable from men's work. Thus, it appears that neither social status within science nor the results of research are actually meant to be neutral or socially impartial. Instead, the discourse of value-neutrality, objectivity, and social impartiality appears to serve projects of domination and control.

The history of science bears this out. The evolution of Western science can be traced back to the period when Europe began its imperial adventure. Science and empire developed and grew together, each enhancing and sustaining the other. In India, for example, European science served as a handmaiden to colonialism. The British needed better navigation so they built observatories and kept systematic records of their voyages. The first sciences to be established in India were, not surprisingly, geography and botany. Western science progressed primarily because of the military, economic, and political power of Europe, focusing on describing and explaining those aspects of nature that promoted European power, particularly the power of the upper classes. The disinterested commitment of European scientists to the pursuit of truths had little to do with the development of science. The subordination of the blacks in the ideology of the black "child/savage" and the confinement of the white women in the cult of "true womanhood" emerged in this period and are both by-products of the Empire. While the blacks were assigned animal and brutish qualities the white women were elevated and praised for their morality. While the blacks were segregated and enslaved, the women were placed in narrow circles of domestic life and in conditions of dependency. Racist and androcentric evolutionary theories were developed to explain human behaviour and canonised in the history of human evolution. The origins of Western, middle class social life, where men go out to do what men have to do, and women tend the babies and look after the kitchen, are to be found in the bonding of "man-the-hunter "; in the early phases of evolution women were the gatherers and men went out to bring in the beef. Now this theory is based on little more than the discovery of chipped stones that are said to provide evidence for the male invention of tools for use in the hunting and preparation of animals. However, if one looks at the same stones with different cultural perceptions, say one where women are seen as the main providers of the group – and we know that such cultures exist even today – you can argue that these stones were used by women to kill animals, cut corpses, dig up roots, break down seed pods, or hammer and soften tough roots to prepare them for consumption. A totally different hypothesis emerges and the course of the whole evolutionary theory changes.

Thus the cultural, racial, and gender bias of modern science can be easily distinguished when it is seen from the perspective of non-Western cultures, marginalised minorities, and women. The kinds of questions science asks when seeking to explain nature's regularities and underlying causal tendencies, the kinds of data it generates and appeals to as evidence for different types of questions, the hypotheses that it offers as answers to these questions, the distance between evidence and the hypothesis in each category, and how these distances are traversed all have the values of white middle class men embedded in them. Put simply, this implies relativism in science as in any other sphere of human knowledge. However, most scientists do not look kindly towards criticism, or sociological, philosophical, historical, and anthropological studies which highlight science's value laden nature. Relativism is anathema to scientists: many believe that they are engaged in revealing nature's absolute truths. Science, they argue, is special and different from any other body of knowledge: it is counter-intuitive and rarely a matter of common sense. Some propagandists for science have even suggested that the entire discipline of the sociology of knowledge is a conspiracy of the academic left against science.

The idolisation and mystification of science, the insistence on its value neutrality and objectivity, is an attempt not only to direct our attention away from its subjective nature but also from the social and hierarchical structure of science. Whenever we think of "the scientists" we imagine white men in white coats: the sort of chaps we see in advertisements for washing powder and skin care preparations, standing in a busy laboratory behind a Bunsen burner and distillation equipment telling us how the application of science has led to a new and improved soap or cold cream. This view of scientists is not far from reality. True power in science belongs to white, middle-aged men of upper classes. Everyone else working in science – women, minorities, black men, white men of lower classes, and third world researchers – are actually basically rank and file laboratory workers. The social hierarchy within science by and large preserves absolute social status, the social status scientific workers hold in the larger society. The people who make decisions in science, who decide what research is to be done, what questions are going to be asked, and how the research is going to be done are a highly selective, tiny minority. These people have the right background, the contacts to get the necessary appointments, and then further contacts to secure funding for their research projects. The actual execution of scientific research, the grinding and repetitive laboratory work, is rarely done by the same person who conceptualises that research; even the knowledge of how to conduct research is rarely possessed by those who actually do it. This is why the dominant (Western) social policy agendas and the conception of what is significant among scientific problems are so similar. This is why the values and agendas important to white, middle class men pass through the scientific process to emerge intact in the results of research as implicit and explicit policy recommendations. This is why modern science has become an instrument of control and manipulation of non-Western cultures, marginalised minorities, and women.

Even if we were to ignore all other arguments and evidence, the very claim of modern science to be value-free and neutral would itself mark it as an ethnocentric and a distinctively Western enterprise. Both claiming and maximising cultural neutrality is itself a specific western cultural value: non-Western cultures do not value neutrality for its own sake but emphasise and encourage the connection between knowledge and values. By deliberately trying to hide its values under the carpet, by pretending to be neutral, by attempting to monopolise the notion of absolute truth, Western science has transformed itself into a dominant and dominating ideology.

Ziauddin Sardar

References

Alvares, Claude. *Science, Development and Violence*. Delhi: Oxford University Press. 1992.

Aronowitz, Stanley. *Science As Power*. London: Macmillan. 1988.

Collins, Harry and Trevor Pinch. *The Golem: What Everyone Should Know About Science*. Cambridge: Cambridge University Press. 1993.

Feyerabend, Paul. *Against Method*. London: NLB. 1975.

Feyerabend, Paul. *Science in a Free Society*. London: Verso. 1978.

Funtowicz, Silvio and J.R. Ravetz. *Uncertainty and Quality in Science for Policy*. Dordrecht: Kluwer. 1990.

Goonatilake, Susantha. *Aborted Discovery: Science and Creativity in the Third World*. London: Zed. 1984.

Gross, Paul R. and Norman Levitt. *Higher Superstition: The Academic Left and Its Quarrels with Science*. Baltimore: Johns Hopkins University Press. 1994.

Harding, Sandra. *The Science Question in Feminism*. Milton Keynes: Open University Press. 1986.

Harding, Sandra, ed. *The "Racial" Economy of Science*. Bloomington: Indiana University Press. 1993.

Harding, Sandra. " Is Science Multicultural? Challenges, Resources, Opportunities, Uncertainties." In *Multiculturalism: A Critical Reader*. Ed. David Theo Goldberg. Oxford: Blackwell. 1994, pp. 344–370.

Jacob, Margaret, ed. *The Politics of Western Science*. Atlantic Highlands, New Jersey: Humanities Press. 1994.

Knorr-Cetina, Karin. *The Manufacture of Knowledge*. Oxford: Pergamon. 1981.

Kuhn, T.S. The *Structure of Scientific Revolutions*. Chicago: University of Chicago Press, 1962; 2nd ed. 1972.

Lakatos, Imre and Alan Musgrove. *Criticism and the Growth of Knowledge*. Cambridge: Cambridge University Press. 1970.

McNeil, Maureen, ed. *Gender and Expertise*. London: Free Association Books. 1987.

Mitroff, Ian. *The Subjective Side of Science*. Amsterdam: Elsevier. 1974.

Moraze, Charles. *Science and the Factors of Inequality*. Paris: UNESCO. 1979.

Nandy, Ashis, ed. *Science and Violence*. Delhi: Oxford University Press. 1988.

Ravetz, J.R. *Scientific Knowledge and Its Social Problems*. Oxford: Oxford University Press. 1971.

Ravetz, J.R. " Science and Values." In *The Touch of Midas: Science, Values and the Environment in Islam and the West.* Ed. Ziauddin Sardar. Manchester: Manchester University Press. 1982. pp. 4353.

Ravetz, J.R. *The Merger of Knowledge With Power*. London: Mansell. 1990.

Rouse, Joseph. *Knowledge and Power*. Ithaca: Cornell University Press. 1987.

Sardar, Ziauddin, ed. *The Touch of Midas: Science, Values and the Environment in Islam and the West*. Manchester: Manchester University Press. 1982.

Sardar, Ziauddin. "Conquests, Chaos, Complexity: The Other in Modern and Postmodern Science." *Futures* 26(6): 665–682. 1994.

Seshadri, C.V. *Equity is Good Science*. Madras: Murugappa Chettier Research Centre. 1993.

Walpert, Lewis. *The Unnatural Nature of Science*. London: Faber and Faber. 1992.

See also: Colonialism and Science – Technology and Culture – Western Dominance Environment and Nature

Varāhamihira

Varāhamihira, who flourished in Ujjain, in Central India, during the sixth century, was perhaps the greatest exponent of the twin disciplines of astronomy and astrology in India. A master of all three branches of the disciplines astronomy, natural astrology, and horoscopic astrology, he was a prolific writer whose works number more than a dozen, some of which are extensive.

Varāhamihira was born in Kāpitthaka, present-day Kapitha, in Uttar Pradesh, known also as Saṅkāśya and mentioned as a great centre of learning by the Chinese pilgrim Yuan Chwang as Kah-pi-t'a. He was the son of Ādityadāsa, and a Śakadvīpī brāhmaṇa of the Maga sect who were sun worshippers. His renown has caused several legends, both Hindu and Jain, being woven round his birth, growth, and predictive propensity. A legend has it that he was one of the nine luminaries of the court of King Chandragupta II Vikramāditya, but the definitively known date of Varāhamihira goes against this identification. Varāhamihira's patron has now been identified as King Mahārājādhirāja Dravyavardhana who ruled over Ujjain during the middle of the sixth century. Varāhamihira also possessed high poetic talents, so that some of the later rhetoricians have extracted verses from his writings to illustrate poetic qualities. Indeed, the fame of Varāhamihira has even induced several places in India to be named after him.

It is a characteristic of Varāhamihira that he produced larger and shorter versions of most of his works. His only work on astronomy is *Pañcasiddhāntikā* (The Five Basic Texts), being a redaction of select topics from the basic texts of five earlier astronomical schools: *Vāsiṣṭha, Paitāmaha, Romaka, Pauliśa* and *Saura*, in 443 verses, in the form of a work manual.

Varāhamihira wrote several astrological works. In his *Bṛhajjātaka* (Large Horoscopy), called also *Horāśāstra* (Science of Hours), in 25 chapters, containing about 400 verses, he treated all conceivable topics on the subject. This work has been a model for later works and is highly popular even today. In this work, Varāhamihira exhibits his understanding of Greek astrology and employs the Sanskritised forms of a number of Greek terms. His *Laghujātaka* (Shorter Horoscopy) is an abridged form of the previous work. On marriage and the prediction of auspicious times to marry, Varāhamihira composed two works, the *Bṛhadvivāha-paṭala* (Larger Treatise on Marriage), and its abridgement, the *Svalpa-vivāha-paṭala*. Prognostication on military marches and domestic journeys were treated in three works: *Bṛhad-Yoga-Yātrā* (Larger Course on Expedition), in 34 chapters, *Yoga-yātrā* (Course on Expedition), and *Svalpa-yātrā* (Shorter Course on Expedition).

On natural astrology, Varāhamihira's major work is the *Bṛhatsaṃhitā* (Large Compendium), the *Svalpa-saṃhitā* (Shorter Compendium) and *Vaṭakaṇikā* (Short Text), the last two known only through profuse quotations in later works. The *Bṛhatsaṃhitā*, in 106 chapters, is an encyclopaedic compendium on numerous subjects relating to life and nature, such as physical astronomy, geography, calendar, meteorology, flora, portents, agriculture, economics, politics, physiognomy, engineering, botany, industry, zoology, erotica, gemmology, hygiene, omens, and prognostication on the basis of asterisms, lunar days, etc. A fund of material on applied science, physical observation, and deduction on the basis of statistics and experimentation went into the production of this work which is perhaps unparalleled in the early literature of the world. Varāhamihira's works, set in precise terminology and in graceful language, have been models for writers of later times, not only in astronomy and astrology, but in other disciplines as well.

K.V. Sarma

References

Bṛhad-yātrā of Varāhamihira. Ed. David Pingree. Madras: Government Oriental Manuscripts Library. 1972.

Bṛhajjātakam of Varāhamihira Ed. and trans. V. Subrahmanya Sastri, 2nd ed. Bangalore: K.R. Krishnamurthy. 1971.

Bṛhat-saṃhitā of Varāhamihira Ed. and trans. N.Ramakrishna Bhat. Delhi: Motilal Banarsidass. 2 Pts. 1982.

Fleet, J.F. *The Topographical List of Bṛhat-saṃhitā*. Ed. Kalyan Kumar Dasgupta. Calcutta: Semushi. 1973.

Pañcasiddhāntikā of Varāhamihira. Ed. and trans. T.S.K. Sastry and K.V. Sarma. Madras: PPST Foundation. 1993.

The Pañcasiddhāntikā of Varāhamihira. Ed. and trans. Otto Neugebauer and David Pingree. Copenhagen: Munksgaard. 2 Pts., 1970, 1971.

Shastri, Ajay Mitra. *India as Seen in the Bṛhat-saṃhitā of Varāhamihira*. Delhi: Motilal Banarsidass. 1969.

Shastri, Ajay Mitra. *Varāhamihira and his Times*. Jodhpur: Kusumanjali Prakashan. 1991.

See also: Astronomy – Astrology

Vaṭeśvara

Vaṭeśvara (b. CE 880), son of Mahadatta, hailed from Ānandapura or Vaḍanagar, in Gujarat in Western India, a great centre of learning of the time. The *Vaṭeśvara-siddhānta,* composed by Vaṭeśvara in CE 904, is one of the largest and most comprehensive works on astronomy, a work which throws much light on the theories, methodologies, and processes of Indian astronomers until the tenth century CE. It was one of the standard works for the study of the discipline, and several of the rules enunciated by Vaṭeśvara were adopted by later astronomers like Śrīpati (eleventh century) and Bhāskara II (b. 1114). It is also noteworthy that the Persian scholar and polymath al-Bīrūnī, who came to India early in the eleventh century, referred to Vaṭeśvara and cited some of his views in his own writings. Vaṭeśvara also wrote a *Karaṇasāra*, known only through references, and a *Gola* (Treatise on Spherics), which is available only in part, the existing portion dealing with a graphical demonstration of planetary motion, construction of the armillary sphere, spherical rationale, and the nature of the terrestrial globe.

The *Vaṭeśvara-siddhānta*, in 1326 verses set out in eight chapters, deals exhaustively with all aspects of Indian astronomy, besides bringing in a number of methodologies, short-cuts, and interpretations. Chapter I, on mean motion, depicts the astronomical parameters, time-measures, calculation of the aeonary days, and computation of the mean planet in three ways: using parameters, using a cut-off date, and by the orbital method. The longitude corrections to be applied to mean longitude are also dealt with here. Chapter II, on true motion, deals with the corrections to be applied to the mean planet by the epicyclic theory, by the eccentric theory, and by the use of the *R* sine table. Different items relevant to the almanac are also set out here. Chapter III treats in detail the three problems in diurnal motion. Chapters IV and V deal with the computation of lunar and solar eclipses. Chapter VI treats of heliacal rising, chapter VII with elevation of lunar horns, and chapter VIII with the conjunction of celestial bodies.

Again, a unique feature of the work, found in no other work, consists in Vaṭeśvara's dividing each chapter into small sections earmarked for different topics. Unlike other works, rules are formulated for all possible alternatives of a theory or practice. It is interesting that sets of problems are posted at the ends of chapters for the student to solve, as in textbooks of modern days. The work is also characterised by the depiction of novel interpretations and methodologies practised by early Indian astronomers. In fact, the invaluable service done by Vaṭeśvara through his *Siddhānta* lies in his presenting the achievements of the Indian astronomers from the sixth century to the tenth century and methodically documenting the astronomical knowledge of the Hindus during those centuries.

K.V. Sarma

References

Sastry, T.S. Kuppanna. "The System of the Vaṭeśvara Siddhānta." *Indian Journal of History of Science* 4 (1–2): 135–43. 1969. Reprinted in *Collected Papers on Jyotisha*. Ed. T.S. Kuppanna Sastry. Tirupati: Kendriya Sanskrit Vidyapeetha. 1989. pp. 76–88.

Shukla, K.S. "Hindu Astronomer Vaṭeśvara and his works." *Ganita* 23(2): 65–74. 1972.

Vaṭeśvara-siddhānta and Gola of Vaṭeśvara, 2 vols. Ed. K. S. Shukla. New Delhi: Indian National Science Academy. 1986.

See also: Śrīpati – Bhāskara – Astronomy – al-Bīrūnī

W

Weights and Measures in the Indus Valley

The golden era of the Indus civilisation in ancient India extended from 2300–1750 BCE. This vast civilisation had significant uniformity and standardisation in its material culture, as reflected in its town planning, building construction, pottery, metallurgy, and system of weights and measures.

The prosperity of the Indus cities depended to a large extent on trade. Many raw materials were brought by land and sea routes from within and outside the Indus valley. Inland trade must have extended beyond the Baluchistan to Afghanistan and the Iranian highlands on the one hand and to the Punjab and Aravalli hills on the other. Overseas trade covered the Makran and Persian Gulf ports on the west and the Gujarat and Konkan ports, if not those of the Malabar Coast in the south. The writing system has not yet been deciphered, and the names of their measuring units are still unknown.

Length

Fragments of linear measures excavated from Mohenjodaro and Lothal are made of shell and ivory. The average length of a unit is estimated to have been 67.6 cm. The linear measures were graduated using the decimal system. Some of the precise linear measures were graduated even to the one-hundredth and one-four hundredth of the unit. A fragment of bronze linear measure, which graduated to one-half (9.34 mm) of the later unit, *aṅgula* (digit) was also discovered in Harappa.

Area and Volume

The use of area measures should have been quite common. There is, however, no piece of evidence which could be connected with an area measure. In the excavations carried out at various sites of the Indus civilisation a variety of pots made of clay, and sometimes of metal, has been discovered. No systematic determinations of the volumes of pottery seem, however, to have been made.

Mass

The oldest known weight in the Indus measuring system was excavated from Dashli Tepe, south Turkmenia in Russia. This era dates back to the fifth millennium BCE. The other three weights belong to the fourth millennium BCE and were discovered in northern Iran. These weights belong to pre-Indus civilisation. In the third millennium BCE the Indus measuring system was further developed in the ancient regions of Iran and Afghanistan.

A total of 558 weights were excavated from Mohenjodaro, Harappa, and Chanhu-daro, not including defective weights. They did not find statistically significant differences between weights that were excavated from five different layers each about 1.5 m in depth. This was evidence that strong control existed for at least a five-hundred-year period. The 13.7 g weight seems to be one of the units used in the Indus valley. The notation was based on the decimal system. Eighty-three per cent of the weights which were excavated from the above three cities were cubic, and sixty-eight per cent were made of chert.

Balance pans were made of copper, bronze, and ceramics. A bronze beam was found with the two pans in Mohenjodaro. The fulcrum was the cord-pivot type (Figure 1).

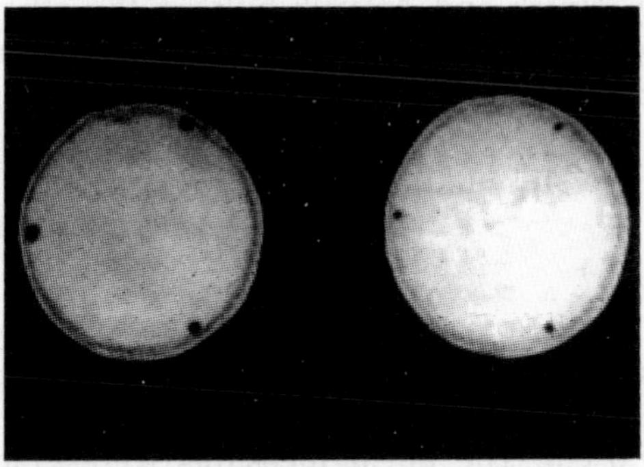

Figure 1 *Balance pans. Photograph by the Mainichi Newspapers, 1961. Used with permission.*

The measuring system used in the Indus valley was different from the Mesopotamian and Egyptian measuring systems, but the sensitivity of precision balances used in these regions is assumed to have been comparable. The weights excavated from Taxila (sixth century BCE – seventh century CE) descend from the system of weights used in the Indus civilisation (Figure 2).

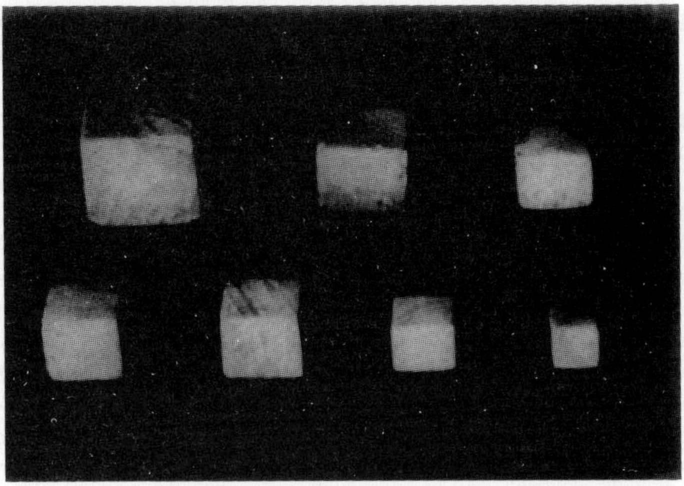

Figure 2 *Weights. Photograph by the Mainichi Newspapers, 1961. Used with permission.*

Shigeo Iwata

REFERENCES

Bhardwaj, H.C. *Aspects of Ancient Indian Technology*. Delhi: Motilal Banarsidass. 1979.

Hori, Akira. "A Consideration of the Ancient Near Eastern Systems of Weights." *Orient* 22: 16–36, 1986.

Iwata, Shigeo. "On the Standard Deviation of the Weights of Indus Civilisation." *Bulletin of the Society for Near Eastern Studies in Japan* 27 (2): 13–26. 1974.

Iwata, Shigeo. "Development of Sensitivity of the Precision Balances." *Travaux Du 1er Congres International De La Metrologie Historique*. Zagreb: Jugoslavenska academija znanosti i umjet-nosti, Historijski Zavod. pp. 1–25 + Fig. 1. 1975.

Mainkar, V.B. "Metrology in the Indus Civilization." In *Frontiers of the Indus Civilization*. Ed. B.B.Lal and S.P. Gupta. New Delhi: Books & Books. 1984. pp. 141–151.

Rao, S.R. *Lothal and the Indus Civilization*. London: Asia Publishing House. 1973.

Western Dominance: Western Science and Technology in the Construction of Ideologies of Colonial Dominance

Scientific curiosity was a major motive behind Western overseas expansion from the fifteenth century onward and technological innovations, particularly in shipbuilding, navigational instruments and firearms, made that expansion possible. But early European explorers and conquistadors did not rely heavily on evidence of scientific or technological achievements as gauges of the worth of the peoples they encountered or as explanations for their growing dominance in the Americas and maritime Africa and Asia. In encounters with the great centres of civilisation in Africa and Asia, European superiority in these endeavours was highly selective, marginal, or in many areas non-existent. In fact, travellers to China and the Indian subcontinent in the early centuries of expansion were as likely to dwell on the technological deficiencies of the West, when compared to these great civilisations, as to boast of European advantages. As in India, China, and Japan, the Europeans were able to make little headway into the heartlands of the Islamic world in this era. That standoff and the fact that they had borrowed so heavily from the scientific learning and technology of Muslim cultures in the centuries of Europe's emergence as a global force, rendered it unlikely that material standards would supplant the long contested religious differences that the Europeans had employed to set themselves off from and above the followers of Islam.

In Africa, disease and geographical barriers and the power of coastal kingdoms prevented the Europeans from translating their technological edge into significant conquests. Failure to move into the African interior also meant that the Europeans had only the vaguest notions about African epistemologies or understandings of the natural world, which were usually dismissed as superstition or fetishism. In sharp contrast to their experience in Africa, European invaders encountered few disease barriers in the Americas. In fact, diseases from smallpox to the measles became powerful allies of the Spanish conquistadors in their assaults on the heavily populated and highly advanced civilisations of Mesoamerica and the Andean highlands. The long isolation of the Amerindian peoples from the Afroeurasian people and cultures left them highly vulnerable to both the microbes borne and the iron-age technology wielded by the European invaders. None the less, the Spanish tended to attribute their startling successes in battle against seemingly overwhelming numbers of Aztec or Inca adversaries and the rapid conquests that followed to supernatural forces and to the superiority of their militant brand of Christianity over the "heathen" faiths of the indigenous inhabitants of the "New" World.

Lacking in-depth knowledge of the epistemologies and scientific learning of most of the peoples they encountered in the early centuries of overseas expansion and often enjoying only very selective (but at times critical) technological advantages over them, European explorers, missionaries, and Crown or Company officials were unlikely to rely on material standards to judge the level of development attained by other cultures or to compare overseas civilisations to Europe itself. Until at least the end of the seventeenth century, religious beliefs, or the Europeans' certitude that they possessed vastly superior understandings of the transcendent world, predominated as the gauge by which other cultures and peoples were assessed and ranked. Additional cultural variables, such as the position and treatment of women, were frequently cited as evidence of advancement or savagery; and physical features, especially skin colour, were sometimes emphasised in attempts to distinguish and rank the peoples encountered overseas.

Scientific and technological gauges of past attainments and present abilities remained peripheral to most evaluations of the peoples and cultures that the Europeans encountered as they expanded across the globe. None the less, signs of material advancement – the existence of large cities, sophisticated techniques of fortress construction, or evidence of complex scientific instruments – were often noted and even cited to support arguments regarding the level of development achieved by different peoples. In two areas in particular, in the perception and measurement of time and in perspectives on space associated with artistic and mathematical advances of the Italian Renaissance, European overseas observers began to see a clear divide between the West and all other civilisations and cultures. In addition, as early as the sixteenth century, European commentators began to rank African cultures beneath those of Asia and the Americas, not so much because of skin colour or other physical features, as has often been argued, but due to what was perceived as a markedly lower level of material culture in Africa than that found by European travellers and traders in India, China, or central Mexico.

Although ideologies justifying overseas expansion and the domination of non-European peoples from the fourteenth to the early eighteenth century were rooted in religious belief and were generally culture- rather than racially-oriented, material accomplishment, including the assumed capacity for invention and scientific thinking, was increasingly associated with racist defences of the enslavement of Africans. Defenders of the slave trade sought to counter the abolitionists' objections with often lurid descriptions of the alleged savagery of African life and the debased level of African material culture. Along with skin colour and other physical differences, racist writers, such as Samuel Estwick and Dominique Lamiral, emphasised material backwardness and ignorance of the workings of the natural world as proof of the subhuman nature of Africans that justified their subjugation as slaves.

By the last decades of the eighteenth century, this rather broad association between racial ideology, material culture, and the defence of slavery, was refined and enhanced by the rise of racist theories allegedly grounded in scientific experimentation and reasoning. Physicians and ethnologists devised a variety of measurements – from skull size and shape to genital configurations – in efforts to prove that there were innate physical, mental and moral differences between human racial groups. The fact that the measurements reflected a priori assumptions and were based on small and suspect samples, and that even the racial categories themselves were hotly contested, did not prevent "scientific" racism from winning widespread support from European scientists, social commentators, and politicians throughout the nineteenth century. The tenets of scientific racism were popularised among the middle and working classes by practitioners of phrenology, whose booths could be found at county fairs and on the promenades of seaside resort towns, and by the pulp press, where allegedly scientific proofs of European racial superiority were linked to social evolutionist arguments for imperialist expansion.

By the first decades of the nineteenth century, scientific and technological gauges of human achievement and worth were clearly in the ascendant. Earlier measures of the level of development achieved by different cultures continued to be cited. Religious belief, for example, remained of paramount importance to missionaries active in overseas lands. But even missionaries increasingly linked Europe's advances in the sciences and invention to the rhetoric of Christian proselytisation. Ignoring past and contemporary tensions between science and religion in Europe itself, prominent missionaries, such as the Abbé Boilat and David Livingstone, argued that Christian culture had been particularly receptive to scientific investigation and technological innovation, and that conversion to Christianity would promote the scientific and material development of colonised peoples. The growing numbers of ethnologists and professionally trained anthropologists, who found in the colonies relatively safe and fertile environments for their research, also privileged non-scientific or technological standards, such as modes of political organisation or gender relations, in assessing the level of development attained by African, Asian, Amerindian, and Pacific Island peoples. But evidence of material culture was increasingly linked to societal advance, and indigenous "superstitions" or at best magical beliefs contrasted with the scientific mindset that was seen as typical of the educated West.

A number of factors account for the dominance of material standards, particularly those linked to science and technological innovation, in nineteenth-century ideologies of European global hegemony. Most critically, the transformations wrought by industrialisation from the middle of the eighteenth century in England and somewhat later in Belgium, Germany, France, and Italy made the gap in scientific and technological capacity and material development between western Europe and non-Western societies increasingly apparent to European and non-European observers alike. Maxim guns, steamboats, and railway lines carried elements of Europe's industrialisation to colonised areas, and champions of imperialist expansion reasoned that these wonders could not help but impress subjugated peoples with the unprecedented degree to which European societies had advanced over their own. Not only was European superiority in science and technology obvious, it could be empirically tested in ways that claims of higher religious understanding or moral probity could not. Europeans had vastly more firepower, could produce incomparably greater quantities of goods much more rapidly, and could move both these products and themselves about the globe with much greater speed and comfort than any other people, including the once highly touted Chinese. In an age when what were held to be scientific proofs were authoritative, attainments that could be measured statistically were viewed as the most reliable gauges of human ability and social development. Mechanical principles and mathematical propositions could be tested; cast iron or steel bridge spans could be compared for size and strength with the stone or wooden structures of non-Western societies; and human skulls could be quantified in seemingly infinite ways to assess the highly variable mental capacity of the "races of man".

The pre-eminence gained in the nineteenth century by scientific and technological standards of human worth and ability not only bolstered proponents of theories of European racial supremacy, but it also proved vital to various formulations of the civilising mission ideology that both inspired and rationalised European imperialist expansion from the early 1800s to 1914. Chauvinistic politicians in the metropoles and imperial proconsuls in the colonies increasingly stressed the importance of the diffusion of Western science and technology to what they viewed as the benighted peoples and backward lands that had come under European control. Proponents of the civilising mission confidently predicted that the world would be remade in the image of industrialising Europe. Given Europe's material advancement, it was seen as appropriate that Europe and North America serve as the sources of capital, both machine and financial; of entrepreneurial, scientific, and managerial expertise; and of manufactured goods for the rest of the globe. In this view, the non-Western world, including both areas that had been formally colonised and those that had come under the informal sway of the Great Powers, were best suited to provide abundant and cheap land, labour, and raw materials that were required to fuel the industrial economies of Europe and North America.

According to the "improvers" or non-racist advocates of the civilising mission, the spread of Western education among colonised peoples – emphasising the inculcation of at least rudimentary Western scientific learning and technological skills – would provide the critical means by which the material level of non-Western societies would gradually be raised. Though they approved of the diffusion of essential Western technology to overseas areas under the paternalist supervision of European colonisers, racist apologists for imperialism had little faith in the ability of subjugated peoples to master the sciences or engineering of the West. Thus, they envisioned the period of European "tutelage" extending for centuries, if not indefinitely, into the future.

The non-Western peoples who were the targets of the European colonial enterprise were very often awed and overwhelmed by their initial encounters with the science and technology of the industrialising West. Whether they were indigenous leaders resisting the growing encroachments of European forces or scribes and merchants who allied themselves with the invaders, the colonised could not help but be impressed with the clear and increasing advantages in power that the Europeans gained through their superior capacity to tap the resources of the natural world, to produce material goods, and to devise more deadly weapons. As surveys taken as late as the post-World War II era, such as those which form the basis for G. Jahoda's *White Man*, the science and technology of the colonisers gave them an aura of magical power among the colonised masses in many areas. Though Western-educated Africans, Asians, or Polynesians were likely to scoff at such expressions of popular admiration, most came to accept that Western science and technology were not only on the whole superior to their own but essential for the future "development" of societies they hoped someday to rule. Therefore, nationalist critiques of imperial domination often deplored the fact that colonialism had severely constricted the flow of science and technology from the West to dominated areas, and demanded that technical education and scientific facilities for indigenous peoples be expanded and improved.

By the last decades of the nineteenth century, however, a number of influential African, Asian, and Caribbean thinkers were mounting cogent challenges both to notions of European racial superiority based on evidence of scientific and technological achievement and to the advisability of the wholesale transformation of non-Western cultures and societies along Western, industrial lines. Much of this resistance to the hegemonic ideologies of the Western colonizers focused on efforts to reassert and revitalise indigenous epistemologies, modes of social organisation, and approaches to the natural world. Thinkers such as Vivekananda and Aurobindo Ghosh contrasted an Indian spiritualism with the deadening abstractions of Western materialism. African writers, such as the Caribbean-born Edward Blyden, deplored the devastating impact of the Atlantic slave trade on African cultures and celebrated the Africans' strong sense of community, reverence for and care of the elderly, and sophisticated artistic creations.

Ironically, these defences of colonised cultures were buttressed by contemporary European anthropological studies, usually carried out under the auspices of colonial administrations; by the intense interest in "Oriental" religions fashionable among European intellectuals in the decades before World War I; and by the "discovery" of the abstract power of

African masks and other forms of "primitive" artistic expression by Picasso, Derain, Matisse, and other avant-garde artists in the early 1900s. Inadvertently, however, the works of these first generations of Indian and African critics of European hegemonic ideologies often validated the very materialistic standards they sought to contest. For example, Indian thinkers, particularly Vivekananda, repeatedly stressed the scientific accomplishments of India's ancient civilisations, while African and West Indian writers, most notably Anetor Firmin, claimed Egypt, with its impressive engineering and architectural feats, as a civilisation that black Africans had done much to build.

With the coming of the First World War, non-Western critics of what had been characterised as the excessively rationalistic, impersonal, and materialistic West found numerous, and highly vocal, European intellectual allies. The horrific trench slaughter on the Western Front and the multitude of ways that Western scientific knowledge and experimentation were harnessed to the war effort as a whole raised profound doubts for noted thinkers, such as Paul Valery, Sigmund Freud, and Georges Duhamel, about the long- assumed progressive nature of Western science and technology.

After such a savage and suicidal war, the tenets of the civilising mission rang hollow, and "scientific" racist thought came under increasing assault in both western Europe and the United States. Collaboration with indigenous elites was increasingly stressed in the governance of the colonies, and a rhetoric of science and technology as agents of development through cooperation with indigenous peoples permeated colonial policy making.

European doubts about the directions taken by the industrial West and its global hegemony gave new impetus to African, Asian, and Caribbean critiques of European global hegemony. René Maran's *Prix Goncourt*-winning novel, *Batouala,* mocked the pretensions of racial superiority held by European colonisers, and idealised village life in French West Africa. The poets of the *négritude* movement, most powerfully L.S. Senghor and Aimé Césaire, mourned the suffering and destruction wrought by European science and technology in Africa and the lands of the slave diaspora, and inverted racist epithets by exulting blackness, intuition, affinity for the natural world, and indifference to inventiveness. Aurobindo Ghosh and Rabindranath Tagore viewed the war as the fulfilment of earlier Indian prophecies of a coming cataclysm in the aggressive and materialist West, and proof of importance of India's spiritual mission in the modern age. Mohandas Gandhi also cited the senseless violence and colossal destructiveness of the war in support of his sweeping assaults on industrial society. In the decades after the war, he sought to formulate for India (and implicitly for other colonised areas) a community-centred, low tech, and conservationist alternative to industrialised society as it had developed in the West. In this same period, Gandhi also worked out a strategy for confrontational but non-violent protest that repeatedly proved an effective antidote to the advanced technologies of repression employed by Western overlords in the decades of decolonisation from the 1920s onward.

Though battered and under assault, the scientific and technological underpinnings of ideologies of Western dominance survived the crisis of two global wars and the powerful critiques of Gandhi and the *négritude* writers largely due to the emergence of the United States as *the* global power from the 1920s onwards. Entering the First World War late and just when new technologies had restored a war of motion and decision, Americans continued their long-standing infatuation with science and technology after the conflict. Faith in the essentially progressive nature and beneficence of science and technology informed the development (later called modernisation) theory that came to dominate both American thinking on colonial issues and that of the European and Japanese imperialist rivals of the United States. After World War II, America's chief rival, the Soviet Union, also championed a rhetoric of development that privileged science and large-scale industrialisation. With modernisation theory (in a number of capitalist and socialist versions) in the ascendant, non-Western alternatives to social development and economic well being, such as that formulated by Mohandas Gandhi, were marginalised or openly spurned by the Western-educated elites that governed the new states that emerged from the collapsing colonial empires. Though alternative approaches have gained significant support in some of these new nations, most notably India and Tanzania, international agencies and Western and non-Western planners continue to rely upon modernisation schemes, based overwhelmingly on Western precedents, to solve the problems of poverty and growing wealth differentials and the demographic and environmental dilemmas that have been building on a global basis for centuries.

Michael Adas

References

Adas, Michael. *Machines as the Measure of Men: Science, Technology and Ideologies of Western Dominance*. Ithaca, New York: Cornell University Press. 1989.

Cohen, William B. *The French Encounter with Africans: White Response to Blacks*, 1530–1880. Bloomington: University of Indiana Press. 1980.

Curtin, Philip.*The Image of Africa: British Ideas and Action,* 1780–1850. Madison: University of Wisconsin Press. 1964.

Gould, Stephen J. *The Mismeasure of Man*. New York: Norton. 1981.

Irele, Abiola. "Négritude or Black Cultural Nationalism." *The Journal of Modern African Studies* 3(3): 321–48. 1965.

Irele, Abiola. "NégritudeLiterature and Ideology. "*The Journal of Modern African Studies* 3(4): 499–526. 1965.

Jahoda, G. *Whiteman: A Study of the Attitudes of the Africans to Europeans in Ghana Before Independence*. Oxford: Oxford University Press. 1961.

Jordan, Winthrop D.*White Over Black: American Attitudes Toward the Negro*, 1550–1812. Chapel Hill: University of North Carolina Press. 1968.

Leclerc, Gérard. their *Anthropologie et Colonialisme: Essai sur l'Histoire de l'Africanisme*. Paris: Fayard. 1972.

Nandy, Ashis.*Science, Hegemony & Violence: A Requiem for Modernity*. Delhi: Oxford University Press. 1988.

Stephan, Nancy.*The Idea of Race in Science: Great Britain 1800–1960*. Hamden, Connecticut: Greenwood. 1982.

See also: Magic and Science – Colonialism and Science – Science as a Western Phenomenon – Technology and Culture

Wind Power

There is little doubt that the first practical use of the wind as an energy source other than as the motive power for sailing ships occurred in the East. Those of Persia were probably the first but precisely when is uncertain. According to a story of al-Ṭabarī writing around CE 850, and later writers, the second orthodox Caliph, ‹Umar ibn al-Khaṭṭāb, was murdered in CE 644 by a captured Persian technician, Abū Lu'lu'a, who claimed to be able to construct mills driven by the power of the wind and was bitter about the taxes he had to pay. This early date cannot be confirmed but Arabic geographers of the tenth century all confirm the existence of windmills in the region of Seistan in north eastern Iran. For example, al-Mas‹ūdī around CE 950 wrote of the wind driving mills and raising water from streams.

Nothing has survived to show how such mills pumped water but later drawings of one type of windmill suggest that it could have been derived from horizontal corn-grinding watermills. In the lower part of such watermills, the water fell down a chute and struck blades placed radially around a vertical shaft. The top of the shaft passed into a higher room and carried the upper millstone and so rotated it above the bedstone which was set on a floor in the middle of the mill. When developed into a windmill, the wind rotor was situated in the bottom of the mill. The building was constructed with four "loop holes" to direct the wind on to the blades from whichever quarter the wind might be coming. There might be eight or ten blades, and the wind was directed by the loop holes onto one side of the rotor. In this way the sails were pushed round on that side away from the direction of the wind while those advancing into the wind on the other side were shielded by the walls around the loop holes. The grinding stones were placed in an upper room with the upper stone turned by an extension of the rotor shaft in the 'underdrift' manner. None of these mills has survived.

The type which may still be in use today had the grinding stones situated in a room below the rotor with the sails above. This necessitated a change to the 'overdrift' method of driving the stones where the spindle passed down through the bedstone to a bearing underneath which had to support the weight of both the upper stone and the sails. It was possible to disconnect the driving shaft at the place where it was connected to the upper stone at the 'rynd' to allow for the stones being separated for dressing. The layout for grinding was similar to other corn mills with a hopper mounted on the wall from which the grain was fed into the central hole of the upper stone through a chute or shoe while the flour was passed out around the circumference of the stones.

The advantage of this second layout was that the rotors with the sails could be greatly enlarged. Mills near Seistan and on the borders of Iran and Afghanistan might have rotors approximately 5 m (16.4 ft) high by 3 m (9.8 ft) in diameter. There could be six or eight sails which had wooden framing that was interlaced with straw or covered with wooden boards. The upper parts of these mills were built with one wall that directed the wind on to one side of the rotor while another wall shielded the other half. Sometimes matting screens were erected to help channel the winds to the sails. There were no brakes, but more screens might be placed across the slots between the walls to regulate the wind reaching the sails; to secure the mill when out of use, the rotor and upper millstone were lowered to rest on the bedstone. A wind speed of 22.40 m/s (50 m.p.h.) was needed to drive these mills at only 30 r.p.m. Such a mill, working with intermittent wind for about four months of the year, would grind enough flour for about fifteen families. Seistan was known as the land of the winds and between mid-June to mid-October the wind regularly blew from the north for a period of 120 days. The mills were built in line to face these winds and the famous example at Neh had a long row of 75 mills.

Because these horizontal mills needed such strong and regular winds to power them, they did not spread in this form much beyond the borders of Iran and Afghanistan. They were invented earlier than the Western vertical type but it is doubtful whether there is any connection between the two. Those in the West, which have been described as the 'full admission, axial flow type', first appeared around CE 1150 possibly either in the south-east of England, the northwest of France, or Flanders.

However these Persian mills were probably the source of the later Chinese mills and possibly the Tibetan wind-powered prayer wheels.

The wind-powered Tibetan prayer wheels were not surrounded by elaborate buildings to guide the wind onto the rotors. The most common form had a vertical axle with horizontal spokes at its top, on the ends of which were fixed sails shaped like cups so that they caught the wind on one side of the rotor, but were smoothed or curved to present less resistance on the other. The wind would be caught in the concave part of the cup, or a curved sheet, during half the circle to turn the rotor, while in the other half, the convex or streamlined side would be advancing into the wind and so present less resistance. Power output was minimal but was sufficient to turn the prayer wheels which was the only use to which they were applied.

Prayer cylinders designed for automatic repetition of the Buddhist mantra are unlikely to have been produced before the reign of K'ri-srong-Ide-brtsan in CE 755 to 797 when Buddhism conquered Tibet. It is unlikely that such cylinders were turned by the wind at this period, although no doubt there were prayer flags fluttering in the wind from around this time. Early in the twelfth century, a new fashion for mechanical piety swept China, but again it seems doubtful whether this included wind-powered prayer wheels, which must therefore be placed later, and certainly therefore after the Persian mills.

More is known about the adoption of the windmill in China. Once again early dates for this have been suggested, but these are doubtful. In about CE 1230, Yelu Chu Zai was captured by Jinghi Khan and became his minister. He was an extremely good scholar, administrator and mathematician. An accurate description of the Persian windmill has been discovered in his memoirs, with a comment on how good it would be if the Chinese used it. A Chinese book of the seventeenth century, the *Zhu Qi Tu Shu*, describes the windmill as if it were a European invention, which could be a mistake for Persia. In China, these mills were used for raising water and evolved into an entirely different form from the ones in Persia or Tibet.

Chinese horizontal windmills are still used today along the eastern sea coast north of the Yangtze and in the region of Thangku and Taku near Tientsin to operate chain pumps through gearing for raising salt water for salt pans or fresh water for irrigation. Once again they have no elaborate structures for directing the wind onto the rotors but are closely linked to the way the sails of junks operate. Canvas-covered sails are mounted at the ends of radial arms, each with its own mast, in such a way that they can be spread to catch the wind on the side of the rotor turning away from the wind and be feathered to present the least resistance on the side turning against the wind. The sails can pivot on these masts, and, as on a junk, the sail extends to the front of the mast. The longer, or driving, side of the sail is tied to the rotor framework by a piece of rope of such a length that, when the sail is turning with the wind, it is held into the wind, but can rotate out of the wind when advancing against the wind. The angle of the sail can be set by the length of rope to a position in which the wind can do useful work on it for more than 180 degrees.

These mills must have been the most efficient of the horizontal types developed before the twentieth century, but all horizontal windmills suffer from the same problem, that only a small part of the wind rotor can be used to its maximum efficiency at any time. The theoretical maximum power coefficient for a simple horizontal windmill is only one third, but in practice it will be much less. These are the reasons why the horizontal windmill has not been developed further and few examples remain at work today.

In the eastern Mediterranean, at some period during the Middle Ages, vertical windmills appeared with a different type of sails from those normally used in the West. The horizontal wind shaft was extended in front of the mill like a ship's bowsprit so that ropes from it could help to stay six or eight radiating spars. The sails were triangular pieces of canvas like the sails of a modern yacht and worked in the same way. The leading edge of each sail was attached to the spar, round which it could be wrapped to reef it in strong winds. The free corner was secured by a rope in much the same way as a ship's boom. These sails were much lighter than those of a conventional western windmill and have found a new application in water pumping mills in some developing countries today.

Richard L. Hills

References

Harverson, Michael. *Persian Windmills.* The Hague: International Molinological Society. 1991.

Hills, Richard Leslie. *Power from Wind. A History of Windmill Technology.* Cambridge: Cambridge University Press. 1994.

Needham, Joseph. *Science and Civilisation in China.* Vol. 1, *Introductory Orientations,* and Vol. IV, *Physics and Physical Technology,* Part II, *Mechanical Engineering.* Cambridge: Cambridge University Press. 1954 and 1965.

White, Lynn. *Medieval Technology and Social Change.* Oxford: Clarendon Press. 1962.

Y

Yavaneśvara

The name Yavaneśvara, meaning Lord of Greeks, referring to one of Greek descent, is said to be the author of a number of Indian astrological works. He is mentioned for the first time by Sphujidhvaja who wrote his *Yavanajātaka* in CE 240. Towards the close of his book, Sphujidhvaja states that his work was based on that of Yavaneśvara. Yavaneśvara, who had been blessed by the Sungod, is stated to have rendered a Greek work on genethlialogy into Sanskrit in CE 150, at the instance of the ruler of the land. This work shows the position and influence of the stars at one's birth. The date mentioned by Sphujidhvaja is a little after the Kṣatrapa dynasty of Greek descent had established itself in the region of Saurashtra and Gujarat and ruled with its capital at Ujjain. The patron of Yavaneśvara has been identified, on the basis of coins and inscriptions, as Rudradāman I. Sphujidhvaja states that he (Rudradāman I) was quite conversant with the said Greek work.

Though the original work of Yavaneśvara is apparently lost, it is possible to get an idea of its extent and contents from *Yavanajātaka* which is its redaction. The *Yavanajātaka*, as it is available now, is an extensive work in 79 chapters, and takes under its purview a large variety of topics on horoscopy and natural astrology, including the delineation of the planets, their lords, their characteristics, the major and minor influences they exert on human beings at different periods, predictions relating to professions, experience of happiness and sorrow on account of planetary combinations, and predictions of the future on the basis of questions, omens, and military astrology. Most of these would have been depicted by Yavaneśvara as well. This is borne out also by several later texts, which are ascribed to Yavaneśvara. Among such works might be mentioned: The *Candrābharaṇahorā*, an extensive work in 101 chapters, prevalent in South India, and a shorter work with the same title prevalent in North India; two texts with the title *Yavanasaṃhitā*; a work called *Bhāvadīpikā* or *Bhāvādhyāya*, making predictions on the basis of the placement of the planets in the horoscope; *Nakṣatracūḍāmaṇi*, being predictions on the basis of the twenty-seven constellations; and, a *Yavanapārijāta* depicting the results of good and evil deeds in life. It is clear from this that a regular school of the Yavana tradition of astrology had developed in India.

What is significant is that in all these texts which ostensibly go under the authorship of 'Yavana', Hellenistic practices have been Indianised, both in the matter of content and presentation. Hindu caste distinctions and social orders are duly taken note of in making predictions; Hindu deities and their descriptions are duly invoked, and suitable modifications are made to their Greek counterparts. In effect, these texts seem to be wholly indigenous but for the ascription of their authorship to Yavana (Greek).

K. V. Sarma

References

Kane, P.V. "Yavaneśvara and Utpala." *Journal of the Asiatic Society* 30(1): 1–8. 1955.

Majumdar, M.R. *Historical and Cultural Chronology of Gujarat*. Baroda: Maharaja Sayajirao University. 1960.

Pingree, David. "The Yavanajātaka of Sphujidhvaja." *Journal of Oriental Research* (Madras) 31(1–4): 16–31. 1961–62.

The Yavanajātaka of Sphujidhvaja. Ed. David Pingree. Cambridge, Massachusetts: Harvard University Press, 1976.

See also: Astrology

Yoga

Yoga is one of the six principal systems of Indian thinking known as *darśanas*. The word *darśana* is derived from the Sanskrit root *dṛś*, meaning "to see". Fundamentally, *darśana* means "view" or "a particular way of viewing". Yoga, as one of the six *darśanas* has its source in the Vedas. In the traditional Indian view these are called *vaidika*, or Vedic *darśanas*. These are: *nyāya, vaiśeṣika, sāṅkhyā, yoga, mīmāṁsā*, and *vedānta* (there are other *Darśanas* that do *not* accept the supremacy of the Vedas, such as Buddhism and Jainism). While the source of Yoga was the Vedas, Patañjali, one of the great Indian sages, formalised Yoga. His classic text is *Yoga Sūtra* (Aphorisms on Yoga). Although there are many other major treatises on Yoga that postdate Patañjali's, his work is the most authoritative.

All the *darśanas* proclaim that it is their aim to help human beings achieve clarity and balance of perception and action. Yoga is unique in as much as it offers practical suggestions and guidelines to achieve this end. According to the tenets of Yoga, human beings are under the influence of *avidyā*, which is what prevents correct perceptive analysis. Sage Patañjali suggests practical ways to reduce and remove *avidyā*. In his *Yoga Sūtra*, three things are suggested to help us explore the meaning of Yoga and therefore feel *avidyā*. These are *tāpas, svādhyaya,* and *īśvara praṇidhāna. Tāpas* is a means by which we keep ourselves fit and clean. Often *tāpas* is defined as penance, mortification, and dietary austerity, but what is meant is the practice of *āsana* (postures), *prāṇāyāma* (control of the breath), and other disciplines. These practices aid in the removal of impurities from our systems. In so doing we gain control of our whole system. It is the same principle as heating gold to purify it.

The next part of Yoga is *svādhyāya*, the study of the self. Where are we? What are we? What is our relationship to the world? It is not enough to keep ourselves fit; we should know who we are and how we relate to others. This is not easy because we do not have an actual mirror for our minds as we do for our bodies. We must use reading, study, discussion, and reflection as a mirror to the mind.

The third means of exploration is *īśvara-praṇidhāna*. It is usually defined as "love of God" but it also means "quality of action". We must carry out our jobs, and all our actions must be done with quality. Since we can never be certain of the fruits of our labours, it is better to remain slightly detached from them and pay more attention to the actions themselves.

Together, these three cover the whole of human action: fitness, inquiry, and quality of action. Taken together, these practices are known as *Kriyā Yoga*, the Yoga of action. Yoga is not passive. We must be involved in life, and preparation is necessary for this involvement.

Patañjali's Yoga is sometimes called *Aṣṭāṅga* Yoga, which literally means Eightfold Yoga. These eight are *yama, niyama, āsana, prāṇāyāma, pratyāhāra, dhārāṇa, dhyāna*, and *samādhi.*

Patañjali considers five different attitudes *(yamas)* or relationships between an individual and "the outside". The first is *ahimsā*. While the word *himsā* means injury or cruelty, *ahimsā* means more than merely the absence of *himsā*. It means kindness, consideration, or thoughtful consideration of people or things. The next *yama* is called *satya,* "to speak the truth". The third *yama* is *asteya. Steya* means "to steal"; *asteya,* the opposite, means if we are in a situation where people trust us, we will not take advantage of them. The next yama is *brahmacarya*. The word is composed of the root *car* (to move) and *brahma* (the truth). If we move towards the understanding of truth, and sensual pleasures get in the way, we must keep our direction and not become lost. The last *yama* is *aparigraha,* "hands off". *Parigraha* is the opposite of the word *dāna,* which means "to give". *Aparigraha* means, "to receive exactly what is appropriate".

Niyamas, like *yamas,* are attitudes and are not to be taken as actions or practices. The five *niyamas* are more intimate in the sense that they are the attitudes we have towards ourselves. The first *niyama* is *śauca*, or cleanliness. There are two parts to this, external and internal. External *śauca* has to do with simply keeping ourselves clean. Internal *śauca* has to do with cleanliness of the internal organs and mind. The practice of *āsanas* or *prāṇāyāma* could be an internal *śauca*. The second *niyama* is *santoṣa*, a feeling of contentment. The next is *tāpas*, a word we have already discussed. With *tāpas* the idea is to bring out *aśuddhi*, "dirt" inside the body. *Svādhyāya* is the fourth *niyama*. As we defined it earlier, *sva* means self; *adhyāya* means study or inquiry. Actually, *adhyāya* means to go near. *Svādayāya* means to go near yourself, that is, to study yourself. Any study, reflection, or contact that helps us understand more about ourselves is *svādhyāya*. The last *niyama* has also been mentioned before. *Īśvarapraṇidhāna* means "to leave all our actions at the feet of the Lord". Since our actions often come from *avidyā* it is possible that they might go wrong. That is why contentment is so important. This attitude suggests that we have done our best, and can leave the fruits of our actions in the hands of something higher than ourselves.

The third *aṇga* is *āsana*. In the theory of *āsana* practice, there are two aspects, *sukha* and *sthira*. We must be comfortable and at ease (*sukta*) and we must be steady and alert (*sthira*). We must be involved and at the same time attentive. Yoga suggests ways to achieve these qualities in *āsana*. The fourth *aṅga* is *prāṇāyāma* which is conscious regulated breathing. *Pratyāhāra*, the fifth *aṅga*, involves the senses. The word *āhāra* means "food". *Pratyāhāra* means "withdrawing from that on which we are feeding". This refers to the senses: when the senses refrain from "feeding" on their objects, that is *pratyāhāra*.

Dhāraṇā comes from the root *dhṛ*, "to hold". *Dhāraṇā* occurs when we create a condition so that the mind, normally going in a hundred different directions, is directed towards one point. *Dhāraṇā* is a step leading towards *dhyāna*. In *dhāraṇā* the mind is moving in one direction; nothing else has happened. In *dhyāna*, when we become involved with a particular thing and we begin to investigate it, there is a link between ourselves and this object; that is, there is a perceptual and continuous communication between the object and our mind. This communication is called *dhyāna*. Further, when we become so involved with an object that our mind completely merges with it, that is called *samādhi*. In *samādhi* we are almost absent; we become one with that object.

There are many varieties of Yoga. Some people say that *dhyāna* is the means to *Jñāna Yoga*. In this context, this means inquiry about the truth, the real understanding that we attain in a state of *samādhi*. Inquiry in which we hear, then reflect, and then gradually see the truth, is *Jñāna Yoga*. In the *Yoga Sūtra* it is said that in the state of mind where there is no *avidyā*, automatically there is *Jñāna*.

Bhakti Yoga comes from the root *bhak* which means "to serve that which is higher than ourselves". This means an attitude of devotion. In *Mantra*

Yoga, a teacher who knows us very well might give us a *mantra* which has a particular connotation because of the way it has been arranged. If that *mantra* is repeated in a certain way, if we are aware of its meaning, and perhaps if we want to use a particular image, *Mantra Yoga* brings about the same effect as *Jñāna* or *Bhakti Yoga*. In *Rāja Yoga*, the word *Rāja* means "the king who is always in a state of bliss, who is always smiling". Any process through which we achieve greater understanding of the mysterious and the obscure is *Rāja Yoga*. In the Vedas there are many references to the word *Rāja* in relation to *īśvara*.

It is best to explain *Laya Yoga* in a context of *samādhi*, when the meditator completely merges with the object of meditation, that is *Laya*. We merge with the object and nothing else exists.

In recent times, much has been written about Hinduism, and a lot of it pertains to and derives from the viewpoint of Vedānta. It is important to see that the viewpoint of Yoga differs in some crucial respects from the viewpoint of Vedānta. Brahman is considered the "Pāramārthika Satya" or ultimate truth, and the world we live in and experience through our senses is granted the status of truth at an operational level, i.e. "Vyāvahārika Satya". In a sense, this carries the implication that the world is false and illusory. According to Yoga, everything we see, experience, and feel is not an illusion but is true and real. This concept is called *satvāda*.

Everything, including *avidyā*, dreams, and even fancy and imagination, is real. However, all these are constantly in a state of flux. This concept of change is called *pariṇāmavāda*. In Yoga, although everything we see and experience is true and real, changes do occur either in character or in content.

A. V. Balasubramanian

References

Aranya, Swami Hariharananda. *Yoga Philosophy of Patañjali*. Trans. P.N. Mukherji. Calcutta: University of Calcutta Press. 1977.

The *Bhagavadgītā*. Trans S. Radhakrishnan. New York: Harper and Row. 1948.

The *Gheraṇḍa Saṃhita*. Trans Sris Chandra Vasu. New York: AMS Press. 1974.

Haṭhayogapradīpikā of Svātmārāma. Trans Srinivasa Iyangar. Madras: Adyar Library and Research Centre. 1972.

Yoga Yājñavalkya. Trans. Sri Prahlad Divanji. Bombay: Royal Asiatic Society. 1954.

Yuktibhāṣā of Jyeṣṭhadeva

Jyeṣṭhadeva (fl. 1500–1610) was a Nambūthiri Brahmin from the Ālattūr village, an important Brahmin settlement near Cochin. He was probably a student of Dāmodara, the son of Parameśvara, who also taught Nīlakaṇṭha Somayāji. His fame rests on the authorship of one of the most important texts of the Kerala school of mathematics and astronomy, the *Yuktibhāṣā* (An Exposition of the Rationale [of mathematics and astronomy]) also called *Gaṇita-nyāya-saṅgraha* (Compendium of Mathematical Rationale). It is a unique work on the rationale of Hindu mathematics and astronomy as it was understood in medieval India. It is unique in the sense that it is neither a textbook nor a commentary, but a work which is wholly devoted to a systematic exposition of mathematical rationale, written in Malayalam, the local language of Kerala.

Born about 1500, Jyeṣṭhadeva probably composed the *Yuktibhāṣā* about 1530, since it is known that a little after 1534, Śaṅkara Vāriyar, another contemporary astronomer, used it in his commentaries. Another work of Jyeṣṭhadeva, the *Dṛkkaraṇam*, also in Malayalam, was composed in CE 1608.

At the outset of the work, the author states that he is attempting "to set out in full the rationale useful for understanding the planetary motion according to the *Tantrasaṅgraha* of Nīlakaṇṭha Somayāji" (b. 1444). But he actually goes much beyond that and subjects to rationalistic analysis the entire gamut of mathematics and astronomy. In fact, he takes up the treatment from the very fundamentals, the concept of numeration and the theory of numbers. The work is made up of two divisions, each one divisible into several sequential chapters.

The first deals with the following subjects: (1) the eight fundamental operations, from simple addition to the roots of sums and differences of squares, wherein several methods, including diagrammatic solutions are offered; (2) algebraic problems; and (3) operations on fractions. The other chapters deal with: (4) the general nature of the Rule of Three (direct proportion) and (5) application of the Rule of Three in the computation of mean planets; (6) elaborate rationalisations of the circumference of the circle; and the last chapter: (7) the rationales of the derivation of the *R* sines, *R* versed sines, and their addition, properties of cyclic quadrilaterals, and the surface area and volume of a sphere. Many of the rationales are demonstrated both algebraically and geometrically.

Part Two is devoted to the exposition of rationales in astronomy, including (1) the computation of mean and true planets by means of two types of epicycles, supplemented by corrections; (2) the celestial sphere, the related great circles and secondaries, the precession of the equinoxes, and the armillary sphere; (3) declination, right ascension, and related matters; (4) problems related to spherical triangles; (5) problems connected with direction and shadow; (6) computation of the rising and setting points of the ecliptic, the ecliptic having constant variation. A direct method, enunciated in Indian astronomy for the first time, is used; (7) eclipses and the attendant parallax corrections; (8) the *Vyatīpāta*, which is the moment when the Sun and the Moon have equal declinations, but in different quadrants; and (9) reduction of computed results to observation and with the phases of the moon.

Some points are of special interest in the *Yuktibhāṣā*. One is the rationale for three or four steps for true planets, although the result can also be obtained in two steps; the derivation of inverse declination and inverse right ascension; novel solutions for some of the problems on spherical triangles and on shadows; refinements for parallax corrections; and an alternate method with a novel correction for the computation of the moment of *Vyatīpāta*. A noteworthy characteristic is its elucidating rationale from the fundamentals, first setting out the axioms and postulates involved, then developing the arguments and methodologies step by step.

It might also be noted here that Śaṅkara Vāriyar provided a valuable service by incorporating rationales from the *Yuktibhāṣā* into Sanskrit, the language of scholars. This he did in two elaborate commentaries called *Kriyākramakarī* on the *Līlāvatī* by Bhāskarācārya, and *Yuktidīpikā* on the *Tantrasaṅgraha*, a work on astronomy by Nīlakaṇṭha Somayāji. There is also a highly corrupt rendering of *Yuktibhāṣā* into Sanskrit.

K. V. Sarma

References

Balagangadharan, K. "Mathematical Analysis in Medieval Kerala." In *Scientific Heritage of India: Mathematics*. Ed. K.G. Poulos. Tripunithutua: Government Sanskrit College. 1991. pp. 29–42.

Līlāvatī of Bhāskarācārya with Kriyākramakarī of Śaṅkara and Nārāyaṇa. Ed. K.V. Sarma. Hoshiarpur: Vishveshvaranand Vedic Research Institute. 1975.

Rajagopal, C.T., and M.S. Rangachari. "On an Untapped Source of Medieval Keralese Mathematics." *Archive for History of Exact Sciences*. 18: 89–101. 1978.

Rajagopal, C.T., and M.S. Rangachari. "On Medieval Kerala Mathematics." *Archive for History of Exact Sciences*. 18: 89–101. 1978.

Sarma, K.V. and S. Hariharan. "Yuktibhāṣā: A Book of Rationales in Indian Mathematics and Astronomy, An Analytical Appraisal." *Indian Journal of History and Science* 26(2): 185–207. 1991.

Tantrasaṅgraha of Nīlakaṇṭha Somayāji with Yuktidīpikā and Laghuvivṛti of Śaṅkara (An Elaborate Exposition of the Rationale of Hindu Astronomy). Ed. K.V. Sarma.

Hoshiarpur: Vishveshvaranand Vishva Bandhu Institute of Sanskrit and Indological Studies. Punjab University. 1977.

Yuktibhāṣā, Pt. I. *Mathematics*. Ed. Ramavarma (Maru) Thampuran and A.R. Akhileswarayyar. Trissivaperur: Mangalodayam Limited. 1948.

See also: Nīlakaṇṭha Somayāji – Parameśvara – Rationale in Indian Mathematics – Śaṅkara Vāriyar

Z

Zero

Mathematics today owes its existence in part to the discovery of zero. For the purpose of calculation it needed a short symbol, which is at present denoted by a small circle in nearly every part of the world. In India in the early period the form of short symbol which represented zero (*śūnya*) was both a dot and a small circle.

In the Vedic literature, in the *Amarakośa*, zero (*śūnyam bindau*) was represented by a dot. The form was also suggested by the word *kṣudra* (very small) in the *Atharvaveda*. The word *randhra*, used in the Hindu *Gaṇita Śastraka*, indicated a small hole. There are many examples of zero being represented as a dot in the Kashmirian *Atharvaveda*, both in the marginal notes and in the text itself.

In the marginal notes, the numbers are as follows:

Symbol	Number	Folio	Page
◇••	100	100b	192
◇•◇	101	101b	194
◇2•	170	170b	310

The symbols in the marginal notes represented the Folio numbers. In most of the pages of the book there are examples of the above type. In the text itself the following numbers are found:

1	◇
2	
3	
4	
5	
6	
7	
8	
9	
10	◇•

Small circles in pairs have been used to denote blank spaces in the text. In Folio No. 178a, page 323 in the last line, pairs of small circles are found as follows: oo oo oo oo oo

The numbers in the marginal notes may be taken from a later period, but the symbols in the text itself must be from the time of the *Atharvaveda* (500 BCE).

In the Bakhshālī Manuscript, the oldest extant manuscript in Indian mathematics (CE 200), the symbol  scan from text 2000 represents the number 4000, while the symbol 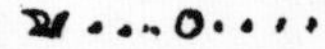 represents 500,000,000. The fourth zero in The manuscript is a small circle; the others are dots. In the Shahpur Stone image Inscription of Aditya Sena Bihar, India (CE 672), the symbol 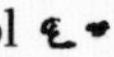 stands for 60, and zero is represented as a dot. In the Malay Inscription at Katakapur (CE 686), symbol 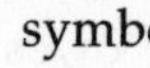 represents the number 608 (Śāka), and zero is represented as a circle with a deep circumference.

In the above examples it is found that zero has been exhibited either as a dot or as a small circle to fit in numbers in the place value scale, or to represent absence. In the inscriptions at Sambor, Palembang, and Kotakapur, which were Indian colonies of the Far East, the numbers have been written to represent the Śaka era. Hence undoubtedly they also represent an Indian origin.

Now the question arises as to why zero has been represented as dot or a small circle and not as a square or rectangle or anything else. Two reasons can be assigned; one is spiritual or metaphysical; the other is physical or atomic.

In the spiritual sphere, *śūnya*, or zero or nothing, the symbol of absence of everything, identifies itself with *Nirguṇa* Brahma, the absence of all qualities. The no-quality *Nirguṇa* in Brahma represents the fact that He is not guided by any of the qualities or constituents of nature. But at the same time Brahma is the source of all qualities, energy, power, and strength. Similarly *śūnya* or zero itself signifies absence when placed independently, but it represents fullness when it is placed in the decimal system of numeration. (Placing zeros on the right side of a number increases the value to an infinite step).

Since the conception of *śūnya* identifies with the conception of Brahma both in absence and fullness the symbol of *śūnya* (zero) must be guided by the symbol of Brahma. The conception of Brahma lies in meditation on a particular point or small circle in the space between the two eyebrows.

Swami Sivananda has given concentration the name 'one- pointedness'. Concentration can also start by fixing one's gaze on a black dot on the wall and later on a bright light first of the size of a pinpoint and later of the size of a Sun coming out from the space in between the two eyebrows. Hence the symbol of Brahma can be taken to be a dot or a small circle, and therefore the symbol of mathematical zero is either a point or a small circle.

There are also physical reasons for the symbol for zero. Planets seen from the earth look just like dots. From Vedic times Indians excelled in astronomical observations as we can see in the *Vedāṅga Jyotiṣa* (1200 BCE). So the physical reason for zero's being represented as a dot or a small circle lies in the fact that the Sun, Moon, and planets were seen as dots or small circles to an observer on the Earth.

Now why should a planet be chosen to represent zero? It is because of the other interpretation of mathematical zero, the absence of atoms. The idea of the atom was mentioned in ancient times in the Buddhist work *Lalita Vistara* (500 BCE), where the diameter of a *paramāṇu* (molecule) was given as 1.32×10^{-7} inches, whereas at present the diameter of an atom is 2×10^{-8} cm. The nucleus of the atom has a radius of 1.37×10^{-7} cm. The quantity is so small that it is a *kṣudra* (minute), a synonym for zero in the *Atharvaveda*. Thus the absence in mathematical zero is guided by the absence inherent in the smallness of an atom or its nucleus. The concept of the fullness of mathematical zero is also present in the infinite motion of the planets and of the electrons around the nucleus of an atom.

The electron has no mass in the material sense, and the mass which it has developed from electrical energy is negligible, having a radius of $1.875 \times 10^{-}$ cm. This guides the concept of mathematical zero. The movement of electrons around the nucleus with a very high velocity and in an infinite motion is identical with the concept of fullness of mathematical zero which guides the numbers to move in an infinite journey like 10, 100, 100, 10,000 to 10^n.

Thus the double interpretations, absence and fullness, of mathematical zero are identical with the double interpretations in a planet, and the symbol of zero is guided by the symbol of a planet which is observed either as a dot or a small circle and whose path is almost circular.

Zero has a double meaning in Vedic literature. Etymologically the word *śūnya* comes from the word *śūna* (*śūna + yat*). The word *śuna* has two meanings. One is the killing of animals or the slaughterhouse, which represents absence; the other is increase, which leads to the conception of fullness.

The synonyms of zero, *randhra, tuccha, kṣudra,* and *rikta* project a concept of nothingness, while the synonyms *vyoma, diba, ākāśa, antarikṣa,* and *jaladhra patha* mean infinite expanse of sky. *Pūrṇa* means full, and *ananta* means infinite. *Drabinam* and *balam* mean strength, vigour, and force.

Zero, though it signifies nothing, has a full voltage battery charge when used in the decimal place value system.

The German mathematician B.L. Van der Waerden opines that the symbol of zero as a small circle came from the first letter 'o' of the Greek word *ouden* meaning nothing. This claim can be compared with a parallel claim (apparently convincing but far from the truth) that 4 in Brahmi Numerals written as ૪ comes from the first letter ચ of the word ચ (our four in English). Similarly in India one may say that the symbol ५ (5) in devanāgari script comes from the first letter of the word પાંચ (five), ६ (6) in devanāgari comes from the first letter of the word છઃ (six). But these are not so. Every number symbol has a heritage and a path through which it has come down to its present form. That 'o' is the first letter of the word is just a coincidence. Rather the symbol of a small circle (o) was used for the numbers ten, seventy, and a hundred in Greece.

R.N. Mukherjee

REFERENCES

Atharvaveda. Ed. Maruice Bloomfield. Strasbourg: K.J. Treubner, 1899, reprinted 1971.

Bag, A.K. "Symbol for Zero in Mathematical Notation in India." *Boletin de la Academia Nacional de Ciencias* 48: 251. 1970.

Bose, D.M., et al. ed. *A Concise History of Science in India*. New Delhi: Indian National Science Academy. 1971.

Datta, B. and A.N. Singh. *History of Hindu Mathematics*, vol. 1. Lahore: Motilal Banarsi Das. 1935.

Kaye, G.R. *Bakhshālī Manuscript. A Study in Mediaeval Mathematics*. Calcutta: Government of India. 1927–1933.

Van der Waerden B.L. *Science Awakening*. Oxford: Oxford University Press. 1961.

See also: Bakhshālī Manuscript

Zij

The term *zīj* is used everywhere in the study of Islamic culture to signify an astronomical handbook. These handbooks consist of a collection of astronomical tables together with such textual material as the reader would need in using the tables. The material is often divided into the following sections.

1. Calendrical conversion;
2. Mean motions of the sun, moon, and planets;
3. Equations of the sun, moon, and planets;

4. Positions of fixed stars;
5. Trigonometrical tables (sine, tangent);
6. Spherical astronomy;
7. Parallax;
8. Eclipses of the sun and moon;
9. Geographical coordinates;
10. Astrological quantities.

The underlying theoretical model of planetary equations was almost invariably Ptolemy's, and to a large extent *zījes* represent a continuation of the *Handy Tables* of Ptolemy (ca. CE 140). However the earliest Arabic *zīj* is that of al-Khwārizmī (ca. CE 830), which was based on procedures derived from the Indian treatise *Brāhmasphuṭasiddhānta* of Brahmagupta (CE 628). A large number of such works survive intact in Arabic and Persian manuscripts, and many others are known by name only. The most notable modern edition (Nallino, 1907) is that of the *Zīj* al- Ṣābiʾ of al-Battānī (fl. CE 880). For the long Islamic period the *zīj*es are a vital repository of data, primarily through the constant improvements in the parameters of mean motions reflecting new observations, and also of improvements in mathematical methods.

In most *zīj*es the textual material was generally restricted to instructions in the use of the tables. Larger astronomical treatises, such as the *Qānūn al-Masʿūdī* of al-Bīrūnī (ca. CE 1030), included all the material which would be found in a *zīj*, but went further in its detailed treatment of the whole subject.

A number of *zīj*es were translated into Latin and Greek, and initially at least, the term *zīj* was transcribed as *ezich*, *ezeig*, etc. (= *al-zīj*), and $\zeta\upsilon\zeta\ell$ respectively. In Latin the term was soon replaced by *tabulae*, while in Greek one finds $\sigma\upsilon\nu\tau\alpha\xi\iota\varphi$, reminiscent of Ptolemy's *Almagest*.

The term *zīj* is originally Middle Persian, where it means 'stretched cord'. The sense 'astronomical handbook' goes back to the early sixth century when Sanskrit works were introduced into Iran. The name *zīj* may have arisen as a literal translation of the Sanskrit term *tantra* (from *tan* 'to stretch') literally 'warp, loom', but which is used in the sense of 'system' or 'text book'. The word is singled out by Varāhamihira (ca. 580) as the name of the branch of astronomy which is concerned with planetary calculations. In a Middle Persian tract of the ninth century, the *Epistles of Manuščihr*, there are references to the *zīg ī hindūg* (Indian Astronomy), and also to a *zīg ī šahriyārān* (Royal Astronomy), a Sasanid compilation probably of the sixth century, referred to later by Arabic authors as the *Zīj al-Shātroyārān*. The former may be one of the works of Āryabhaṭa (ca. CE 520), which were referred to as Tantras by an early commentator.

Raymond Mercier

References

Kennedy, E.S. "A Survey of Islamic Astronomical Tables." *Transactions of the American Philosophical Society* 46: 123–176, 1956.

Kennedy, E.S. "The Sasanian Astronomical Handbook Zīj-i Shāh and the Astrological Doctrine of 'Transit' (mamarr)." *Journal of the American Oriental Society* 78: 246–262, 1958.

Mercier, R.P. "Astronomical Tables in the Twelfth Century." In *Adelard of Bath, an English Scientist and Arabist of the Early Twelfth Century*. Ed. Charles Burnett. London: The Warburg Institute, 1987, pp. 87–118.

Nallino, C.A. *Al-Battānī sive Albatenii Opus Astronomicum*. 3 vols., Milan: Reale Observatorio, 1899–1907.

See also: Astronomy in Islam

Zodiac

The signs of the zodiac originated in Mesopotamia. In the first stage of development, twelve constellations along the ecliptic (i.e. the apparent course of the Sun in the sky) were roughly marked out and each was named after the animal whose shape it resembled. Later, with the need for a rigid coordinate system for planetary positions, the zodiacal sign (Table 1) assumed a new meaning: the length of 30 degrees along the ecliptic, so that twelve equal signs comprised a complete circuit (360 degrees) of the ecliptic. The change from the older irregular constellations to the signs of regular spacing took place somewhere around 500 BCE.

In the cuneiform texts the ecliptic coordinates were side-really fixed, and the vernal equinox was several degrees off "the first point of Aries": at the tenth degree of Aries in System A and at the eighth in System B. The Mesopotamian idea of twelve zodiacal signs of equal length was transmitted to Greece about 300 BCE, where the iconography of the signs was modified by their mythology.

With the discovery of the precession of equinoxes by Hipparchus in about 150 BCE, the significance of the zodiacal signs changed drastically. The first point of Aries was equated with the vernal equinox. Since this is in constant retrograde motion relative to the fixed stars (the shift being about 51 minutes of arc per year), the original relation between the constellations and signs was completely severed and the zodiacal signs became a purely mathematical reference system.

With the development of astrology, which preserved the old association of zoomorphic shape with zodiacal signs (except Libra), the zodiacal signs assumed new meanings. They were classified in various ways: by sex, the ownership of the house of planets, seasons, tastes, four humours, four elements, the governorship over the parts of body, plants, animals, geographical regions, etc.

All these ideas were transmitted to India in the second century of the Christian era. The very Sanskrit names of the zodiacal signs show that they were translated from Greek. In some texts even phonetic translations of Greek words are found. The earliest Sanskrit text which contains a list of these names is the *Yavanajātaka* (ca. CE 269), a Sanskrit version of a Greek book on horoscopic astrology.

Table 1 *The signs of the zodiac*

Degrees	English	Sanskrit
0	Aries	meṣa
30	Taurus	vṛṣa
60	Gemini	mithuna
90	Cancer	karkaṭa
120	Leo	siṃha
150	Virgo	kanyā
180	Libra	tulā
210	Scorpio	vṛścika
240	Sagittarius	dhanus or dhanvin
270	Capricorn	makara or mṛga
300	Aquarius	kumbha
330	Pisces	mīna

Three iconographic modifications in the process of transmission are worth mentioning. (1) While Gemini are the twin boys in Western iconography, *mithuna* in Sanskrit is a couple consisting of a male and a female, and this was interpreted as "husband and wife" in Chinese texts on Buddhist astrology. (2) The Sanskrit word *makara* stands for a kind of sea monster, and *mṛga* for a "forest animal" such as a deer. Thus Capricorn was divided into two separate animals. (3) The word *dhanvin* (one who has a bow) is a better translation of Sagittarius (archer), but the simpler *dhanus* (bow) without a human figure is more frequently used in Sanskrit texts.

In spite of the similarity of the names, the astronomical meaning of Indian zodiacal signs is different from that of the Western ones, because the precession of the equinoxes was not taken into account in India, and the first point (*meṣādi*) of the ecliptic coordinates was sidereally fixed some time in the third or fourth century CE. The difference (*ayanāṃśa)* between the vernal equinox and the *meṣādi,* which has accumulated in the present day (1994), is about 23°40′; thus the Sun's entry into *meṣa* now falls on the 14th of April, and the *makarasaṃkrānti,* originally a winter solstice festival, on the 15th of January.

This seemingly conservative attitude is closely related to the Indian system of naming the lunar month. A year is divided into twelve solar months by the Sun's entry (*saṃkrānti*) into a new zodiacal sign. The lunar month is named after the *saṃkrānti* which falls during that month. For example, the lunar month Caitra is defined as the month during which the Sun's entry into *meṣa* occurs. The full moon of that month has to be located near the diametrically opposite point on the ecliptic, that is, at the lunar mansion *citrā.* Thus, in order to keep the relation of the month name and the constellation name, they had to stick to the sidereal (*nirayana*) system even at the sacrifice of the correspondence between the seasons and month names.

The Western system has ignored the original association between constellations and zodiacal signs, and the word Aries, for example, has two meanings, one as an actual constellation and the other as the first thirty degrees in the ecliptic longitude. The former is used in astronomy and the latter in astrology. This is not the case in the traditional Indian system.

In south India and Nepal the solar month is still used in the civil calendar. Since a solar month is the time during which the true sun stays in one zodiacal sign, the length of a month varies from 28 to 32 days.

As mentioned above, Indian zodiacal signs were transmitted to China in the eighth century by Buddhist astrology and ultimately to Japan in the ninth century. The iconography of the Indian zodiacal signs was preserved in the star *maṇḍalas* (especially in the temples belonging to the Shingon sect) which were used in the ritual of worshipping the planetary deities.

Yano Michio

References

Neugebauer, Otto. *A History of Ancient Mathematical Astronomy*, 3 vols. New York: Springer. 1975.

Pingree, David. *Yavanajātaka of Sphujidhvaja*, 2 vols. Cambridge, Massachusetts: Harvard University Press. 1976–1978.

See also: Precession of the Equinoxes – Lunar Mansions – Astrology

List of Contributors

George Abraham, India	Gnomon
Michael Adas, USA	Western Dominance
Claude Alvares, India	Irrigation in India and Sri Lanka Technology and Culture
Gene Ammarell, USA	Astronomy in the Indo-Malay Archipelago
R.K. Arora, India	Agriculture
A.V. Balasubramanian, India	Knowledge Systems Metallurgy: Iron and Steel Yoga
S. Balachandra Rao, India	Bhāskara I
Bruce C. Berndt, USA	Ramanujan
A K Chakravarty, India	Calendars
D. P. Chattopadhyaya, India	Environment and Nature
Bhagwan Dash, India	Alchemy Ātreya Caraka Medicine: Āyurveda, Suśruta
Deepak Kumar, India	Colonialism and Science
Prakash N. Desai, USA	Medical Ethics
Ashok K Dutt, USA	Geography
S.D.Gomkale, India	Salt
Susantha Goonatilake, Sri Lanka	East and West: India in the Transmission of Knowledge from East to West
Paul Gregory, Canada	City Planning

R.C. Gupta, India	Āryabhaṭa Baudhāyana Bhāskara II Brahmagupta Mādhava of Saṅgamagrāma Mahāvīra pi in Indian Mathematics Sridhara
Takao Hayashi, Japan	Algebra: Bījagaṇita Arithmetic: Pāṭīgaṇita Bakhshālī Manuscript Combinatorics in Indian Mathematics Magic Squares in Indian Mathematics Number Theory
Richard L. Hills, England	Wind Power
Shigeo Iwata, Japan	Weights and Measures in the Indus Valley
S.K. Jain, India	Ethnobotany
William T. Johnson, USA	Physics
Karen Louise Jolly, USA	Magic and Science
George Gheverghese Joseph, England	Geometry Mathematics
M.S. Khan, India	Medieval Science and Technology
Takanori Kusuba, Japan	Nārāyaṇa Paṇḍita
Murdo J. Macleod, USA	Dyes
Bala V. Manyam, USA	Epilepsy
William J. Mcpeak, USA	Military Technology
Raymond Mercier, England	*Zīj*
R.N. Mukherjee, India	Zero
Roddam Narasimha, India	Atomism Rockets

Vijaya Narayan Tripathi, India	Astrology
Yukio Ohashi, Japan	Astronomical Instruments
Arnold Pacey, England	Rainwater Harvesting
Deepa Pande, India	Forestry
Marvin A. Powell, USA	Sexagesimal System
Baldev Raj, India	Metallurgy: Bronzes of South India; Metallurgy: Zinc and its Alloys: Ancient Smelting Technology
A.S. Ramanathan, India	Meteorology
Vijaya Ramaswamy, India	Textiles
S R Rao	Bricks
Roshdi Rashed, Japan	Science as a Western Phenomenon
Abdul Latif Samian, Malaysia	Al-Bīrūnī
Ziauddin Sardar, England	Values and Science
K.V. Sarma (late), India	Armillary Spheres Astronomy Calculus Candraśekhara Sāmanta Decimal notation Deśāntara Devācārya Haridatta Jagannātha Samrāṭ Jayadeva Kamalākara Lalla Lunar Mansions in Indian Astronomy, Mahādeva Mahendra Sūri Makaranda Munīśvara Pakṣa Parameśvara Paulisa

	Precession of the Equinoxes Putumana Somayājī Rationale in Indian Mathematics Śaṅkara Vāriyar Śatānanda Sphujidhvaja Śulbasūtras Śuryasiddhānta Vākyakaraṇa Varāhamihira Vaṭeśvara Yavaneśvara Yuktibhāṣā of Jyeṣṭhadeva
Nataraja Sarma, India	Time
Joseph E. Schwartzberg, USA	Maps and Mapmaking
Virendranath Sharma, India	Jai Singh Observatories
Akhtar H. Siddiqi, USA	Al-Bīrūnī: Geographical Contributions
Kripanath Sinha, India	Śrīpati
F. Richard Stephenson, England	Eclipses
B.V. Subbarayappa, India	Atomism
M.A. Tolmacheva, USA	Navigation
David Turnbull, Australia	Knowledge Systems: Local Knowledge
Edwin J Van Kley (late), USA	East and West
B. Venkatraman, India	Metallurgy: Bronzes of South India; Metallurgy: Zinc and its Alloys: Ancient Smelting Technology
D.M. Vijalakshmi, India	Metallurgy: Bronzes of South India; Metallurgy: Zinc and its Alloys: Ancient Smelting Technology
Michio Yano, Japan	Trigonometry Zodiac

Index